Superconductors: Emerging Technologies

Superconductors: Emerging Technologies

Editor: Stone Frankenstein

STATES
ACADEMIC PRESS
www.statesacademicpress.com

Published by States Academic Press
109 South 5th Street,
Brooklyn, NY 11249, USA
www.statesacademicpress.com

Superconductors: Emerging Technologies
Edited by Stone Frankenstein

International Standard Book Number: 978-1-63989-501-4 (Hardback)

Cataloging-in-Publication Data

Superconductors : emerging technologies / edited by Stone Frankenstein.
p. cm.
Includes bibliographical references and index.
ISBN 978-1-63989-501-4
1. Superconductors. 2. Electronics--Materials. 3. Solid state electronics.
4. Materials science. I. Frankenstein, Stone.
TK7872.S8 S87 2022
621.3--dc23

Contents

Preface

Superconductivity is a set of physical properties observed in certain materials where electrical resistance vanishes and magnetic fields are expelled from the material. The materials that exhibit these properties are called superconductors. In ordinary metallic conductors, resistance decreases gradually as its temperature is lowered down to near absolute zero. In a superconductor, the resistance drops to absolute zero below its characteristic critical temperature. Electric current that passes through a loop of superconducting wire can remain without any power source indefinitely. Superconductors are characterized by the Meissner effect, where the magnetic field lines are ejected completely from the interior of the superconductor during its transitions into the superconducting state. This book is a compilation of chapters that discuss the most vital concepts and emerging trends in the field of superconductors. The various studies that are constantly contributing towards advancing technologies and evolution of this field are examined in detail. This book includes contributions of experts and scientists which will provide innovative insights into this field.

This book is the end result of constructive efforts and intensive research done by experts in this field. The aim of this book is to enlighten the readers with recent information in this area of research. The information provided in this profound book would serve as a valuable reference to students and researchers in this field.

At the end, I would like to thank all the authors for devoting their precious time and providing their valuable contribution to this book. I would also like to express my gratitude to my fellow colleagues who encouraged me throughout the process.

Editor

1

Ultra-Low-Power Embedded SRAM Design for Battery-Operated and Energy-Harvested IoT Applications

Arijit Banerjee

Abstract

Internet of Things (IoT) devices such as wearable health monitors, augmented reality goggles, home automation, smart appliances, etc. are a trending topic of research. Various IoT products are thriving in the current electronics market. The IoT application needs such as portability, form factor, weight, etc. dictate the features of such devices. Small, portable, and lightweight IoT devices limit the usage of the primary energy source to a smaller rechargeable or non-rechargeable battery. As battery life and replacement time are critical issues in battery-operated or partially energy-harvested IoT devices, ultra-low-power (ULP) system on chips (SoC) are becoming a widespread solution of chip makers' choice. Such ULP SoC requires both logic and the embedded static random access memory (SRAM) in the processor to operate at very low supply voltages. With technology scaling for bulk and FinFET devices, logic has demonstrated to operate at low minimum operating voltages (V_{MIN}). However, due to process and temperature variation, SRAMs have higher V_{MIN} in scaled processes that become a huge problem in designing ULP SoC cores. This chapter discusses the latest published circuits and architecture techniques to minimize the SRAM V_{MIN} for scaled bulk and FinFET technologies and improve battery life for ULP IoT applications.

Keywords: IoT, SoC, ULP, SRAM, FinFET, assists, canary sensor SRAM

1. Introduction

The revolutionizing Internet of Things (IoT) devices connect us to a new horizon of smart wearable gadgets, home appliances, health monitors, home automation controllers, etc. According to a growth projection of IoT devices by CISCO in 2013, the number of these connected IoT devices could reach 50 billion by the year 2020 [1]. Among these IoT devices, a significant amount of products would be of wearable or portable categories. Thus, the portability and form factor of such smaller devices restrict the use of power source to smaller batteries. Besides, these mobile IoT devices could harvest energy from ambient light, body heat, etc. energy sources. Based on the power consumption of these IoT devices, the battery life would vary for different applications [2]. However, power storage capacity of smaller batteries, both non-rechargeable and rechargeable, is limited. Therefore, all of these so-called battery-operated portable devices are limited by the battery life, and battery replacement of millions of IoT devices per year could result in millions of dollars in replacement cost.

On the other hand, energy harvesters could transform light, radio wave, and vibration energy to electrical energy that could be a potential solution for battery life and replacement issues. However, the limited harvested power [2] from various energy sources may be insufficient to power IoT devices for applications requiring milliwatt or even hundreds of microwatts of power with the constraints of a smaller form factor. Also, guaranteed availability of energy sources may not be available for long-term application usage. Therefore, batteries remain the primary power source and choice for most of the IoT applications. However, due to self-leakage and energy consumption in IoT applications, the battery life and replacement time of IoT devices are major concerns, which last much less than the shelf-life of batteries of about 10 years. As battery energy density doubles every 10 years [3], which is much slower than Moore's law of the number of transistors doubling every 2 years [3], the low-power circuit solutions show great promises to empower IoT devices for longer battery life.

Every modern-day electronic gadget that has a digital processor in its circuit board, starting from the household micro-oven to Apple's iPhone and the commercial Amazon's cloud servers, uses a fast and power-efficient on-chip memory called the static random access memory (SRAM). The SRAM has three operations: one can write some desired data into a particular memory address location or read some data from a specific memory address or hold the written data to access in the future. Hence, the usual metrics to evaluate an SRAM are (1) the ability to write (write-ability), (2) ability to read (readability), and (3) ability to retain data (data retention) without any operation. Also, there is another metric called read stability that evaluates the stability of unselected bitcell columns while writing a data in selected columns. A simple architecture of the SRAM is given in **Figure 1(a)**, which shows it has an address bus to select an address for a write or read operation. The other pins are a data input bus DIN, a data output bus DOUT, a chip enable signal EN, a synchronous clock signal Clk, and a write and read select signal WRRD. More advanced SRAMs can have additional pins, such as test pins, write and read margin control pins, power management pins, etc. SRAMs are nonvolatile: disconnecting the power supply from the SRAM would result in loss of memory data stored previously. The SRAM typically shares the power rail with the microprocessor's digital circuits (logic core). The SRAM and microprocessor logic core can also have a separate supply rail at the cost of power rail routing, silicon area of the DC-DC converter, chip design time, and overheads. **Figure 1(b)** and **(c)** shows the two usual topologies used in system on chip (SoC) integrated circuits. The advantage of SRAM power rail topology shown in **Figure 1(b)** over **Figure 1(c)** is that it saves silicon area required by the additional on-chip DC-DC converter blocks; those are usually very large compared to the other blocks in the SoC. Due to the square-law dependency of power with supply voltage, one of the best ways for

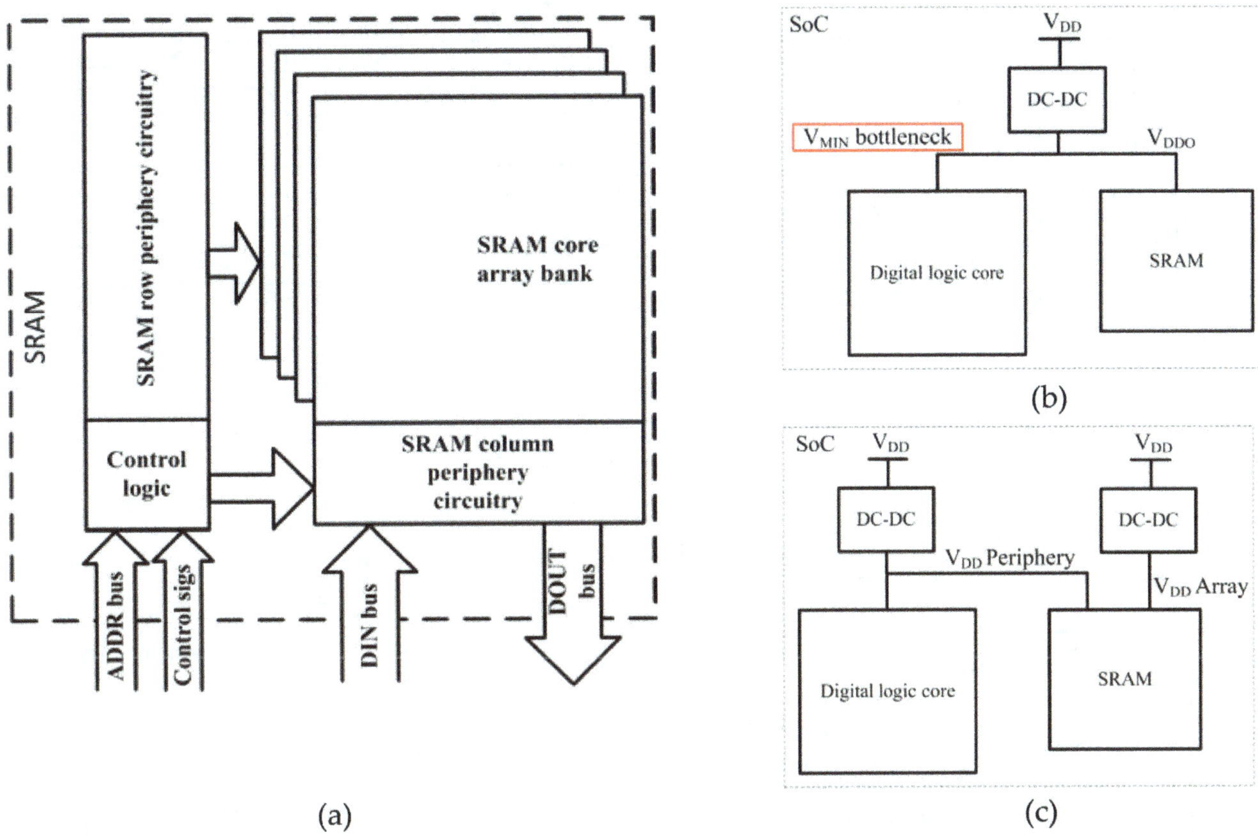

Figure 1. (a) SRAM architecture, (b) digital core and SRAM sharing the same rail, and (c) digital core and SRAM having dual-rail architecture.

low-power operation of an SoC is to lower the supply voltage (V_{DD}) and operate the entire digital microprocessor block at the scaled V_{DD}. Digital logic has been demonstrated to work at subthreshold [4] supply voltages [4–6] (100 mV and lower) that is lower than the threshold voltage (V_T) of bulk MOSFETs, as shown in **Figure 2**, in a MOSFET I_D-V_{GS} curve. However, the conventional 6T SRAM bitcell (**Figure 3(a)**) being a ratioed logic, which shares the same M5 and M6 transistors

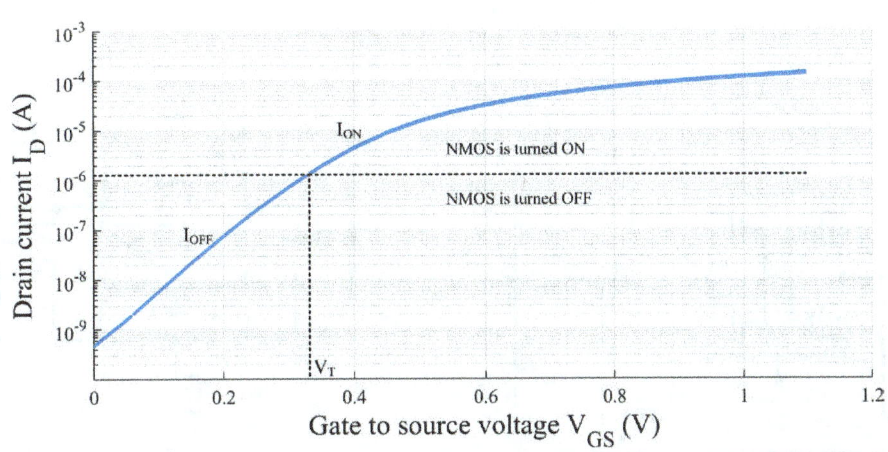

Figure 2. Drain current (I_D) vs. gate-to-source voltage (V_{GS}) plot for an NMOS transistor showing on and off states in 130 nm bulk predictive technology model from Arizona State University. Below the threshold voltage (V_T) of the MOSFET, the transistor is still operable.

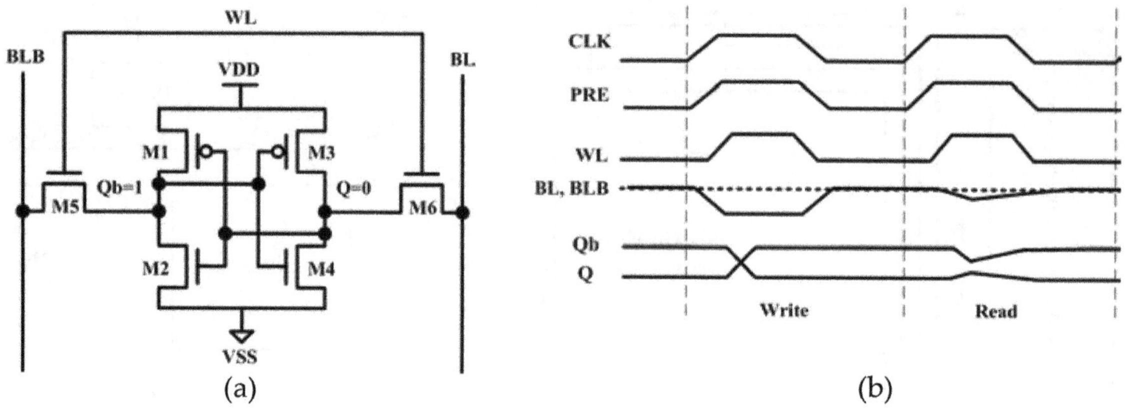

Figure 3. (a) Conventional 6T SRAM and (b) its write and read waveforms.

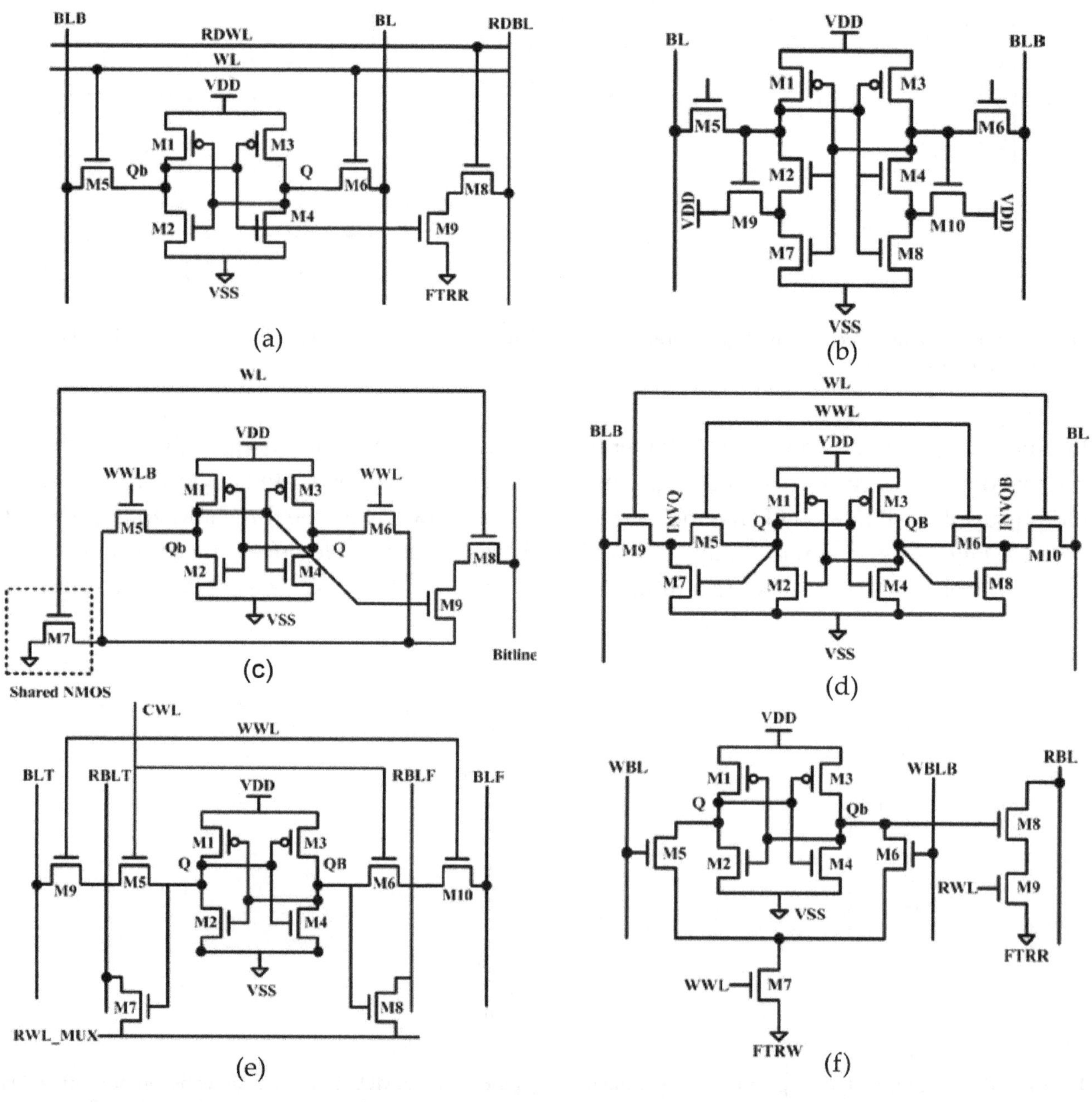

Figure 4. (a) Conventional 8 T SRAM bitcell, (b) Kulkarni's 10 T bitcell, (c) Chiu's 8 T bitcell, (d) Chang's 10 T bitcell, (e) Feki's bitcell, and (f) Arijit's 9 T bitcell.

for write as well as read operation, faces write-ability and read stability challenges across process variation, and the minimum operating voltage (V_{MIN}) of SRAM increases. Thus, sharing the same power rail of the logic core with SRAMs limits the voltage scaling of the SRAM with logic core for low-power operations. Additionally, with technology scaling in nanometer domain, the 14-nm FinFETs experience huge process variation [7], and the conventional high-density (HD) 6T FinFET bitcell (**Figure 3(a)**) with 1:1:1 M1:M2:M5 beta ratios has insufficient write-ability and read stability across process variation. With further technology scaling in 7 nm and smaller processes, it will be very challenging to make the conventional 6T SRAM memory to work, which has been there for decades.

There are mainly two available solutions to address these challenges of 6T SRAM by trading off SRAM area such as alternative bitcells and a write-read peripheral assist (PA) to improve V_{MIN} of 6T SRAM bitcell. The alternative bitcells are a class of bitcells that has lower V_{MIN} or lower energy consumption than the conventional 6T SRAM bitcell. A very popular alternative bitcell is 8 T bitcell, as shown in **Figure 4(a)**. Here, the write and read path are decoupled to improve the write-ability, readability, and read stability of the 8 T SRAM compared to the 6Ts. However, after a certain V_{DD}, even alternative bitcells are inoperable, and one of the popular SRAM schemes comes into the play for further V_{MIN} lowering: peripheral write and read assist techniques. Although the PAs reduce the worst-case SRAM V_{MIN}, it does not remove the SRAM V_{MIN} guardbanding across process variation, which costs additional area and energy penalty in the typical and best case dies. A couple of recently published works address this V_{MIN} guardbanding issue by tracking it using canary sensor SRAM. The canary SRAM extends the SRAM operating range by reducing V_{MIN} guardbanding across process variation, which promises to enable a multitude of IoT applications. This chapter will discuss aforementioned three major techniques that could enable ULP low-V_{MIN} SRAMs for IoT applications as follows. Before delving details into these topics, we need to understand the SRAM design metrics as follows.

2. SRAM write and read design metrics

As discussed earlier SRAM has four different categories of design metrics such as write-ability, readability, read stability, and data retention. The first three categories of design metrics can have static and dynamic measures. Here the static measures are obtained using DC SPICE simulations, and dynamic measures are obtained using transient simulations. Static measures for write and read metrics are easy to evaluate and are widely being used to quantify the SRAM static V_{MIN} across process and temperature corners. On the other hand, dynamic measures for write and read metrics are more accurate to represent an actual SRAM write or read operation; however, they are harder to evaluate and time-consuming. The static measures for write-ability are called write margin (WM) and write static noise margin (WSNM). Both WM and WSNM assume an infinitely long wordline pulse. The WM during a write is defined in two ways: the margin between V_{DD} and WL while BL and BLB are fixed at V_{SS} and V_{DD} and the margin between V_{DD} and BL while WL is fixed at V_{DD}. On the other hand, WSNM has a single definition for measuring the SRAM static noise margin (SNM) when the wordline is turned on. The static measure of readability is the DC read current (I_{read}) drawn from the bitline while reading a bitcell. The static measure of read stability is read static noise margin (RSNM), which assumes an infinitely long wordline pulse too. The measurement technique of RSNM and WSNM using the SNM measurement technique shown in [8] is widely used across industry and academia. The wordline is turned off during a standby operation, and the corresponding hold metric is called hold static noise margin (HSNM). During the read operation, the internal nodes of the 6T SRAM bitcell are read stressed, and thus, RSNM is the worst-case SNM among the RSNM and HSNM. On the other hand, the quantifiable metric to retain data at the lowest supply voltage is called the data retention voltage (DRV) or at the supply voltage at which the HSNM is almost zero. Due to the reason that the static metrics assume an infinitely long wordline pulse, the measurement of WM is optimistic, and the measurement of RSNM is pessimistic. Moreover, the static metrics does not represent the true nature of the SRAM write and read operations, which has a finite wordline pulse-width. Thus, dynamic metrics play an important role to accurately determine the write-ability, readability, and read-stability metrics and their corresponding V_{MIN} of SRAM. There can be many measures of dynamic metrics, such as dynamic write-ability and readability margins; the critical wordline pulse-width [9] for write-ability, readability, etc.; the failure rate of write-ability, readability, or read stability for a given wordline pulse-width; etc. Among these measures, the measurements of failure rates are the more popular choice to determine the V_{MIN} of SRAM. This section concludes the discussion of SRAM write and read design metrics, which paves the path for discussion to alternative bitcell in the next section.

3. Alternative bitcells for low-power IoT applications

As broad categories of ULP and mid-high performance IoT applications demand to run on modern IoT SoCs, the SoC must be operable throughout a wide range of supply voltages. The SRAM in the SoC for such IoT application is no exception. However, at a lower supply voltage, the conventional 6T high-density (smallest area) bitcell has poor write-ability, readability, and read-stability metrics, such as WM, I_{read}, and RSNM. Across process and temperature variation, these metrics degrade even more, and the conventional 6T SRAM becomes inoperable at lower supplies. Device sizing for write improvement hampers the read stability and vice versa due to shared write and read path and thus is not an option for ULP IoT applications. Moreover, near and below the subthreshold supplies, sizing does not work well to improve WM and RSNM metrics. On the other hand, at lower supply voltage, the soft error rate (SER) [10] increases. The SER can cause soft error disturb (SED) caused by high-energy particle strike that can flip the internal content of the bitcells in an SRAM. Error-correcting codes (ECC) [10] are essential to fix the SED errors; however, it requires additional ECC hardware and memory row or column to fix single-bit single-word errors. Detecting and correcting a multi-bit single-word (MBSW) error is expensive regarding ECC hardware and layout area. An MBSW error is usually lowered using bitline interleaving scheme, which is also known as column muxing. However, in a column mux scenario, selecting a 6T bitcell row for a write using the so-called wordline boosting-type peripheral write assist for V_{MIN} lowering degrades the read stability of the half-selected bitcells, which is known as the half-select issue [16]. The root of the problem in the conventional 6T is the shared write and read path that degrades both the write and read operations in a column mux scenario. Thus, separating the shared path for write and read operation is the desired solution for low-V_{DD} operation of SRAMs.

State-of-the-art alternative bitcells' [11–16] innovations in the last decade, having separate write and read path, show promises for low-V_{DD} operations. Among these bitcells the 8 T

(**Figure 4(a)**) bitcell is very popular and widely used in register files. Here, the write operation is performed using the 6T part of the 8 T bitcell, which is exactly the same as the conventional 6T operation. According to the data, one of the WBIT or WBITB lines goes high while the write wordline (WWL) is turned on. The read operation uses the two transistor read buffers M8 and M9. During a read, the read bitline (RBL) is initially precharged to V_{DD}, and after the read wordline (RWL) turns on the RBL discharges if the internal node Qb is holding a logic "1," else not. This RBL discharge directly drives an inverter or logic gate or a single-ended sense amplifier to generate the read-out signal. Although the 8 T bitcell separates the write and read paths, it suffers from read-stability issues in column mux scenario due to the half-select problem. Thus, an ultra-low voltage (ULV) operation using 8 T may not be viable in scaled technology across process and temperature variation. On the other hand, some of the other alternative bitcells that are shown in **Figure 4**, which includes Kulkarni's [12], Chiu's [13], Chang's [14], Feki's [15], and Arijit's [16] bitcells, show promise for ULV operation. Kulkarni's bitcell (**Figure 4(b)**) uses Schmidt-trigger type topology to have higher read stability and shown to operate down to 160 mV. However, due to feedback in the Schmidt-trigger-type topology, the write and read energy, as well as leakage current of the bitcell, is higher than the other state-of-the-art alternative bitcells. It also suffers from the half-select issue. Chiu's and Wang's bitcell has a unique data-aware cross-point selection in the topology itself, which not only avoid the half-select issue but also serve as a lower energy bitcell. On the other hand, Feki's bitcell has two wordlines (**Figure 4(e)**) that separated the write from read operations and has lower leakage numbers. All of these ULV alternative bitcells show improvement in V_{MIN} or dynamic energy or leakage numbers. However, it does not necessarily mean that any capacity ULV SRAM using any of these alternative bitcell would be suitable for all the ULV application. Where the battery life is extremely important, such as invasive or noninvasive ECG, EEG, or EMG monitoring for patients for a long time, a careful selection of bitcells is required based on total energy per cycle consumption and the duty cycle of the active IoT device.

With the voltage scaling in subthreshold supplies although the dynamic energy per cycle decreases, the cycle time increases due to the exponential relationship of MOSFET drain current with gate supply voltage. Thus, with voltage scaling the leakage energy per operation increases in SRAMs, and there can be a minimum energy point (MEP) [16]. Hence, arbitrary scaling down supply voltage for alternative bitcell arrays using different methods may not be fruitful from the standpoint of energy consumption or battery life. Authors in [16] compare Kulkarni's, Feki's, Chiu's, and Chang's bitcells with Arijit's 9 T that shows the MEP contours are best for Arijit's 9 T bitcell for low-energy biomedical applications due to its lower read, write, and leakage energy per operation. **Figure 5(a)** and **(b)** shows across design knobs (word width and size) the MEP comparison of the abovementioned bitcells as described in prior work [16], which is useful for selecting greener bitcells for low-energy consumption for extending battery life of biomedical devices. Note that all of the alternative bitcells have area penalty and energy tradeoffs compared to the high-density 6T SRAMs. Although alternative bitcells allow us to somewhat lower the V_{MIN} of SRAMs for low-energy operation, there is another widely used design knob, called peripheral assists (PAs), for achieving a low-V_{MIN} in SRAMs. Without the V_{MIN} lowering PAs, even for alternative bitcells, below some V_{DD} doing write and read operation is challenging, such as subthreshold V_{DD}s. The V_{MIN} lowering PAs are discussed next.

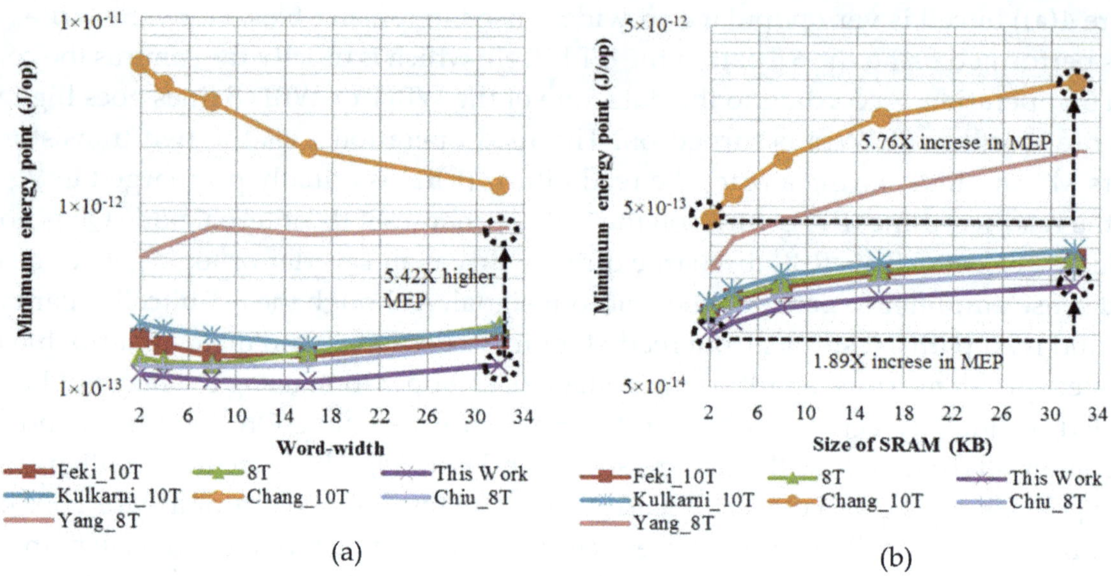

Figure 5. Minimum energy point (MEP) vs. (a) word width and (b) size of SRAM [16].

4. Write and read peripheral assist techniques for low-V_{MIN} applications

In moderate- to high-speed IoT applications, such as 100 MHz–1GHz, an alternative bitcell may not be the choice of SRAM designer due to high timing as well as area penalty. Thus, the lowest area 6T bitcell is still a popular choice for mid- to high-speed IoT applications. However, the 6T V_{MIN} is heavily guardbanded due to process and temperature variation. Thus, lowering V_{MIN} requires write and read peripheral assist [17] (PA) techniques. Moreover, alternative bitcells in ULV application involves the help of PA to have correct write and read functionality across process variation. We define the PAs as a class of circuit techniques used in SRAM periphery that improves the write-ability, readability, and read stability of SRAM bitcells. Mainly, a PA technique would either bump up or lower the wordline or bitline control voltages of the SRAM to make the write or read operation successful, as shown in **Figure 6(a)**. A PA can also decrease the SRAM cycle time by shortening the write or read operations. The PAs are transient in nature and can be classified into write-ability, readability, and read-stability PAs. For the conventional 6T SRAM bitcell, the control signals are mainly wordline and bitline. Thus, an example of write-ability PA would be wordline boosting (WLB) [17] or negative bitline (NBL) [17]. Although V_{DD} and ground (V_{SS}) signals are usually static, they can serve as control signals for SRAM write operation. Thus, V_{DD} lowering and V_{SS} rising [17] are also write assist techniques. On the other hand, to improve the readability or differential development or shorten the differential development time, one can apply a small percentage of WLB (as bigger percentages could induce read-stability issues in half-selected bitcells in column mux scenario) or negative V_{SS} (NVSS). Applying a suppressed wordline in write improves the read stability in half-selected bitcells, which have better RSNM numbers; however, it degrades the write-ability in selected bitcells. Additionally, column-wise boosting the V_{DD} or making the V_{SS} negative during the write operation in half-selected

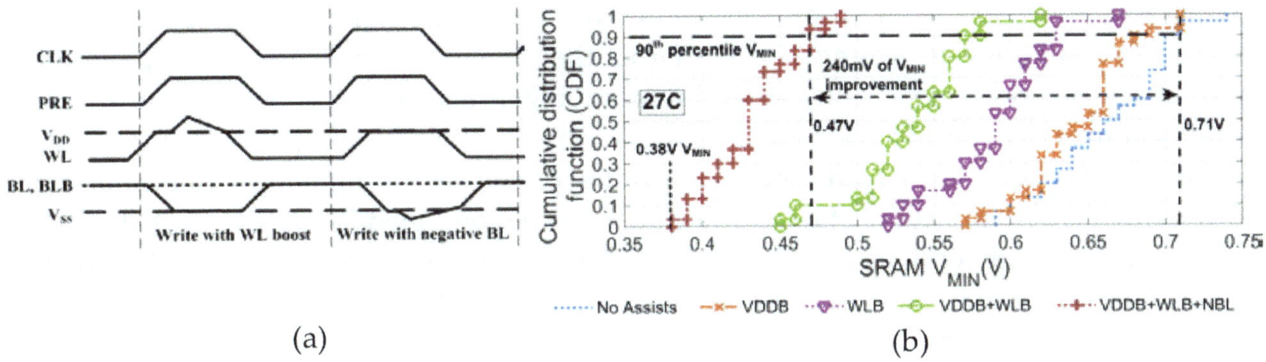

Figure 6. (a) Example of write assists using wordline boost and negative bitline techniques and (b) measured CDF of 256 kb SRAM V_{MIN} showing 90th percentile V_{MIN} improvement of 240 mV using combined assists [V_{DD} boosting (VDDB), WL boosting (WLB), negative bitline (NBL)] [20].

bitcells also improves the read stability. Note that usually increasing the percentage of assist further enhances the write-ability, readability, and read stability but has a limit called assist line contour [17], which is controlled by the V_{MAX} of the process technology. The list of possible PAs for write-ability, readability, and read stability can be found in [17]. PAs can affect the V_{MIN} and yield of SRAMs differently in different technology. Thus, evaluations of PAs are necessary for new scaled-technologies, as past technology trends may not hold true in newer ones.

More than a decade ago, when bulk CMOS technology scaling at 65 nm and lower was facing challenges of higher process variation, the single write or read PAs showed enormous promises to improve the V_{MIN} and yield of 6T SRAMs. However, with the introduction of scaled 28 nm technology, the process variation was so high that the HD 6T bitcell was not writeable in all process corners, especially for the worst case. Post 28 nm bulk the FinFETs become a device fabrication option, and the trend of write-ability issues in the HD 6T bitcell persisted due to huge process variation. Thus, from 28 nm onward applying a particular single write or a read assist may not lower the SRAM V_{MIN} across process variation anymore. Authors in [18] show the use of dual write and read PAs that reduces the V_{MIN} and improves the yield. Moreover, authors in [19] discussed some appropriate combination of PAs (CPAs) that could lower the V_{MIN} further for FinFETs at near-subthreshold supplies, such as a combination of negative bitline with boosting the V_{DD}, etc. One could employ different CPAs based on V_{MIN} lowering application needs. Because write-ability and read stability are more important metrics in FinFET SRAM design, and they often contradict the use of certain assists, such as wordline boosting for write improvement, the SRAM designer must make a careful selection of CPA. Usually, a widely used CPA combination for FinFETs nowadays is V_{DD} underdrive with wordline underdrive [18] schemes.

Moreover with technology scaling the metal width and pitch scale. Thus, there exist challenges of electro-migration, IR drop, and cross talk issues, which could restrict the use of a specific assist or limit the size of an SRAM bank. With the explosion of IoT application needs, ULP SoCs are targeted to run ultra-low energy as well as high-speed applications from time to time. Thus, voltage scaling down to near-subthreshold or deep-subthreshold supplies for SoC is a need nowadays. As logic V_{MIN} easily scales down to lower V_{DD}s, but SRAM V_{MIN} is comparatively higher due to process and temperature variation, the overall V_{MIN} of IoT SoCs increase that has a logic core and SRAM sharing the same power rail. Note that splitting the logic and SRAM power rail, called dual-rail technology, requires additional silicon area and power routing costs due to additional DC-DC converters. Thus, lowering SRAM V_{MIN} is essential for wide-range ULP IoT applications depending on the speed requirements of the applications. Authors [20] show for the first time the use of three combined PAs (NBL + VDDB + WLB) that lowered the conventional 6T V_{MIN} from 0.71 V (90%) to 0.47 V as shown in **Figure 6(b)** using a measured cumulative distribution function. The work reports the total V_{MIN} improvement as 240 mV in a commercial bulk 130 nm technology. This work shows more than 300X active power lowering using the triple CPA technique.

However, it is not imperative that always lowering the V_{MIN} would help to reduce the SRAM energy consumption. Lowering V_{MIN} requires energy penalty due to the use of assists, which could lead to the cause of overall SRAM energy could increase in some case. The total SRAM energy could increase with certain higher assist percentages and lower V_{MIN} is not always an intended requirement for low-energy applications. However, if the SRAM shares the same power rail with the logic core, as shown in **Figure 1(b)**, the energy savings from voltage scaling in the logic core could be much higher than the energy increase in SRAM, and thus, in this scenario only it might help.

Although write and read PAs usually improve the V_{MIN} guardbanding, it does not remove it entirely. Moreover, due to design and application of PAs, we trade off additional silicon area and energy for SRAM and overall SoC sharing the same power rail with SRAM. Additionally, with the design for the worst-case approach, the nominal and the best-case corner dies suffer from additional area and energy overhead. Thus, an important research question emerges: how to minimize this additional V_{MIN} guardbanding of SRAMs across process and temperature variation? The answer lies in tracking the SRAM V_{MIN} using in situ canary sensor SRAMs that helps to apply CPA for individual dies differently across process and temperature variation. The next section describes the canary SRAM techniques.

5. Canary sensor SRAMs for V_{MIN} tracking and guardband lowering

The story of canary SRAMs ties to the story of the canary in a coalmine. Eighteenth-century coal miners used to carry this beautiful yellow canary bird for poisonous gas, such as methane detection. A moderate presence of such gases could be fatal to human beings. If there is a significant level of methane being present in the mines, the canaries used to feel sick. By observing the canaries, thus, the miners get enough time to evacuate the coalmines. The Canaries from the standpoint of a circuit could mean a weak circuit that fails earlier than the main circuit. The canary circuits are first being introduced as canary flip-flops [21]. Later in the year 2007, the authors in [22] show canary techniques could reduce the data retention voltage (DRV) of an SRAM, thus saving a huge amount of leakage power for ULP applications. This work indicates that canary SRAM bitcells are a modified version of the SRAM cells; those use an additional bias control knob to weaken the DRV of sets of canaries to tune to fail them earlier than the population of the core SRAM bitcells [22]. A bias generator circuit is used to generate the bias voltage for the canaries in a row. A failure detector senses the canary retention failures in a closed loop. Thus, canaries could achieve a failure point before the SRAM DRV in each dies fabricated that lower the SRAM DRV and leakage current. This technique avoids the design for the worst case using canary-based DRV tracking.

In the year 2014, the authors in [23] demonstrated a theory for dynamic write V_{MIN} tracking for the conventional 6T SRAM. This work introduces the term reverse assists (RA) as one of the canary design knobs. As discussed earlier, the peripheral assists (PA) improve the write-ability or readability of the SRAMs. On the contrary, the RA **Figure 7(a)** degrades the write-ability or readability of the canaries to fail earlier than the population of core SRAM, as shown in **Figure 7(b)**. Thus, with the increase of the RA percentage, the canary distribution of write V_{MIN} would shift to the right-hand side from distribution A to B to C. A user can tune the failure point of the canaries by selecting the proper reverse assist percentages or settings [23]. Another input design knob that helps to tune the canaries at a desired V_{MIN} failure point is the failure threshold condition (F_{th}) [22], which defines the no. of canary failures that correspond to a threshold failure point. The work also derives a mathematical formulation for dynamic write V_{MIN} tracking as shown in Eqs. (1) and (2) [23]. Here the meanings of the variables of the equations are described in [23]. Here, the two equations relate the input and output SRAM design knobs and metrics to the canary design knobs and metrics. The work explains how to calculate the output metric named canary chip failure probability. The intended SRAM bit failure rate vs. V_{MIN} data is calculated first. Then from the canary failure rate vs. V_{MIN} data, the corresponding canary bit failure rate p_f is calculated. This serves as the input data to Eq. (2) [23] for the calculation of the output metric of canary chip failure probability P_{fc}. The authors show that one can achieve the desired canary-chip-failure-probability either by selecting a smaller no. of canaries with larger reverse assist voltage strength or the vice versa, as shown in **Figure 8** [23]. The work further shows that for a fixed reverse assist voltage to track the V_{MIN} of a bigger SRAM, more number of canaries are required. Moreover, with the same reverse assist voltage, increasing the SRAM yield requires more number of canary bitcells to track the corresponding SRAM's V_{MIN} and so on.

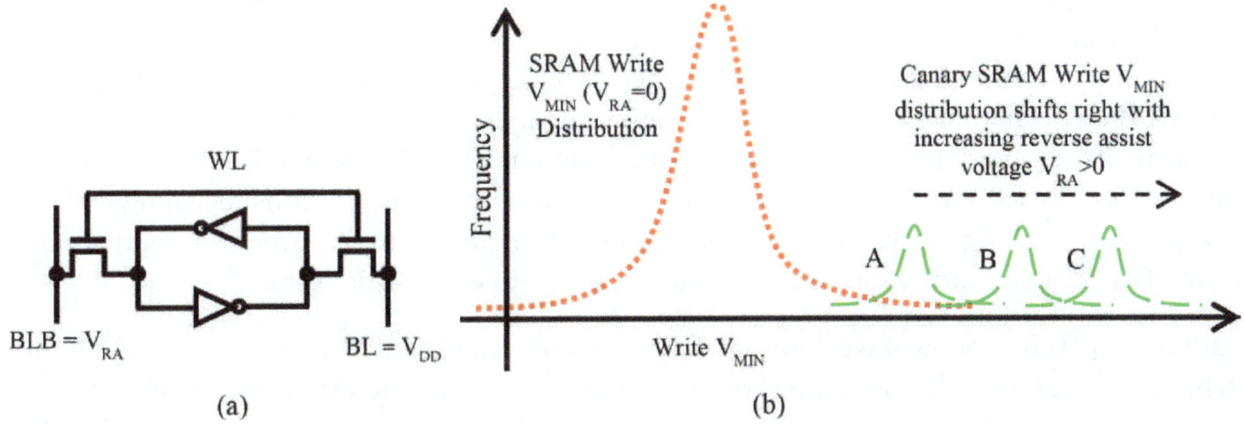

Figure 7. (a) SRAM write operation using bitline-type reverse assist and (b) write V_{MIN} distributions with a reverse assist (A, B, Cs are canary V_{MIN} distributions) [23].

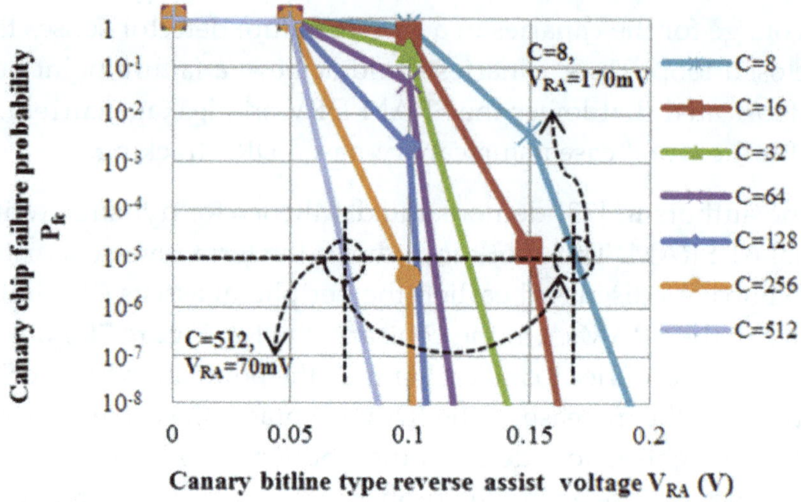

Figure 8. Canary chip failure probability vs. reverse assist voltage for 1 million SRAM bitcells with 95% yield @ TT_85C [23].

The work [23] shows the proposed bitline-type peripheral reverse assist circuit, as shown in **Figure 9**. The peripheral RA circuit uses a configurable NMOS-NMOS voltage divider to pass the generated voltage using an analog demultiplexer to the BL or BLB lines controlled by data D and data-bar Dbar. The reverse assist voltage generation can be disabled for normal write mode by asserting the AON signal to logic "0." The proposed block diagram of the integrated canary SRAM architecture is shown in [23], which is physically adjacent to the SRAM itself that shares the power rails. However, for independent write and read operations, at the canary and SRAM boundary, the bitlines are disconnected. The advantage of canary being an independent memory permits simultaneous operation to track voltage droops occurring at the SRAM-canary power rails to take actions if the canary SRAM fails. Such actions include either stopping the SRAM operation or lowering the SRAM clock frequency to prevent voltage scaling further or selecting an apt PA to lower the V_{MIN} further. The proposed algorithm in this work starts with an initial V_{RA} voltage and writes and reads canary rows to compare if the data written is correct. If the canary write operation is successful, the V_{RA} is increased else lowered gradually to reach the minimum V_{RA} settings. Unless the minimum V_{RA} setting is reached, the dynamic voltage and frequency scaling (DVFS) is allowed else the DVFS has to be stopped, as reaching the minimum V_{RA} would indicate the SRAM V_{MIN} is reached. The minimum V_{RA} setting would vary with the SRAM and canary input design knobs. The work also shows the area and power tradeoffs for SRAM and canary design knobs. It shows that for an increase in the number of canary bits, the normalized canary area and power overhead are amortized in bigger SRAM and increase with smaller capacity and so on. The work [23] showcases interesting results revealing that due to write V_{MIN} tracking, the canaries can save a minimum of 31% in SS corner dies and a maximum of 51.5% in FS corner dies compared to the worst-case SF corner dies.

Authors in [24] first show a working prototype of the canary SRAM in a commercial 130 nm technology that reveals the necessary properties of canary SRAM to track SRAM V_{MIN}. The work further shows a proof of concept V_{MIN} tracking canaries that fail earlier than the SRAM starts to fail, which is controllable using the canary design knobs (F_{th} and RAS) post-fabrication. The architecture of the SRAM is shown in **Figure 10**, which is similar to the [23].

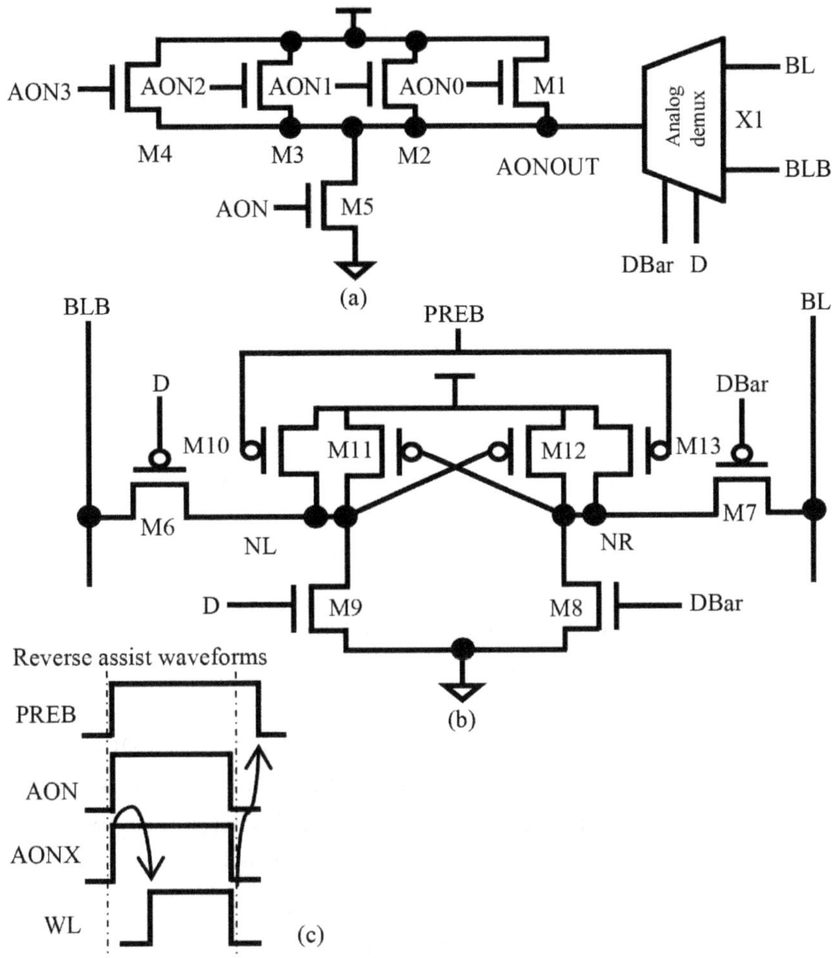

Figure 9. (a) Canary SRAM reverse assist circuit. (b) Canary write driver. (c) Reverse assist waveforms [23].

The testchip includes an 8kb core SRAM, 512 kb canary SRAM, a memory BIST (MBIST), a canary BIST (CBIST), and boundary scan chain blocks. The 6T bitcells used in both canary and core SRAM are same; it uses an external BLVRA voltage to apply as reverse assists to the canary SRAM. Both the MBIST and CBIST architecture are similar to a traditional MBIST [25]; however, they are specialized in measuring the number of bit failures in the core and canary SRAM. This work characterizes some important properties of canary SRAM that helps to track the core SRAM write V_{MIN} (WV_{MIN}). The authors show that using BLVRA and WLVRA reverse assists across different voltage, frequency, and temperatures (VFT), the canary failure curve shifts distinctly compared to each other. Without this distinction in shifting of failure curves, canary SRAM would not work, as there will be no way to tell if the input design knobs are changed, such as VFT. As discussed earlier, this work shows the first silicon proof that canaries can be tuned to fail earlier than the core SRAMs.

With the intuition presented in [23, 24] the authors in [20] show a closed-loop 256 kb self-tuning SRAM that can automatically track the SRAM V_{MIN} using canaries and apply apt write-read PAs to improve the V_{MIN} based on frequency needs for ULP IoT application. This work shows a 67% extension of operating voltage from 1.2 to 0.38 V deep into subthreshold supplies. Reverse assists are used to track the core SRAM V_{MIN} using canaries to allow

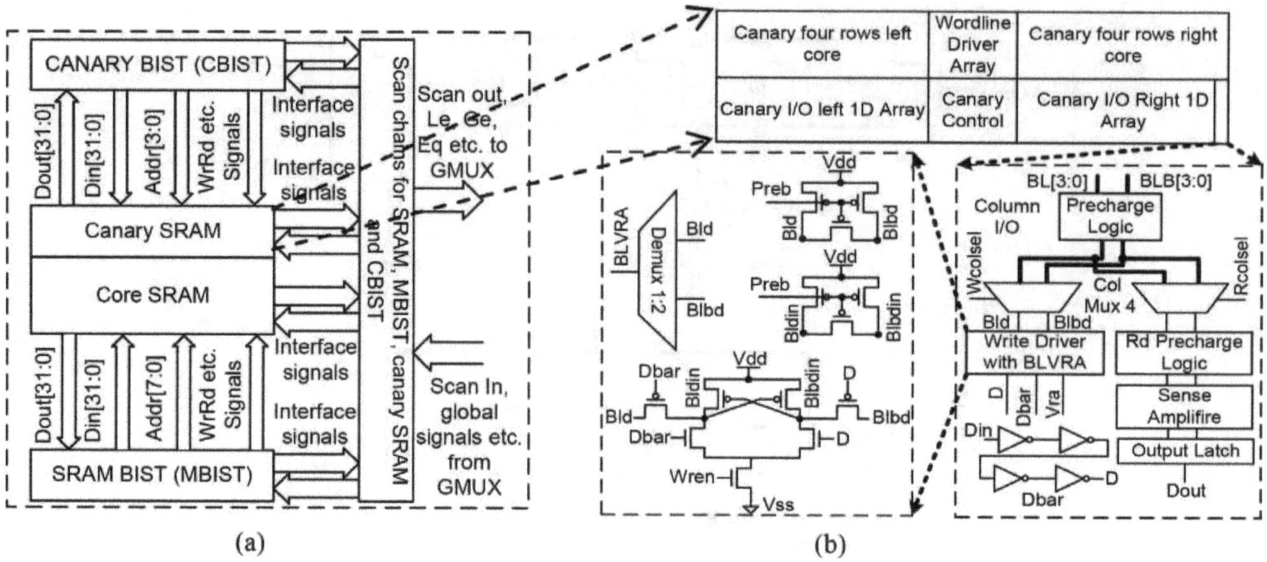

Figure 10. (a) Block diagram (not in scale) of the memory block and (b) block diagram (not in scale) of the canary SRAM column periphery (I/O) and BL-type reverse assist [24].

a closed-loop control of the system supply voltage at an intended operating frequency. The system uses write and read combined PA (CPA) along with in situ canary sensor SRAM-based V_{MIN} tracking to maximize the operating range of the SRAM into subthreshold supplies. This work meets the design needs for SRAMs of highly variable IoT applications while retaining the density of the conventional 6T bitcells. As the battery-operated or harvested energy IoT devices have an operating range of 10kHz to 10 MHz [26, 27], it is needed as a highly versatile feature to expand the 6T SRAM operating range to ULV supply voltages for low power operation. PAs can lower SRAM V_{MIN}; however, selecting the best CPA depends on the supply voltage that could influence the power-performance tradeoff.

This work uses write assists NBL and WLB along with read stability assist VDB to achieve a 90% V_{MIN} of 0.47 V compared to the other assist combinations as well as a no-assist case (shown in the CDF plot in **Figure 6** [20]). However, CPA alone requires a V_{MIN} guardbanding that ensures all chips functioning across PVT variation, hampering potential power savings. Canaries play a vital role to extend the power saving achieved using CPA using runtime determination of V_{MIN} that allow us to reduce the guardbanding. The block diagram of the proposed system is shown in **Figure 11**. The SRAM testchip comprises a 256 kb SRAM with 2 kb integrated canaries, a PA controller (ASC), a frequency to digital converter (FDC), an MBIST, and a CBIST. This architecture shares the SRAM periphery with canary sensors, such as write drivers, sense amplifiers, pre-charge circuits, etc. The RA circuit uses a wordline slope degrading programmable control for canaries.

The work [20] employs a self-tuning algorithm described in [20] that tracks SRAM V_{MIN} dynamically, which also adjust the supply voltage and the selection of PAs. The algorithm uses the FDC to measure and convert the input clock frequency to a digitized output and to initialize the off-chip low dropout (LDO) regulator to an initial V_{DD} programmed by a given look-up table (LUT). Using the ASC the algorithm chooses required PAs based on the LUT,

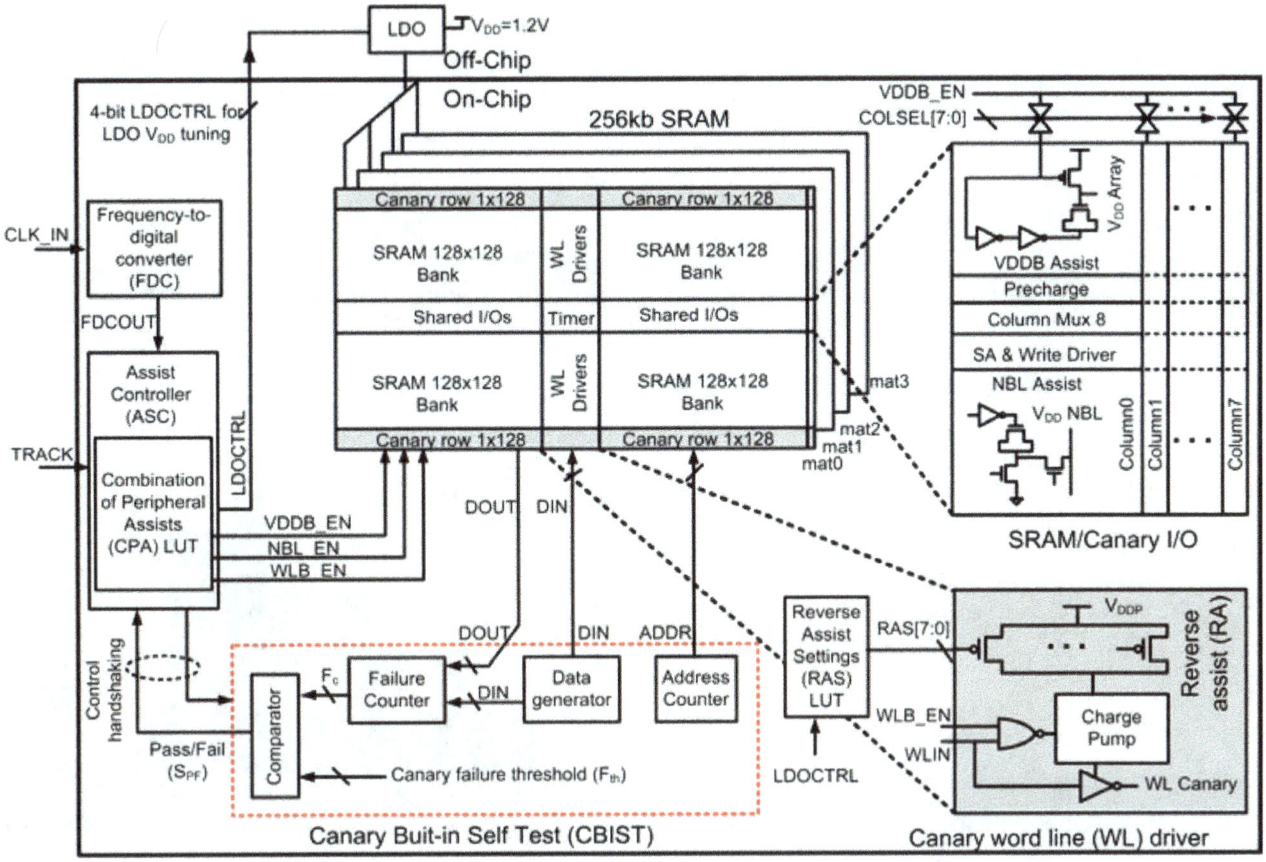

Figure 11. System-level block diagram for the 256 kb 6T self-tuning SRAM subsystem showing subcomponents [20].

and the CBIST iterates the canary write and read operations across all canary addresses to compare and determine if the canary failure (Fc) crosses the canary failure threshold condition (F_{th}). Based on the comparison, the CBIST generates a control signal for the ASC to increase or decrease the LDO supply voltage accordingly. Therefore, the closed loop tracking using self-tuning completes once the canary failure point is reached, which indicates the approaching SRAM V_{MIN}. Once the canary V_{MIN} is tuned to the SRAM V_{MIN} using F_{th} and RA settings, the worst-case SRAM bitcells are mapped into canaries, and the canary sensors track properties of the worst-case SRAM bitcells across a range of voltage, frequency, and temperature (VFT) variations. The authors show measured tracking of SRAM V_{MIN} across VFT variation as shown in **Figure 12** [20]. The canary sensors, system components without the BISTs, and CPA have reported overheads of 0.77, 1.8, and 2.8%, respectively. The work allows V_{DD} scaling using CPA at the 90th percentile worst-case V_{MIN} of 0.47 V with guardbands that reduces 337X active power. Moreover, enabling canary-based V_{MIN} tracking provides a 4.3X power savings by removing the V_{MIN} guardbanding to achieve up to 1444X active power savings at 0.38 V [20]. The authors show using CPA and in situ canary-based tracking down to 0.38 V gives a 12.4X leakage savings, too. The canary-based V_{MIN} tracking is scalable to lower technologies such as 45 and 32 nm, which shows promise to reduce the effect of process variation in FinFET SRAM in the highly-variant 7 nm and beyond technology nodes for a wide range of IoT applications.

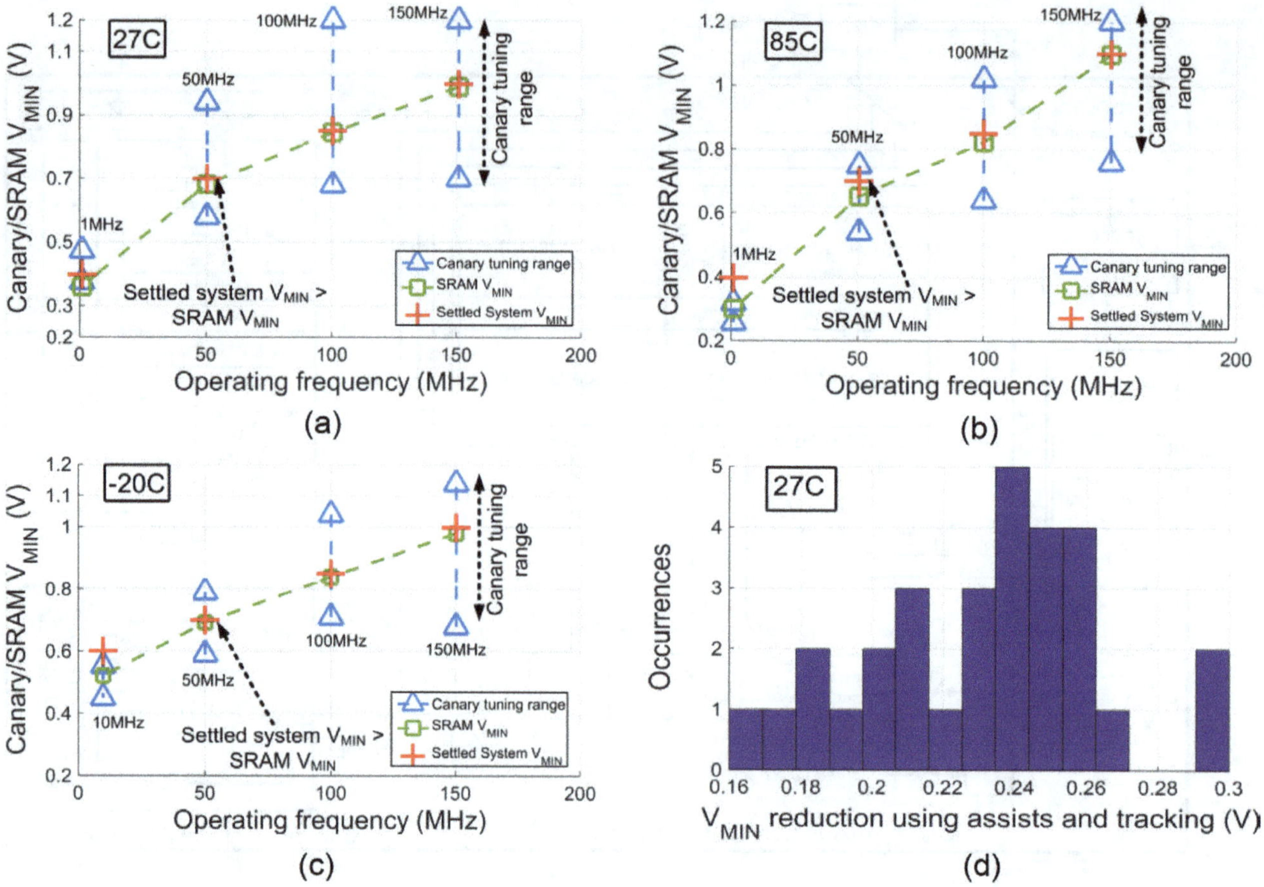

Figure 12. Measured canary V_{MIN} tracking across clock frequencies [1 or 10, 50, 100, and 150] MHz and temperatures (a) 27°C, (b) 85°C, and (c) −20°C, showing V_{MIN} tuning range, and (d) the distribution of overall V_{MIN} reduction using assist and tracking [20].

6. Discussion

Energy consumption in billion node IoT networks is expected to increase, as the total no. of IoT devices may reach 50 billion by the year 2020. A portion of these massive numbers of IoT devices will be plugged into the outlets in homes, factories, and outdoor settings. On the other hand, a huge number of IoT devices will be battery-operated or energy-harvested portable systems. The billions of IoT devices plugged into the outlets will draw power from the energy grid resulting in millions of dollars in energy bills and will increase the carbon footprint of this planet. Moreover, with a shorter battery life and replacement time, supporting billions of battery-operated IoT devices will require a massive production of portable batteries increasing the carbon footprint of Earth, too. Reducing the carbon footprint of these IoT devices requires reduction of power consumption, usage of low voltage operation for quadratic energy savings, and harvesting energy from the environment, which will require ULP IoT SoCs to reduce the energy cost and improve the battery life for a greener IoT electronics. However, technology scaling in the latest 7 nm FinFET and beyond will become a hindrance to lower operating voltage of the widely used embedded SRAMs, which shares the same power line with the digital core in ULP IoT SoCs. This chapter reviews some of the state-of-the-art SRAM design techniques, which are promising candidates for reducing power consumption in greener IoT applications, such as alternative bitcell topologies, a combination of peripheral assists, and in situ canary-based V_{MIN} tracking for guardband lowering.

7. Conclusions

Technology scaling in FinFET devices 7 nm node and beyond is going to experience a higher degree of process variation, which could affect the design and production of so-called lowest area 6T SRAM memory cells used in modern IoT system on chips. Based on the latest published works, there are three key directions to solve this issue. One of the directions is to use appropriate alternative bitcells for SRAMs trading off core array area that will enable ultra-low energy and lower leakage memory operation to sustain a longer battery life for portable home automation, wearable, and biomedical IoT applications. For low-cost system on chips using 6T SRAMs supporting low-power and mid- to high-speed applications, the use of appropriate combined peripheral assists is essential for a low-V_{MIN} application. Although the combined assist lowers the V_{MIN} and improves the SRAM yield, it does not eliminate the costly V_{MIN} guardbanding due to process and temperature variation. To remove or minimize this V_{MIN} guardbanding, the in situ canary sensor SRAM shows great promises for V_{MIN} tracking across voltage, frequency, and temperature variation. Combined peripheral assists along with canary sensor SRAM show promise for improvement in the power consumption of IoT systems by more than 1000X supporting a wide range of IoT application in a single SoC. Hence, to support a wide range of greener IoT applications, SRAM designers need to choose appropriate design techniques, such as alternative bitcells, combined peripheral assist, and in situ canary sensor SRAMs to enable technology scaling for SRAMs in 7 nm node and beyond.

Conflict of interest

The author has no conflict of interest.

Author details

Arijit Banerjee

Address all correspondence to: ab9ca@virginia.edu

University of Virginia, Charlottesville, Virginia, USA

References

[1] Evans D. The Internet of Things, How the Next Evolution of the Internet Is Changing Everything. CISCO, IBSG. April 2011. http://www.cisco.com/c/dam/en_us/about/ac79/docs/innov/IoT_IBSG_0411FINAL.pdf

[2] Vullers RJM, van Schaijk R, Doms I, Van Hoof C, Mertens R. Micropower energy harvesting, Solid-State Electronics, Volume 53, Issue 7, 2009, Pages 684-693. DOI: 10.1016/j.sse.2008.12.011 ISSN 0038-1101

[3] Rabey J. Low Power Design Essentials. USA: Springer; 2009. 6 p. DOI: 10.1007/978-0-387-71713-5

[4] Bryant A et al. Low-power CMOS at Vdd=4kT/q. Device Research Conference. 2001. pp. 22-23

[5] Hwang ME. Supply-voltage scaling close to the fundamental limit under process variations in nanometer technologies. IEEE Transactions on Electron Devices. Aug 2011;**58**(8):2808-2813

[6] Roy K, Kulkarni J, Hwang M. Process-tolerant ultralow voltage digital subthreshold design. In: IEEE Topical Meeting on Silicon Monolithic Integrated Circuits in RF Systems. Jan 2008. pp. 42-45

[7] Wang X, Cheng B, Brown AR, Millar C, Asenov A. Statistical variability in 14-nm node SOI FinFETs and its impact on corresponding 6T-SRAM cell design. In: 2012 Proceedings of the European Solid-State Device Research Conference (ESSDERC), Bordeaux. 2012. pp. 113-116

[8] Seevinck E, List FJ, Lohstroh J. Static-noise margin analysis of MOS SRAM cells. IEEE Journal of Solid-State Circuits. Oct 1987;**22**(5):748-754. DOI: 10.1109/JSSC.1987.1052809

[9] Nalam S, Chandra V, Aitken RC, Calhoun BH. Dynamic write limited minimum operating voltage for nanoscale SRAMs. In: 2011 Design, Automation & Test in Europe, Grenoble. 2011. pp. 1-6. DOI: 10.1109/DATE.2011.5763081

[10] Tsiligiannis G et al. SRAM soft error rate evaluation under atmospheric neutron radiation and PVT variations. In: 2013 IEEE 19th International On-Line Testing Symposium (IOLTS), Chania. 2013. pp. 145-150

[11] Verma N, Chandrakasan AP. A 256 kb 65 nm 8T subthreshold SRAM employing sense-amplifier redundancy. IEEE Journal of Solid-State Circuits. 2008;**43**:141-149

[12] Kulkarni JP, Kim K, Roy K. A 160 mV robust schmitt trigger based subthreshold SRAM. IEEE Journal of Solid-State Circuits. 2007;**42**:2303-2313

[13] Chiu Y-W, Lin J-Y, Tu M-H, Jou S-J, Chuang C-Y. 8T single-ended sub-threshold SRAM with cross-point data-aware write operation. In: 2011 Proceedings of the International Symposium on Low Power Electronics and Design (ISLPED); 1-3 August; Fukuoka, Japan. 2011. pp. 169-174

[14] Chang I-J, Kim JJ, Park SP, Roy K. A 32 kb 10T sub-threshold SRAM array with bit-interleaving and differential read scheme in 90 nm CMOS. IEEE Journal of Solid-State Circuits. 2009;**44**:650-658

[15] Feki A, Allard B, Turgis D, Lafont J, Ciampolini L. Proposal of a new ultra low leakage 10T sub threshold SRAM bitcell. In Proceedings of the International SoC Design Conference (ISOCC); 4-7 November 2012; Jeju Island, Korea. 2012. pp. 470-474

[16] Banerjee A, Calhoun BH. An ultra-low energy subthreshold SRAM bitcell for energy constrained biomedical applications. Journal of Low Power Electronics and Applications. 2014;**4**(2):119-137

[17] Mann RW, Nalam S, Wang J, Calhoun BH. Limits of bias based assist methods in nano-scale 6T SRAM. In: 2010 11th International Symposium on Quality Electronic Design (ISQED); San Jose, CA. 2010. pp. 1-8

[18] Karl E et al. 17.1 A 0.6V 1.5GHz 84Mb SRAM design in 14nm FinFET CMOS technology. In: 2015 IEEE International Solid-State Circuits Conference–(ISSCC) Digest of Technical Papers; San Francisco, CA. 2015. pp. 1-3

[19] Banerjee A, Kamineni S, Calhoun BH. Multiple Combined Write-Read Peripheral Assists in 6T FinFET SRAMs for Low-VMIN IoT and Cognitive Applications. In: 2018 ACM/IEEE International Symposium on Low Power Electronics and Design (ISLPED); Bellevue, WA; 23-25 July 2018:1-6

[20] Banerjee A, Liu N, Patel HN, Calhoun BH, Poulton J, Gray CT. A 256kb 6T self-tuning SRAM with extended 0.38V–1.2V operating range using multiple read/write assists and VMIN tracking canary sensors. In: 2017 IEEE Custom Integrated Circuits Conference (CICC); Austin, TX. 2017. pp. 1-4

[21] Calhoun BH, Chandrakasan AP. Standby power reduction using dynamic voltage scaling and canary flip-flop structures. IEEE Journal of Solid-State Circuits. Sept. 2004; **39**(9):1504-1511

[22] Wang J, Calhoun BH. Techniques to extend canary-based standby V_{DD} scaling for SRAMs to 45 nm and beyond. IEEE Journal of Solid-State Circuits. Nov. 2008; **43**(11):2514-2523

[23] Banerjee A, Sinangil ME, Poulton J, Gray CT, Calhoun BH. A reverse write assist circuit for SRAM dynamic write VMIN tracking using canary SRAMs. In: Fifteenth International Symposium on Quality Electronic Design; Santa Clara, CA. 2014. pp. 1-8

[24] Banerjee A, Breiholz J, Calhoun BH. A 130nm canary SRAM for SRAM dynamic write VMIN tracking across voltage, frequency, and temperature variations. In: 2015 IEEE Custom Integrated Circuits Conference (CICC); San Jose, CA. 2015. pp. 1-4

[25] Kincel A, Balaz M. MBIST for LEON3 processor core cache. In: 2013 IEEE 16th International Symposium on Design and Diagnostics of Electronic Circuits Systems (DDECS). 2013. pp. 287-288

[26] Kwong J, Chandrakasan AP. An energy-efficient biomedical signal processing platform. IEEE Journal of Solid-State Circuits. July 2011;**46**(7):1742-1753. DOI: 10.1109/JSSC.2011.2144450

[27] Roy A, Grossmann PJ, Vitale SA, Calhoun BH. A 1.3μW, 5pJ/cycle sub-threshold MSP430 processor in 90nm xLP FDSOI for energy-efficient IoT applications. In: 2016 17th International Symposium on Quality Electronic Design (ISQED); Santa Clara, CA. 2016. pp. 158-162

2

Biomolecules and Pure Carbon Aggregates: An Application Towards "Green Electronics"

Ruby Srivastava

Abstract

"Green electronics" is a novel scientific term which aims to identify the compounds of natural origin (economically safe and biodegradable) and establish economically efficient route for production of synthetic materials. The purpose of green electronics is to create path for the production of human and environmental friendly electronics and the integration of electronics with living tissue in particular. These researches may help to fulfill not only the organic electronics to deliver low cost energy efficient materials and devices, but also achieve unimaginable functionalities for electronics. In this chapter we have considered the molecular electronic devices biomolecules: deoxyribonucleic acid (DNA) and pure carbon aggregates: (carbon nanotubes (CNTs)/graphene), their properties and applications.

Keywords: biosensing, carbon nanotubes (CNTs), graphene, nucleobases, sensors

1. Introduction

Nanobiotechnology is gaining tremendous impetus in this era due to its ability to modulate metals into their nanosize and further interaction to the biological complexes, which efficiently changes their physicochemical and optical properties. Accordingly, considerable attention is being given to the development of novel strategies for the different nanoparticles of specific composition and size using biological sources. As the currently available techniques are expensive, environmentally harmful, and inefficient with respect to materials and energy use, so the emphasis is given to design the user friendly, non-toxic complexes, which can be used in biomedical and environmental applications. The major key prerequisite for achieving sustainability in the electronics industry is the usage of materials and technologies that have low embodied energy. Numerous efforts have been made throughout the world to develop environmentally benign technologies with nontoxic products using green nanotechnology and biotechnological tools. The synthesis of these nanoparticles using biological methods or green technology has diverse nature, greater stability and appropriate dimensions. So it is the current demand of the entire world to use specific techniques to characterize the potential of these hybrid complexes as an application towards drug delivery and biomedical fields. As we are focused towards Molecular electronics, so *bottom up* approach is included in this chapter, in which the individual molecular devices are built by synthesizing molecules with the desired electronic properties which are interconnected into an electronic circuits using attachment techniques like self-assembly. The main advantages of these complexes are their natural nanoscale structures, which can be created as absolutely identical in vast quantities and low fabrication cost [1–8] (**Figure 1**).

The reason for green electronics arises as the elements used previously in the field of green electronics such as Pb, Cd, Zn, Cu, Cr and As are potential bio-accumulative toxins in the production system of milk and dairy products [9]. Cadmium cause bone demineralization, either through direct bone damage or indirectly as a result of renal dysfunction [10].

In contrast to the above, the application of nanoscale materials for electrochemical biosensors has been grown exponentially due to high sensitivity and fast response time [11, 12]. So we can say that there evolved a parallel study as nano-bio-complexes in the field of green electronics, blending the biomolecules and nanotechnology. These nano-biosensor designs have created revolution in the field which can become a pioneering research work. Recently hierarchical cluster analysis (HCA) and principal component analysis (PCA) [13] were used to assay the receptor signaling mechanisms which can be used in biosensors or nano-biosensors. The findings above lead to amperometric biosensor [14] based on enzyme from *Brassica napus* hairy roots to determine ochratoxin, is a colorless crystalline compound that is classified as pentaketides [15]. All biological molecules and cell organelles are chemo mechanically controlled systems known to every biologist. It is an interdisciplinary art to activate them to work as an electronic device [16]. Above all, developing an Immunosensor depends on immobilization of

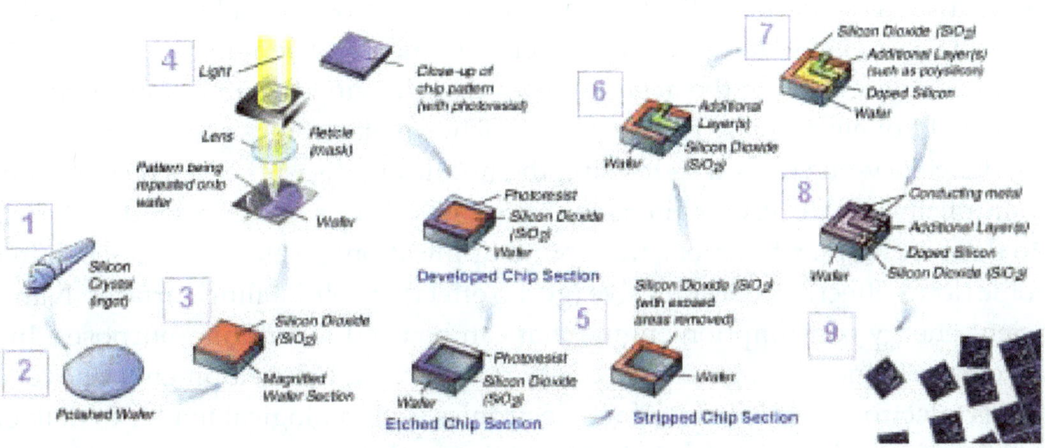

Figure 1. Schematic sequence of processes in the construction of a commercial chip (image from official INTEL website http://www.intel.com).

antibody molecules, which becomes an important factor for successful fabrication of immunosensors [17], which involves screen printing technology. Photoacoustic imaging technique has been developed which works on ultrasound spatial resolution and intrinsic rich optical contrast, penetrates deep into the tissues [18, 19] with acoustic ultra sounds and act as a detection tool in diagnostic medicine [20]. Thomas et al. [21], demonstrated all aspects of overhand throwing, using a 12 camera Vicon motion analysis system in a general motor program using bio-electronic signals. The synthesis from bioprocess or the material used as biomaterial, or the organism used for "GE" i.e. "Genetics Engineering," or the sensor and the receptor in the field, there is a relation between each research studies; that is nothing but the *field of biotechnology*.

Molecular electronics (ME) deals intensively with a long term alternative for increasing the device density in an IC and continuing Moore's law down to the nanometer scale. The basic idea of ME is to use individual molecules to act as wires, switches and memories. Till date, no detailed investigations have been carried out for various complexes using green technology, their utilization, or their analyses have been published. Therefore, this chapter is conducted to highlight the use of various nano-bio complexes, their use in green technology and different techniques for characterization of nanoparticles to provide a better understanding of the these sources to improve their uses in modern technology.

The ME can be further divided to (a) small covalently bonded organic molecules ((aromatic chains, conjugated polymers); b) large biomolecules (DNA, nucleoside-based aggregates, proteins); c) pure carbon aggregates (carbon nanotubes, graphenes).

The organic electronics based on conjugated polymers or small molecules as the core semiconductor element hold the high promise of delivering low-cost and energy-efficient materials and devices, yet the performance and stability of organic semiconductors remain major hurdles in their development as solid competitors of the inorganic counterparts. So the "soft" nature of carbon-based materials are considered enabling fabrication of extremely flexible, highly conformable and even imperceptibly thin electronic devices. In recent years, the graphene based nanomaterials have received considerable attention owing to their distinguished electronic and transport properties and act as promising candidates for electronics and spintronics. So green materials can act as emerging concept with carbon based class and integration of electronics into living tissue with the aim of achieving biochemical monitoring, diagnostic drug delivery tasks or generating human and environmentally benign technologies. Now the green technologies are carving the avenues towards achieving the ambitious goal of sustainability in the field of electronics. The quest is to achieve electronics sustainability by solving the energy deficiency puzzle and redressing the unfolding disaster, for which we look to the apparent simplicity of nature. Our aim is to create a novel class of engineered materials which are able to deliver complex functions that found applications in electronics; designing superhydrophobic (lotus effect), super-adhesive (gecko effect) or self-healing surfaces. Nature is the most efficient energy consumption engine that can be used for infinite purposes. In the last 10 years, we have witnessed a great deal of effort towards the development of novel conductive materials (electrodes) able to interface electronics with biological matter to deliver recognize events (i.e. biosensing, bio-recognition) and the modulate events (i.e. tissue engineering). Here we have selected the larger biomolecules and pure carbon aggregates and their role in the field of green electronics.

2. DNA biomolecular electronics

The role of ME is to provide reproducible well-structured architectures, easy to wire in a programmed manner. Supramolecular chemistry seems to fulfill these needs [22]. The two properties which are attractive for this purpose is the (a) molecular recognition and (b) self-assembly. Molecular recognition is the capability of a molecule to form selective bonds with other molecules or with substrates, which rest on the information stored in the structural features of the interacting partners. Molecular recognition processes (a) the building up of the devices from their components (b) incorporate them into supramolecular arrays; (c) allow selective operations on given species (e.g. ions, dopants), and (d) control the response to external perturbations (e.g. external fields, light, electrons, other molecules, etc.).

The self-assembly has the capability of molecules to spontaneously organize in supramolecular aggregates under well-defined experimental conditions. Self-organization may occur both in solution and in the solid state, and make use of hydrogen bonding, electrostatic donor-acceptor effects (Van der Waals, dipolar, etc.) or metal-ion coordination as basic interactions between the components. Due to these two properties, DNA molecules seem particularly suitable to be used as components for the construction of nanometer scale devices [23–26]. The idea of using DNA in molecular devices is its natural function of storing and coding the genetic information. DNA transmits well-defined chemical information through the pairing properties of the bases. In addition, it occurs in a large variety of structures and display physiochemical stability and mechanical rigidity.

2.1. Electron transfer through DNA

The deoxyribonucleic acid (DNA) is a biopolymer in a double helical form, which is constituted by an extended array of aromatic π-stacked base pairs adenine-thymine (AT) and guanine-cytosine (GC) within a polyanionic sugar-phosphate backbone (**Figure 2**). Due to the biological implications, the studies about the charge migration in DNA were related to physiological processes: the possibility and efficiency of charge transfer is significant, because the migration of the radical cation is a critical issue to understand problems related to radiation damage and mutation [27, 28].

The role of π-π interactions between stacked base pairs in double-stranded DNA could provide a pathway for rapid one-dimensional charge separation. Various experiments were performed

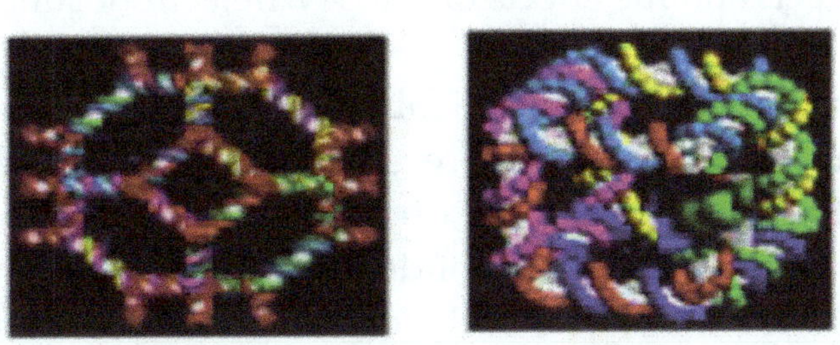

(a) **(b)**

Figure 2. Representations of left (a) truncated octahedron containing six squares and eight hexagons. (b) Each edge of the truncated octahedron contains two double helical turns of DNA a DNA cube containing six different cyclic strands. Their backbones are shown with different colors.

to understand whether DNA facilitates the charge transfer over long distances and whether the base pair stack can act as a conducting medium. The issue of charge migration in DNA has recently become a hot topic with solution chemistry (in particular) after the first reports by Jacqueline Barton's group [29–33] in the early '90s. Although the answer to the question: Is DNA a molecular wire? is still elusive.

The recent achievement is the construction of few DNA-hybrid devices, which requires the application of some state-of-the-art nanotechnologies are: electron beam lithography for the fabrication of metallic nanocontacts, trapping techniques to compel the molecules into the desired device scheme, Atomic Force Microscopy (AFM) or Scanning Tunneling Microscopy (STM) for imaging and probing samples.

The achievement of DNA-based devices requires [33]:

- Construction of nucleic-acid networks.
- Conversion of nucleic-acid (or DNA-protein) network into electron conducting system.

The DNA-based materials may be used either as conductive wires or as a template for other conductive materials. By exploiting the molecular recognition of its functional groups, it is possible to synthetize branched DNA-motifs that may be assembled into periodic arrays. Though a lot of research conducting in this field is giving the contrasting results.

The most important ones are:

1. The intrinsic properties of the different DNA molecules employed in the experiments, the length of the DNA (from a few nanometers to some microns) and the structural conformation of the double helix.
2. The properties of the buffer solution in which the DNA is kept and the presence and the concentration of counter ions;
3. The experimental conditions in which measurements are realized: in air, in vacuum, different humidity, and different temperatures (from 1 to 300 K);
4. The structural aggregation forms of DNA (films, network bundles, single molecules);
5. The presence of contacts and the effects of the DNA/electrode junction.

Apart from experimental difficulties in the fabrication of DNA-based device, several fundamental questions are still open: what are the interactions which control the electrical properties of DNA? How do they depend upon the sequence? What are the mechanisms for charge transport? What are the effects of dopants or defects? How does DNA attach to a metal electrode? What are the effects of the contacts on the conduction properties of the device?

2.2. Application of DNA in green electronics

DNA can be used in various applications in green electronics, which is discussed herein.

2.2.1. *G4-wires DNA as nanowire*

Nanowires known as G4-wires [34] (or quadruplexes), consist of stacked guanine (G) tetrads (G4). These one-dimensional polymers act as prospective candidates for bio-molecular electronics because, due to the low ionization potential of guanine (the lowest among nucleic-acid bases), they might be suitable to mediate charge transport by hole conduction along the helix, and have even been suggested as nano-mechanical extension-contraction machines [35]. In the presence of appropriate metal cations (especially K^+ and Na^+), solutions of homoguanylic strands in water [36, 37] as well as lipophilic guanosine monomers in organic solvents [38], self-assemble in right-handed quadruple helices. Recent investigations has been carried out for the G4-nanowires, but the conduction properties of these nanowires are basically unknown and a direct measurement of electrical properties of G4-wires is still missing.

2.2.2. *DNA as sensors and imaging agents for metal ions*

Sensing and imaging of metal ions have attracted much attention by scientists and engineers because of the important roles of metals in many fields such as environmental, biological, and medical sciences. Significant progress has been made in developing sensors and imaging agents for the detection of metal ions, mostly based on organic molecules, peptides, proteins, or cells [39–45]. Prof. Yi Li has given a significant insight to the role of DNA as sensors and imaging agents for metal Ions [46], the structure of these systems is illustrated and described in **Figure 3**. DNA does not appear to be a good candidate for sensing metal ions with high selectivity because the negatively charged phosphodiester backbones of DNA are known to be capable of binding cationic metal ions with poor selectivity for any particular metal ion. While the four DNA bases can also serve as ligands for metal ions [47–49], many of these DNA–metal ion interactions are nonspecific and weak, making the use of DNA as sensors for metal ions very challenging because selectivity and sensitivity are required for the successful detection of a specific metal ion in the presence of other potentially interfering metals in

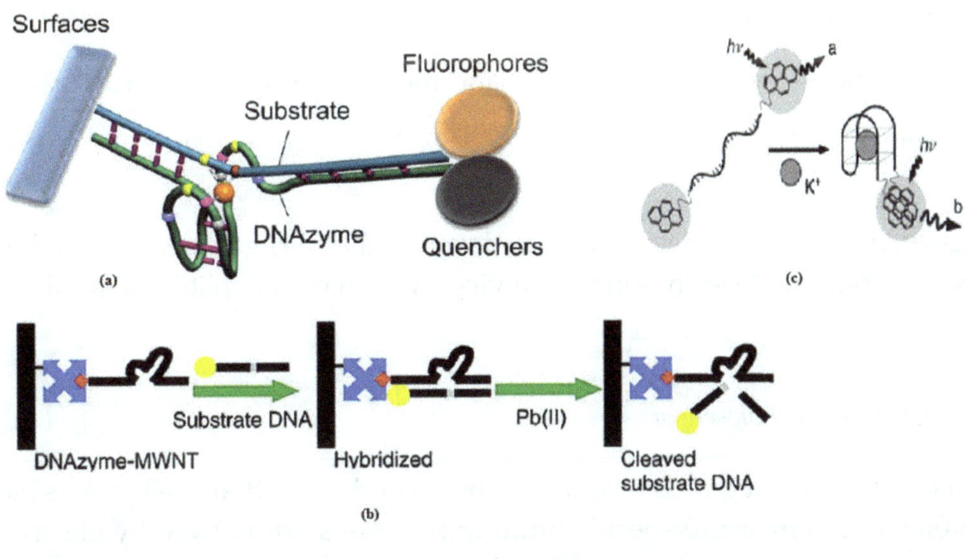

Figure 3. a. General sensor design based on nucleic acid cleavage of DNAzymes for metal-ion detection. b. Fluorescent Ag^+ sensor based on C–Ag^+–C. c. Sensors based on G-quadruplex DNA stabilized by K^+. Adapted with permission from Ref. [46].

complex samples. By incorporation of signal reporters such as chromophores, fluorophores, electrochemical tags, and Raman tags, these metal-ion-specific DNA sequences have been transformed into colorimetric, fluorescent, electrochemical, and Raman sensors and imaging agents for a broad range of metal ions with high sensitivity and selectivity [50–55]. DNAzymes that is highly selective to use specific metal ions as cofactors to catalyze reactions can be obtained. In this way, DNAzymes that are dependent on bivalent metals for various chemical and biological reactions have been successfully discovered. One report of DNAzyme sensor was a fluorescent sensor for Pb^{2+} based on DNAzyme [56–58], which showed much higher specificity to Pb^{2+} over other metal ions in catalyzing the cleavage of DNA substrates with a single RNA linkage (rA) at the cleavage site.

These sensors can be further classified into different parts:

1. Fluorescent sensors based on metal ion-dependent DNAzymes
2. Fluorescent sensors labeled with fluorophores and quenchers [59]
3. Surface-immobilized fluorescent sensors [59]
4. Label-free fluorescent sensors [60–62]
5. Colorimetric sensors based on metal ion-dependent DNAzymes
6. Colorimetric sensors based on gold nanoparticles [63, 64]
7. Colorimetric "dipstick" tests using lateral-flow devices [65, 66]
8. Electrochemical and Raman sensors based on metal ion-dependent DNAzymes [67]
9. Sensors based on metal binding structures [68, 69]
10. Hg^{2+} sensors based on T–Hg^{2+}–T-containing DNA
11. Ag^{+} sensors based on C–Ag^{+}–C-containing DNA
12. Sensors for K^{+}, Pb^{2+}, Cu^{2+}, and Ag^{+} based on G-quadruplex DNA
13. Combination based sensors (DNAzymes and metal-binding DNA structure) [70]
14. Portable sensors [71]

In this category, new technologies have been developed to design sensors based on commercialized devices, compatible with portable devices, which could enable to monitor metal ions by all.

2.2.3. DNA-electrochemical biosensors

DNA biosensors are the integrated receptor-transducer devices that use DNA as biomolecular recognition element to measure specific binding processes with DNA, by electrical, thermal or optical signal transduction methods. The characteristics of DNA probes with the capacity of direct and label-free electrochemical detection find applications in rapid monitoring of pollutant agents or metals in the environment, investigation and evaluation of DNA-drug interaction mechanisms, detection of DNA base damage in clinical diagnosis, or detection of specific DNA sequences in human, viral and bacterial nucleic acids [72–75].

2.2.4. *DNA oxidative biomarker*

Oxidative DNA damage caused by oxygen-free radicals lead to multiple modifications in DNA, including base-free sites and oxidized bases. The damage caused to DNA bases is potentially mutagenic [76] and can be enzymatically repaired. The interest lies in the sensitive determination and full characterization of the mechanism involved in oxidative damage to DNA bases. Electrochemical methods are used to study the DNA oxidative damage and in the investigation of the mechanisms of DNA-drug interactions. In recent study, it has been anticipated that the mispairs-coinage metal complexes can also be used as a biomarker [77].

2.2.5. *Electrochemical biosensors for detection of DNA damage*

The DNA-electrochemical biosensor enables pre-concentration of the hazard compounds which are investigated onto the sensor surface and *in situ* electrochemical generation of radical intermediates, which cause damage to the DNA immobilized on the electrode surface and can be electrochemically detected.

2.2.6. *DNA as radiation sensors*

DNA has an optical response towards temperature, magnetic field, radiation and others. The flexibility of DNA can be modified by the radiations. When irradiated using gamma rays and neutrons (non-ionizing radiation), the dynamics of DNA macromolecules [78] changes its configuration when involved in environmental interactions with other components of the living cells [79]. Whenever any radiation passes through a semiconductor device, different effects are observed which depends on the range of energy of the particle (proton, alpha, neutron and both types of beta) and rays, such as gamma radiation [80]. These include defects as: vacancies, defect clusters, dislocation loops near the surface and adjustment of band gaps [81]. Electrical properties of DNA molecules can be understood by the electrical conduction mechanism, namely: thermionic emission, tunneling and hopping [82]. These all properties can be applied for DNA as radiation sensors.

3. Pure carbon aggregates

Carbon Nanotubes (CNTs) has interesting physicochemical properties as electrical conductance, high mechanical stiffness, light weight, transistor behavior, piezoresistance, thermal conductivity, luminescence, electrochemical bond expansion as well as their versatile chemistry make them superb materials for a broad spectrum of applications ranging from energy storage devices, nanosensors and drug/gene delivery vehicles.

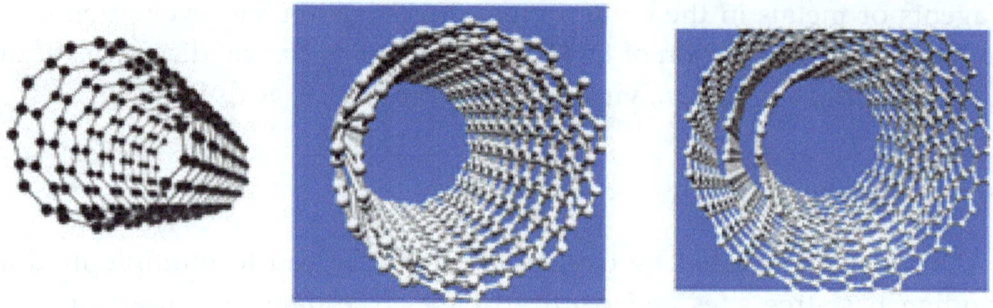

(a) (b) (c)

Figure 4. Representation of SWCNT and MWCNTs (armchair, zigzag and chiral).

3.1. Carbon nanotubes (CNTs)

Carbon nanotubes (CNTs) are allotropes of carbon with a cylindrical nanostructure, a member of the fullerene structural family. Due to their extraordinary thermal conductivity, mechanical, and electrical properties, carbon nanotubes find applications as additives to various structural materials. The various representations of CNTs are given in **Figure 4**.

3.1.1. Applications of CNTs in green electronics

Single-walled carbon nanotubes (SWCNTs) also have unique properties which make them suitable for applications in a variety of imaging modalities, such as magnetic resonance, near-infrared fluorescence, Raman spectroscopy, photoacoustic tomography, and radionuclide-based imaging.

The various features of carbon nanotubes are:

1. Drug delivery and cancer treatment [83] (**Figure 5**)
2. Physical cancer therapies delivered by CNTs [84]
3. Biosensing
4. Optical [85]
5. Electronic and electrochemical sensors [86]
6. Biomedical imaging
7. Molecular imaging [87, 88]
8. MRI with SWCNTs [89, 90]
9. Optical imaging with SWCNTs [91, 92]
10. Raman spectroscopy with SWCNTs [93, 94]
11. Photoacoustic tomography [95, 96]

12. Radionuclide-based imaging with SWCNTs [97, 98]
13. Scaffolds in tissue engineering
14. CNTs used as scaffolds in bone regeneration [99]
15. CNTs for neural applications

3.2. Graphene

The two-dimensional carbon material graphene find application for sensors, electronics and catalysis applications due to its exceptional electrical and mechanical properties. Some of these applications require the adsorption of metal clusters onto graphene and metal-graphene systems which are now become a subject of intense investigation. It show many interesting properties as the observable quantum Hall effect at room temperature [100, 101], existence of two-dimensional gas of massless Dirac fermions [102], ballistic transport properties on the sub micrometer scale [103], etc. As graphene is unique in nature, so researchers explore its unique properties in storage [104], spintronics [105], microelectronics [106], etc. A number of theoretical and experimental work have been carried out for the electronic and magnetic behaviors of dimers [107] and adatoms of different elements [108] adsorbed on graphene system, which have been found to yield many interesting results. Graphene also has immense potential to act as a key ingredient for new devices as single molecule gas sensors, ballistic transistors, and spintronic devices. Bilayer graphene, which consists of two stacked monolayers, has a quadratic low-energy band structure which generates very different scattering properties from those of the monolayer. It also presents the unique

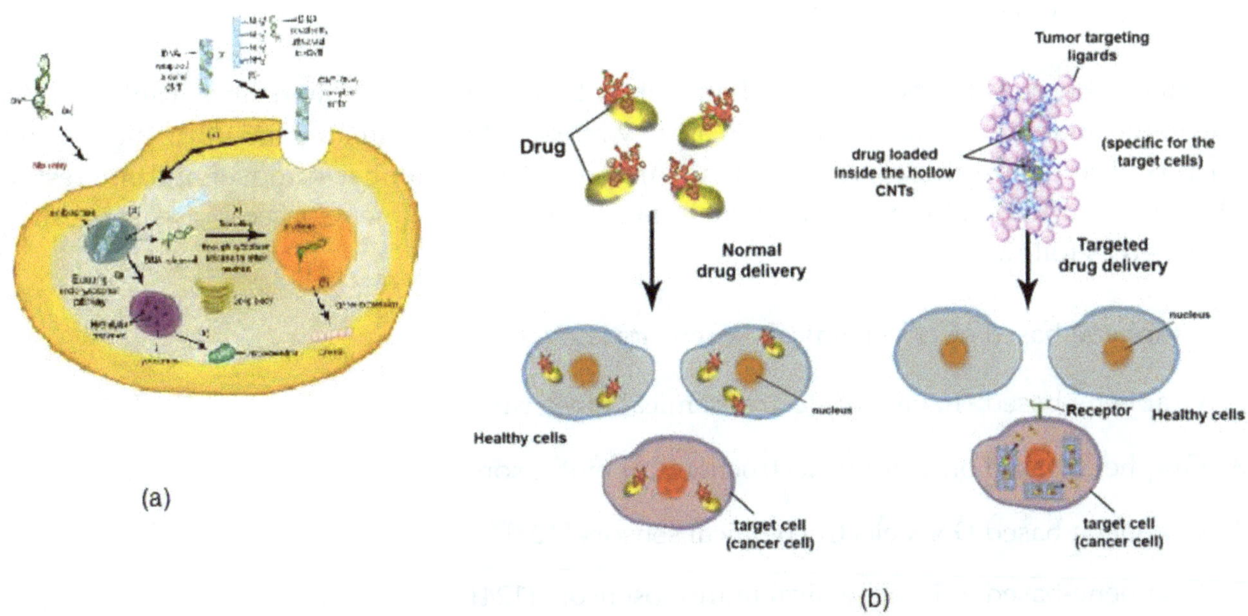

Figure 5. Representation of the role of CNTs in drug delivery. (a) Gene delivery by CNTs, (b) Normal drug delivery versus the efficient target drug delivery using CNTs.

property i.e. *the tunable band gap can be opened and controlled easily by a top gate*. Another property of graphene is the high electronic mobility, which is crucial for many of its potential applications [109]. So, understanding the mechanism, which limit the mobility of carriers in graphene is extremely important. It is also of a high conceptual interest, since transport properties of chiral massless fermions are essentially different from those of conventional charge carriers in metals and semiconductors [110]. Charged impurity scattering has received the most attention [102], with the majority of studies modeling the impurities as point-like objects (1/r potential). Recently, it has been revealed in a theoretical study that the physical structure of the charged impurities and clusterization of charged impurities might be the an important factor, which influence their scattering properties [111]. Graphene nanoribbons (GNRs) present reactive edges which make GNRs not only more accessible to doping and chemical modification, but also more susceptible to structural defects [112]. In particular, for zigzag graphene nanoribbons (ZGNRs) terminated with one hydrogen atom on each zigzag edge, there are quite a few localized edge states near the Fermi energy level on both edges. Such localized edge states can lead to a spin induced energy gap [113] providing a significant effect on the electronic and transport properties [114]. The electronic and transport properties are thus very sensitive to the atomic structures and chemical modification of the edges. One of the natural ways of chemically modifying graphene is to include metal adatoms [115–117]. Motivated by the special transport properties of atomic wires of Al, Ag, Au and Cu, and inspired by the localized edge states of ZGNRs that greatly enhance the binding energy of adatoms, we also conducted the studies on (Al, Au and Cu) adatomed-objects on edge chlorinated nano graphenes to investigate the electronic, magnetic and adsorption and transport properties of these (C1 ($C_{42}CI_{18}$)/C2($C_{48}CI_{18}$)/C3($C_{60}CI_{22}$))–metal systems (**Figure 6**). Metals adsorbed on nanoscale carbon surfaces have been reported experimentally and theoretically to form a variety of structures, such as continuous coatings or discrete clusters [118] and novel interesting phenomena were observed to occur through suitable modification.

Metallic nanowires (namely, linear chains of metal atoms) have drawn significant attention due to the quantum confinement effect, as they represent the ultimate miniaturization of conductors. The nanoscales of metallic nanowires result in a number of novel and interesting phenomena different from their bulk materials [119–121]. Graphene based biosensors can be classified as follows:

1. Graphene-based electrochemical biosensors [122]
2. Graphene-based enzymatic electrochemical biosensors
3. Graphene-based bioaffinity electrochemical biosensors
4. Graphene-based DNA electrochemical sensors [123]
5. Graphene-based electrochemical immunosensors [124]
6. Graphene-based field-effect transistor (FET) biosensors [125]
7. Graphene-based optical biosensors [126]

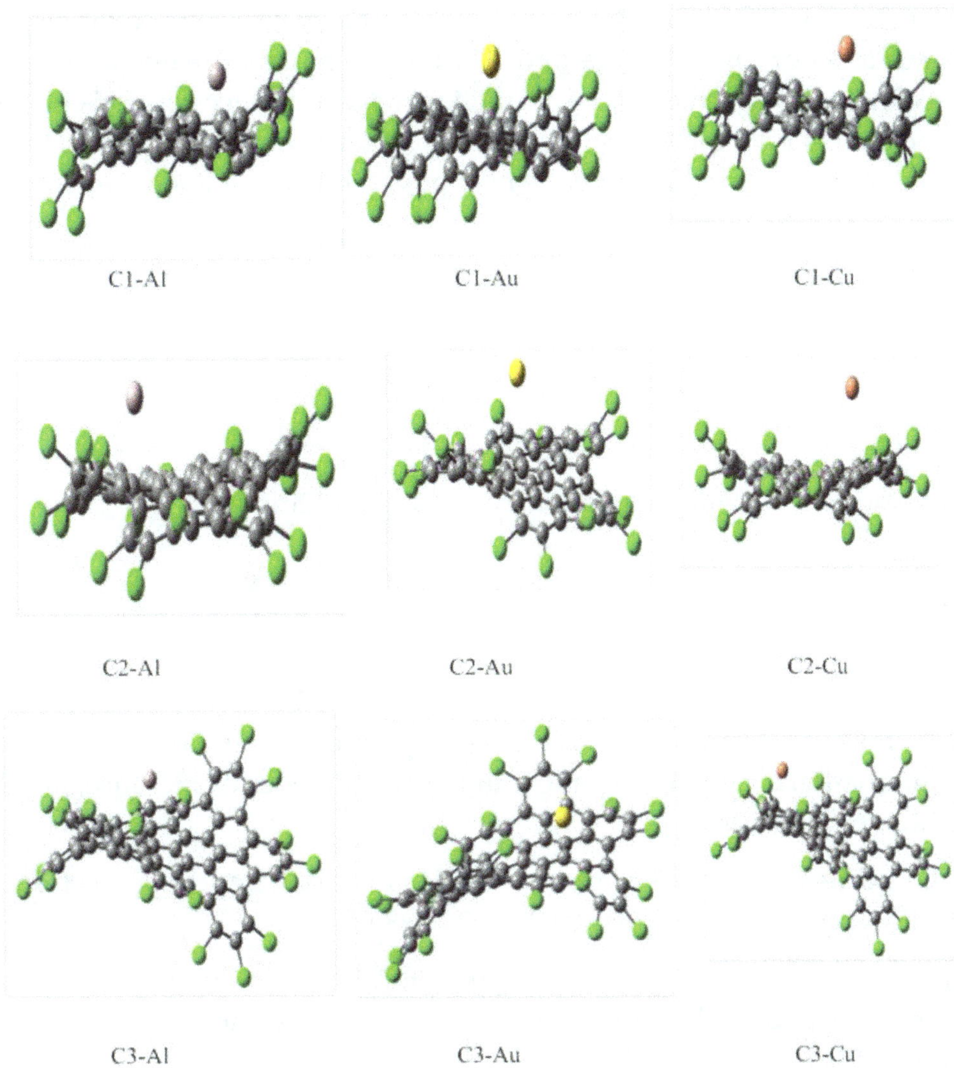

Figure 6. Potential adsorption sites and the most stable structures for Al, Au and Cu (dimers) on C1 ($C_{42}CI_{18}$), C2 ($C_{48}CI_{18}$) and C3 ($C_{60}CI_{22}$). Adapted from Ref. [117].

4. Conclusions

In this chapter, we have discussed the role of DNA and CNTs in the field of green electronics. Metal nanoparticles and its interaction to the nucleobases and pure carbon aggregates are also described in detail. Current and future investigations of green nanotechnology will provide a more complete knowledge regarding various factors that influence green synthesis of nanoparticles and the most sophisticated technology that can be used for characterization of the synthesized nanoparticles for its more efficient future applications in environmental, optoelectronic and biomedical field.

"United Nations World Commission on Environment and Development" has stated that the sustainable development is established when humanity ensures its present needs without compromising the ability of future generations to meet their own needs. So the time has come when we have to bear the responsibility for the shape and type of environment our future generations will live in; a healthy environment, a non-toxic world. Right now the world is not what we have set up for the mankind. Since electronics has now become an indispensable part of our life- for us and our future generations. The natural and nature inspired materials allow the "green" technologies to achieve the substantial goals in the electronics field: they embody low energy and have biodegradable and biocompatible materials as their backbone. Now the time has started to imagine and explore the new possibilities of "Green" organic electronics.

Author details

Ruby Srivastava

Address all correspondence to: amitruby1@gmail.com

CSIR-Indian Institute of Chemical Technology, Hyderabad, India

References

[1] Ravariu C, Ionescu-Tirgoviste C, Ravariu F. Glucose biofuels properties in the bloodstream in conjunction with the beta cell electro-physiology. In: Proceedings of 2-nd IEEE - ICCEP International Conference on Clean Electrical Power Conference; Jun. 09-11, 2009; Capri, Italy. pp. 124-127

[2] Moore GE. Cramming more components onto integrated circuits. Electronics. 1965;**38**: 114-117. Publisher Item Identifier S 0018-9219(98)00753-1

[3] Naskar A, Khan H, Jana S. Cobalt doped ZnO–graphene nanocomposite: Synthesis, characterization and antibacterial activity on water borne bacteria. Advanced Nano-Bio-Materials and Devices. 2017;**1**(4):182-190

[4] Joachim C, Gimzewski JK, Aviram A. Electronics using hybrid-molecular and mono-molecular devices. Nature. 2008;**408**:541-548. DOI: 10.1038/35046000

[5] Hadadi A, Bolverdi AR, Jamshidi M, Rezaei O. Ab initio study of Heme complex as a nano-biosensor for CO and O_2 gases. Advanced Nano-Bio-Materials and Devices. 2017;**1**(1): 90-99

[6] Srivastava R. The optical phenomena of interplay between nanobio complexes: A theoretical insight into their biomedical applications. In: Pyshkin SL, Ballato J, editors. Optoelectronic Devices. Chapter 12. "Optoelectronics - Advanced Device Structures". ISBN: 978-953-51-3370-4. DOI: 10.5772/67527

[7] Ratner MA. Introducing molecular electronics. Materials Today. 2002;**5**(2):20-22. DOI: 10.1016/S1369-7021(02)05226-4

[8] Joachim C, Ratner MA. Molecular electronics: Some views on transport junctions and beyond. PNAS. 2005;**102**(25):8801-8808. DOI: 10.1073/pnas.0500075102

[9] Li Y, McCrery DF, Powell JM, Saam H, Jackson-Smith D. A survey of selected heavy metal concentration in Wisconsin dairy feeds. Journal of Dairy Science. 2005;**88**:2911-2922. DOI: 10.3168/jds.S0022-0302(05)72972-6

[10] European Food Safety Authority. Cadmium in food: Scientific opinion of the Panel on Contaminants in the Food Chain. The EFSA Journal. 2009;**980**:1-139

[11] Musameh M, Wang J, Merkoci A, Lin Y. Low-potential stable NADH detection at carbon-nanotube-modified glassy carbon electrodes. Electrochemistry Communications. 2002;**4**:743-746. DOI: 10.1016/S1388-2481(02)00451-4

[12] Dequaire M, Degrand C, Limoges B. An electrochemical metallo immunoassay based on a colloïdal gold label. Analytical Chemistry. 2000;**72**:5521-5528. PMID: 11101226

[13] Tateishi A, Cauchi M, Tanoue C, Migita S, Coleman SK, et al. Discerning data analysis methods to clarify agonistic/antagonistic actions on the ion flux assay of ligand-gated ionotropic glutamate receptor on engineered post-synapse model cells. Journal of Biosensors and Bioelectronics. 2011;**2**(104):1-5. DOI: 10.4172/2155-6210.1000104

[14] Ramírez EA, Granero AM, Zón MA, Fernández H. Development of an amperometric biosensor based on peroxidases from *Brassica napus* for the determination of ochratoxin a content in peanut samples. Journal of Biosensors and Bioelectronics. 2011;**S3**:001-006. DOI: 10.4172/2155-6210.S3-001

[15] Steyn PS. Ochratoxins and related dihydro-isocoumarins. In: Bettina V, editor. Mycotoxins: Production, Isolation, Separation and Purification. New York: Elsevier; 1984. p. 183. DOI: 10.4172/2155-6210

[16] Henry R, Durai, Net S, Balraj A, Priya WS. Modeling a micro tubule as a diode. Journal of Biosensors and Bioelectronics. 2011;**2**:106-110. DOI: 10.4172/2155-6210.1000106

[17] Lien TTN, Viet NX, Chikae M, Ukita Y, Takamura Y. Development of label-free impedimetric Hcg-immunosensor using screen-printed electrode. Journal of Biosensors and Bioelectronics. 2011;**2**:107-113. DOI: 10.4172/2155-6210.1000107

[18] Brecht H, Su R, Fronheiser M, Ermilov SA, Conjusteau A, et al. Whole body three-dimensional optoacoustic tomography system for small animals. Journal of Biomedical Optics. 2009;**14**:064007-064015. DOI: 10.1117/1.3259361

[19] Sun Y, Sobe ES, Jiang H. First assessment of three-dimensional quantitative photoacoustic tomography for in vivo detection of osteoarthritis in the finger joints. Medical Physics. 2011;**38**:4009-4017. DOI: 10.1118/1.3598113

[20] Sun Y, Jiang H, O'Neill BE. Photoacoustic imaging: An emerging optical modality in diagnostic and theranostic medicine. Journal of Biosensors and Bioelectronics. 2011;**2**:108. DOI: 10.4172/2155-6210.1000108

[21] Thomas JR, Alderson JA, Thomas KT, Campbell AC, Edwards WB, et al. Is there a general motor program for right versus left hand throwing in children? Journal of Biosensors and Bioelectronics. 2011;**S1**:001. DOI: 10.4172/2155-6210.S1-001

[22] Lehn JM. Perspectives in supramolecular chemistry—From molecular recognition towards molecular information processing and self-organization. Angewandte Chemie International Edition in English. 1990;**29**:1304-1319. DOI: 10.1002/anie.199013041

[23] Di Mauro E, Hollenberg CP. DNA technology in chip construction. Advanced Materials. 1993;**5**:384-386. DOI: 10.1002/adma.19930050512

[24] Korri-Youssou H, Garnier F, Srivastava P, Godillot P, Yassar A. Toward bioelectronics: Specific DNA recognition based on an oligonucleotide-functionalized polypyrrole. Journal of the American Chemical Society. 1997;**119**:7388-7389. DOI: 10.1021/ja964261d

[25] Herrlein MK, Nelson JS, Letsinger RL. A covalent lock for self-assembled oligonucleotide conjugates. Journal of the American Chemical Society. 1995;**117**:10151-10152. DOI: 10.1021/ja00145a042

[26] Niemeyer CM. Nanoparticles, proteins, and nucleic acids: Biotechnology meets materials science. Angewandte Chemie International Edition in English. 2001;**40**:4128-4158. DOI: 10.1039/c2cs35265f

[27] Armitage B. Photocleavage of nucleic acids. Chemical Reviews. 1998;**98**:1171-1200. DOI: 10.1021/cr960428

[28] Boom EM, Ceres DM, Drummond TG, Hill MG, Barton JK. Mutation detection by electrocatalysis at DNA-modified electrodes. Nature Biotechnology. 2000;**18**:1096-1100. DOI: 10.1038/80301

[29] Murphy CJ, Arkin MR, Jenkins Y, Ghatlia ND, Bossmann SH, Turro NJ, Barton JK. Long range photoinduced electron transfer through a DNA helix. Science. 1993;**262**:1025-1030. PMID: 8202486

[30] Murphy CJ, Arkin MR, Ghatlia ND, Bossmann SH, Turro NJ, Barton JK. Fast photoinduced electron-transfer through DNA intercalation. Proceedings of the National Academy of Sciences of the United States of America. 1994;**91**:5315-5319. PMID: 19159277

[31] Hall DB, Holmlin RE, Barton JK. Oxidative DNA damage through long-range electron transfer. Nature. 1996;**382**:731-735. DOI: 10.1038/382731a0

[32] Erik Holmlin R, Dandliker PJ, Barton JK. Charge transfer through the DNA base stack. Angewandte Chemie International Edition in English. 1997;**36**(24):2714-2730. DOI: 10.1002/anie.199727141

[33] Kelley SO, Barton JK. Electron transfer between bases in double helical DNA. Science. 1999;**283**:375-381. PMID: 9888851

[34] Alberti P, Mergny JL. DNA duplex–quadruplex exchange as the basis for a nanomolecular machine. Proceedings of the National Academy of Sciences of the United States of America. 2003;**100**:1569-1573. DOI: 10.1073/pnas.0335459100

[35] Li JJ, Tan W. A single DNA molecule nanomotor. Nano Letters. 2002;**2**:315-318. DOI: 10.1021/nl015713

[36] Saenger W. Principles of Nucleic Acids. NY: Springer-Verlag; 1984. DM 79. ISBN: 3-540-90761-0

[37] Gottarelli G, Spada GP, Garbesi A. Comprehensive supramolecular chemistry. In: Atwood JL, Davies JED, MacNicol DD, Vögtle F, Lehn JM, Sauvage JP, Wais Hosseini M editors. Pergamon: Oxford; 1996. pp. 483-506

[38] Kotch FW, Fettinger JC, Davis JT. A lead-filled G-quadruplex: Insight into the G-quartet's selectivity for Pb^{2+} over K^+. Organic Letters. 2000;**2**:3277-3280. DOI: 10.1021/ol0065120

[39] Domaille DW, Que EL, Chang CJ. Synthetic fluorescent sensors for studying the cell biology of metals. Nature Chemical Biology. 2008;**4**:168-175. DOI: 10.1038/nchembio.69

[40] Nolan EM, Lippard S. Tools and tactics for the optical detection of mercuric ion. Chemical Reviews. 2008;**108**:3443-3480. DOI: 10.1021/cr068000q

[41] Que EL, Domaille DW, Chang CJ. Metals in neurobiology: Probing their chemistry and biology with molecular imaging. Chemical Reviews. 2008;**108**:1517-1549. DOI: 10.1021/cr078203u

[42] Chen PR, He C. Selective recognition of metal ions by metalloregulatory proteins. Current Opinion in Chemical Biology. 2008;**12**:214-221. DOI: 10.1016/j.cbpa.2007.12.010

[43] Nolan EM, Lippard S. Small-molecule fluorescent sensors for investigating zinc metallo-neurochemistry. Accounts of Chemical Research. 2009;**42**:193-203. DOI: 10.1021/ar8001409

[44] McRae R, Bagchi P, Sumalekshmy S, Fahrni CJ. In situ imaging of metals in cells and tissues. Chemical Reviews. 2009;**109**:4780-4827. DOI: 10.1021/cr900223a

[45] Kikuchi K. Design, synthesis and biological application of chemical probes for bio-imaging. Chemical Society Reviews. 2009;**39**:2048-2053. DOI: 10.1039/b819316a

[46] Xiang Y, Lu Y. DNA as sensors and imaging agents for metal ions. Inorganic Chemistry. 2014;**53**:1925-1942. DOI: 10.1021/ic4019103

[47] Srivastava R. Complexes of DNA bases and Watson–Crick base pairs interaction with neutral silver Agn (n = 8, 10, 12) clusters: A DFT and TDDFT study. Journal of Biomolecular Structure & Dynamics. 2017;1-14. DOI: 10.1080/07391102.2017.1310059

[48] Sigel RKO, Sigel H. A stability concept for metal ion coordination to single-stranded nucleic acids and affinities of individual sites. Accounts of Chemical Research. 2010;**43**:974-984. DOI: 10.1021/ar900197y

[49] Lippert B. Progress in inorganic chemistry. Coordination Chemistry Reviews. 2000;**200**:487-516

[50] Lee J, Lin L, Li Y. In: Demchenko AP, editor. Advanced Fluorescence Reporters in Chemistry and Biology iii: Applications in Sensing and Imaging. Vol. 10. 2011. pp. 201-221

[51] Jiang Y, Fang XH, Bai C. Signaling aptamer/protein binding by a molecular light switch complex. Analytical Chemistry. 2004;**76**(17):5230-5235. DOI: 10.1021/ac049565u

[52] Ali MM, Aguirre SD, Mok WWK, Li Y. Detection of DNA using bioactive paper strips. Methods in Molecular Biology. 2012;**848**:395-418. DOI: 10.1007/978-1-61779-545-9

[53] Tram K, Kanda P, Li Y. Lighting up RNA-cleaving DNAzymes for biosensing. Journal of Nucleic Acids. 2012:958683-958683. DOI: 10.1155/2012/958683

[54] Lee JH, Wang Z, Lu Y. In: Tiquia-Arashiro SM, editor. Molecular Biological Technologies for Ocean Sensing. New York: Springer; 2012

[55] Ellington AD, Szostak JW. In vitro selection of RNA molecules that bind specific ligands. Nature. 1990;**346**:818-822. DOI: 10.1038/346818a0

[56] Liu JW, Lu Y. FRET study of a trifluorophore-labeled DNAzyme. Journal of the American Chemical Society. 2002;**124**:15208-15216. DOI: 10.1021/ja027647z

[57] Brown AK, Li J, Pavot CMB, Lu Y. A lead-dependent DNAzyme with a two-step mechanism. Biochemistry. 2003;**42**:7152-7161. DOI: 10.1021/bi027332w

[58] Kim HK, Li J, Nagraj N, Lu Y. Probing metal binding in the 8-17 DNAzyme by Tb(III) luminescence spectroscopy. Chemistry: A European Journal. 2008;**14**:8696-8703. DOI: 10.1002/chem.200701789

[59] Dalavoy TS, Wernette DP, Gong MJ, Sweedler JV, Lu Y, Flachsbart BR, Shannon MA, Bohn PW, Cropek DM. Immobilization of DNAzyme catalytic beacons on PMMA for Pb^{2+} detection. Lab on a Chip. 2008;**8**:786-793. DOI: 10.1039/b718624j

[60] Wang FL, Wu Z, Lu YX, Wang J, Jiang JH, Yu RQ. A label free DNAzyme sensor for lead (II) detection by quantitative polymerase chain reaction. Analytical Biochemistry. 2010;**405**:168-173. PMID: 20599653

[61] Xiang Y, Wang ZD, Xing H, Wong NY, Lu Y. Label-free fluorescent functional DNA sensors using unmodified DNA: A vacant site approach. Analytical Chemistry. 2010;**82**: 4122-4129. DOI: 10.1021/ac100244h

[62] Liu M, Zhao HM, Chen S, Yu HT, Zhang YB, Quan X. Controllable oxidative. DNA cleavage-dependent regulation of graphene/DNA interaction. Chemical Communications. 2011;**47**:7749-7751. DOI: 10.1039/c1cc00107h

[63] Liu JW, Lu Y. A colorimetric lead biosensor using DNAzyme-directed assembly of gold nanoparticles. Journal of the American Chemical Society. 2003;**125**:6642-6643. DOI: 10.1021/ja034775u

[64] Liu JW, Lu Y. Colorimetric Cu^{2+} detection with a ligation DNAzyme and nanoparticles. Chemical Communications. 2007;**46**:4872-4874. DOI: 10.1039/B712421J

[65] Mazumdar D, Liu JW, Lu G, Zhou JZ, Lu Y. Easy-to-use dipstick tests for detection of lead in paints using non-cross-linked gold nanoparticle-DNAzyme conjugates. Chemical Communications. 2010;**46**:1416-1418. DOI: 10.1039/b917772h

[66] Fang ZY, Huang J, Lie PC, Xiao Z, Ouyang CY, Wu Q, Wu YX, Liu GD, Zeng LW. Lateral flow nucleic acid biosensor for Cu^{2+} detection in aqueous solution with high sensitivity and selectivity. Chemical Communications. 2010;**46**:9043-9045. DOI: 10.1021/ac100244h

[67] Ono A, Togashi H. Highly selective oligonucleotide-based sensor for mercury(II) in aqueous solutions. Angewandte Chemie International Edition. 2004;**43**:4300-4302. DOI: 10.1002/anie.200454172

[68] Ono A, Cao S, Togashi H, Tashiro M, Fujimoto T, Machinami T, Oda S, Miyake Y, Okamoto I, Tanaka Y. Specific interactions between silver(I) ions and cytosine-cytosine pairs in DNA duplexes. Chemical Communications. 2008;**39**:4825-4827. DOI: 10.1039/b808686a

[69] Ueyama H, Takagi M, Takenaka S. A novel potassium sensing in aqueous media with a synthetic oligonucleotide derivative. Fluorescence resonance energy transfer associated with Guanine quartet-potassium ion complex formation. Journal of the American Chemical Society. 2002;**124**:14286-14287. DOI: 10.1021/ja026892f

[70] CNieh CC, Tseng WL. Thymine-based molecular beacon for sensing adenosine based on the inhibition of S-adenosyl homocysteine hydrolase activity. Biosensors and Bioelectronics. 2014;**61**:404-409. DOI: 10.1016/j.bios.2014.05.031

[71] Xiang Y, Lu Y. Using personal glucose meters and functional DNA sensors to quantify a variety of analytical targets. Nature Chemistry. 2011;**3**:697-703. DOI: 10.1038/nchem. 1092

[72] Oliveira Brett AM, Matysik FM. Voltammetric and sonovoltammetric studies on the oxidation of thymine and cytosine at a glassy carbon electrode. Journal of Electroanalytical Chemistry. 1997;**429**:95-99. DOI: 10.1016/S0022-0728(96)05018-8

[73] Thevenot DR, Toth K, Durst RA, Wilson GS. Electrochemical biosensors: Recommended definitions and classification. Pure and Applied Chemistry. 1999;**71**:2333-2348. DOI: 10.1351/pac199971122333

[74] Mascini M, Palchetti I, Marrazza G. DNA electrochemical biosensors. Fresenius' Journal of Analytical Chemistry. 2001;**369**:15-22. DOI: 10.1007/s002160000

[75] Oliveira-Brett AM, Vivan M, Fernandes IR, Piedade JAP. Electrochemical detection of in situ adriamycin oxidative damage to DNA. Talanta. 2002;**56**:959-970. DOI: 10.1016/ s0039- 9140(01)00656-7

[76] Halliwell B, Gutteridge JMC. Biologically relevant metal ion-dependent hydroxyl radical generation. FEBS Letters. 1992;**307**:108-112. DOI: 10.1016/0014-5793(92)80911-Y

[77] Srivastava R. Theoretical studies on the electronic and optoelectronic properties of [A.2AP (w)/A*.2AP(WC)/ C.2AP(w)/C*.2AP(WC)/C.A(w)/C*.A(WC)]–Au8 mismatch nucleobase complexes. Molecular Physics. 2017;**116**(2):263-272. DOI: 10.1080/00268976.2017.1382737

[78] Takada T, Fujitsuka M, Majima T. Single molecule observation of DNA charge transfer. Proceedings of the National Academy of Sciences. 2007;**104**(27):11179-11183. DOI: 10.1073/pnas.0700795104

[79] Tsoulou E, Kalfas C, Sideris E. Changes in DNA flexibility after irradiation with γ rays and neutrons studied with the perturbed angular correlation method. Radiation Research. 2003;**159**(1):33-39. PMID: 12492366

[80] Finn RS, Aleshin A, Slamon DJ. Targeting the cyclin-dependent kinases (CDK) 4/6 in estrogen receptor-positive breast. Cancer. 2016;**18**(17):17-28. DOI: 10.1186/s13058-015-0661-5

[81] Sönmezoğlu S, Sönmezoğlu ÖA, Çankaya G, Yıldırım A, Serin N. Electrical characteristics of DNA based metal-insulator semiconductor structures. Journal of Applied Physics. 2010;**107**(12):124518. DOI: 10.1063/1.3447985

[82] Bixon M, Giese B, Wessely S, Langen bacher T, Michel-Beyerle ME, Jortner J. Long range charge hopping in DNA. Proceedings of the National Academy of Sciences. 1999;**96**(21):11713-11716

[83] Dhar S, Liu Z, Thomale J, et al. Targeted single-wall carbon nanotube-mediated Pt (IV) prodrug delivery using folate as a homing device. Journal of the American Chemical Society. 2008;**130**:11467-11476. DOI: 10.1021/ja803036e

[84] Liu X, Tao H, Yang K, et al. Optimization of surface chemistry on single-walled carbon nanotubes for *in vivo* photothermal ablation of tumors. Biomaterials. 2011;**32**:144-151. DOI: 10.1016/j.biomaterials.2010.08.096

[85] Kim JH, Ahn JH, Barone PW, et al. A luciferase/single walled carbon nanotube conjugate for near infrared fluorescent detection of cellular ATP. Angewandte Chemie International Edition. 2010;**122**:1498-1501. DOI: 10.1002/anie.200906251

[86] Wang J, Liu G, Jan MR. Ultrasensitive electrical biosensing of proteins and DNA: Carbon-nanotube derived amplification of the recognition and transduction events. Journal of the American Chemical Society. 2004;**126**:3010-3011. DOI: 10.1016/j.nano.2009.02.006

[87] Rudin R, Weissleder R. Molecular imaging in drug discovery and development. Nature Reviews. Drug Discovery. 2003;**2**:123-131. DOI: 10.1038/nrd1007

[88] Cai W, Rao J, Gambhir SS, Chen X. How molecular imaging is speeding up anti angio genic drug development. Molecular Cancer Therapeutics. 2006;**5**:2624-2633. DOI: 10. 1158/1535-7163

[89] Sitharaman B, Kissell KR, Hartman KB, Tran LA, Baikalov A, Rusakova I, et al. Superparamagnetic gadonanotubes are high-performance MRI contrast agents. Chemical Communications. 2005;**31**:3915-3917. DOI: 10.1039/b504435a

[90] Hartman KB, Laus S, Bolskar RD, Muthupillai R, Helm L, Toth E, et al. Gadonanotubes as ultrasensitive pH-smart probes for magnetic resonance imaging. Nano Letters. 2008;**8**:415-419. DOI: 10.1021/nl0720408

[91] Frangioni JV. In vivo near-infrared fluorescence imaging. Current Opinion in Chemical Biology. 2003;7:626-634. PMID: 14580568

[92] Cai W, Shin DW, Chen K, Gheysens O, Cao Q, Wang SX, et al. Peptide-labeled near-infrared quantum dots for imaging tumor vasculature in living subjects. Nano Letters. 2006;**6**:669-676. DOI: 10.1021/nl052405t

[93] Keidar M, Levchenko I, Arbel T, Alexander M, Waas AM, Ostrikov KK. Increasing the length of single-wall carbon nanotubes in a magnetically enhanced arc discharge. Applied Physics Letters. 2008;**92**:043129. DOI: 10.1063/1.2839609

[94] Itkis ME, Perea DE, Jung R, Niyogi S, Haddon RC. Comparison of analytical techniques for purity evaluation of single-walled carbon nanotubes. Journal of the American Chemical Society. 2005;**127**:3439-3448. DOI: 10.1021/ja043061w

[95] Kim G, Huang SW, Day KC, O'Donnell M, Agayan RR, Day MA, et al. Indocyanine-green-embedded PEBBLEs as a contrast agent for photoacoustic imaging. Journal of Biomedical Optics. 2007;**12**:044020. DOI: 10.1117/1.2771530

[96] Ku G, Wang LV. Deeply penetrating photoacoustic tomography in biological tissues enhanced with an optical contrast agent. Optics Letters. 2005;**30**:507. DOI: 10.1364/OL. 30.000507

[97] De la Zerda A, Zavaleta C, Keren S, Vaithilingam S, Bodapati S, Liu Z, et al. Carbon nanotubes as photoacoustic molecular imaging agents in living mice. Nature Nanotechnology. 2008;**3**:557-562. DOI: 10.1038/nnano.2008.231

[98] Cai W, Chen X. Multimodality imaging of vascular endothelial growth factor and vascular endothelial growth factor receptor expression. Frontiers in Bioscience. 2007;**12**:4267-4279. DOI: 10.2741/2386

[99] Abarrategi A, Gutiérrez MC, Moreno-Vicente C, et al. Multiwall carbon nanotube scaffolds for tissue engineering purposes. Biomaterials. 2008;**29**:94-102. DOI: 10.1016/j.biomaterials.2007.09.021

[100] Zhang Y, Tan YW, Stormer HL, Kim P. Experimental observation of the quantum Hall effect and Berry's phase in graphene. Nature. 2005;**438**:201-204. DOI: 10.1038/nature04235

[101] Zhang Y, Jiang Z, Small JP, Purewal MS, Tan YW, Fazlollahi M, Chudow JD, Jaszczak JA, Stormer HL, Kim P. Landau-level splitting in graphene in high magnetic fields. Physical Review Letters. 2006;**96**:136806. DOI: 10.1103/PhysRevLett.96.136806

[102] Novoselov KS, Geim AK, Morozov SV, Jiang D, Katsnelson MI, Grigorieva IV, Dubonos SV, Firsov AA. Two-dimensional gas of massless Dirac fermions in grapheme. Nature. 2005;**438**:197-200. DOI: 10.1038/nature04233

[103] Areshkin DA, Gunlycke D, White CT. Ballistic transport in graphene nanostrips in the presence of disorder: Importance of edge effects. Nano Letters. 2006;**7**:204-210. DOI: 10.1021/nl062132h

[104] Kim KS, Zhao Y, Jang H, Lee SY, Kim JM, Kim KS, Ahn JH, Kim P, Choi JY, Hong BH. Large-scale pattern growth of graphene films for stretchable transparent electrodes. Nature. 2009;**457**:706-710. DOI: 10.1038/nature07719

[105] Miura Y, Kasai H, Diño W, Nakanishi H, Sugimoto T. First principles studies for the dissociative adsorption of H_2 on grapheme. Journal of Applied Physics. 2005;**93**:3395-3400. DOI: 10.1063/1.1555701

[106] Yazyev OV, Katsnelson MI. Magnetic correlations at graphene edges: Basis for novel spintronics devices. Physical Review Letters. 2008;**100**:047209. DOI: 10.1103/PhysRevLett.100.047209

[107] Nakada K, Fujita M, Dresselhaus G, Dresselhaus MS. Edge state in graphene ribbons: Nanometer size effect and edge shape dependence. Physical Review B. 1996;**54**:17954. DOI: 10.1103/PhysRevB.54.17954

[108] Li G, Liu XD, Zhang H, Wang X, Bu H, Chen M, Li F, Zhao M. Adsorption and diffusion of gold adatoms on boron nitride nanoribbons: A first-principles study. Journal of Applied Physics. 2012;**112**:104305. DOI: 10.1063/1.4766411

[109] Fujita M, Wakabayashi K, Nakada K, Kusakabe K. Peculiar localized state at zigzag graphite edge. Journal of the Physical Society of Japan. 1996;**65**:1920-1923. DOI: 10.1143/JPSJ.65.1920

[110] Wang G. Density functional study on the increment of carrier mobility in armchair graphene nanoribbons induced by Stone-Wales defects. Physical Chemistry Chemical Physics. 2011;**13**:11939-11945. DOI: 10.1039/c1cp20541b

[111] Yan QM, Huang B, Yu J, Zheng FW, Zang J, Wu J, Gu BL, Liu F, Duan WH. Intrinsic current-voltage characteristics of graphene nanoribbon transistors and effect of edge doping. Nano Letters. 2007;**7**:1469-1473. DOI: 10.1021/nl070133j

[112] Li J, Li ZY, Zhou G, Liu ZR, Wu J, Gu BL, Ihm J, Duan WH. Dirac fermions in strongly bound graphene systems. Physical Review B: Condensed Matter and Materials Physics. 2010;**82**:115410. DOI: 10.1103/PhysRevB.82.115410

[113] Son YW, Cohen ML, Louie SG. Energy gaps in graphene nanoribbons. Physical Review Letters. 2006;**97**:216803. DOI: 10.1103/PhysRevLett.98.089901

[114] Wu TT, Wang XF, Zhai MX, Liu H, Zhou LP, Jiang YJ. Negative differential spin conductance in doped zigzag graphene nanoribbons. Applied Physics Letters. 2012;**100**:052112. DOI: 10.1063/1.3681775

[115] Valencia F, Romero A, Ancilotto F, Silvestrelli P. Lithium adsorption on graphite from density functional theory calculations. The Journal of Physical Chemistry. B. 2006;**110**:14832. DOI: 10.1021/jp062126

[116] Mao YL, Yuan JM, Zhong JX. Density functional calculation of transition metal adatom adsorption on graphene. Journal of Physics: Condensed Matter. 2008;**20**:115209. DOI: 10.1088/0953-8984/20/11/115209

[117] Srivastava R. Theoretical studies of optoelectronic, magnetization and heat transport properties of conductive metal adatoms adsorbed on edge chlorinated graphene nanoribbons. (submitted to RSC advances)

[118] Zanella I, Fagan SB, Mota R, Fazzio A. Electronic and magnetic properties of Ti and Fe on graphene. Journal of Physical Chemistry C. 2008;**112**:9163-9167. DOI: 10.1021/jp711691r

[119] Amft M, Lebègue S, Eriksson O, Skorodumova NV. Adsorption of Cu, Ag, and Au atoms on graphene including van der Waals interactions. Journal of Physics: Condensed Matter. 2011;**23**:395001. DOI: 10.1088/0953-8984/23/39/395001

[120] Varns R, Strange P. Stability of gold atoms and dimers adsorbed on graphene. Journal of Physics: Condensed Matter. 2008;**20**:225005-225013. DOI: 10.1088/0953-8984/20/22/225005

[121] Granatier J, Lazar P, Prucek R, Šafářová K, Zboril R, Otyepka M, Hobza P. Interaction of graphene and arenes with noble metals. Journal of Physical Chemistry C. 2012;**116**: 14151-14162

[122] Liu Y, Liu Y, Feng H, Wu Y, Joshi L, Zeng X, Li J. Layer-by layer assembly of chemical reduced graphene and carbon nanotubes for sensitive electrochemical immunoassay. Biosensors & Bioelectronics. 2012;**35**(1):63-68. DOI: 10.1016/j.bios.2012.02.007

[123] Zhu L, Luo L, Wang Z. DNA electrochemical biosensor based on thionine-graphene nanocomposite. Biosensors & Bioelectronics. 2012;**35**(1):507-511. DOI: 10.1016/j.bios.2012.03.026

[124] Chen X, Qin P, Li J, Yang Z, Wen Z, Jian Z, Zhao J, Hu X, Jiao X. Impedance immunosensor for bovine interleukin-4 using an electrode modified with reduced graphene oxide and chitosan. Microchimica Acta. 2015;**182**(1):369-376. DOI: 10.1007/s00604-014-1331-5

[125] Ohno Y, Maehashi K, Matsumoto K. Label-free biosensors based on aptamer-modified graphene field-effect transistors. Journal of the American Chemical Society. 2010; **132**(51):18012-18013. DOI: 10.1021/ ja108127r

[126] Zhang M, Yin BC, Tan W, Ye BC. A versatile graphene-based fluorescence "on/off" switch for multiplex detection of various targets. Biosensors & Bioelectronics. 2011;**26**(7): 3260-3265. DOI: 10.1016/j.bios.2010.12.037

3

Environmental Application of High Sensitive Gas Sensors with Tunable Diode Laser Absorption Spectroscopy

Xiaojuan Cui, Fengzhong Dong, Zhirong Zhang,
Hua Xia, Tao Pang, Pengshuai Sun, Bian Wu,
Shuo Liu, Luo Han, Zhe Li and Runqing Yu

Abstract

Due to the fact of global warming, air quality deterioration and health concern over the past few decades, great demands and tremendous efforts for new technology to detect hazard gases such as CH_4, CO_2, CO, H_2S, and HONO have been performed. Tunable diode laser absorption spectroscopy (TDLAS) is a kind of technology with advantages of high sensitivity, high selectivity, and fast responsivity. It has been widely used in the applications of greenhouse gas measurements, industrial process control, combustion gas measurements, medicine, and so on. In this chapter, we will briefly summarize the most recent progress on TDLAS technology and present several kinds of gas sensors developed mainly by our group for various field applications. These could expand from energy, environment, and public safety to medical science.

Keywords: TDLAS, wavelength modulation, hazard gases, HONO, $\delta^{13}C$

1. Introduction

Over the past few decades, environmental pollution problem has occurred to different degrees in the whole world, such as atmospheric pollution, marine pollution, and urban environmental problems. With the globalization of economy and trade, environmental pollution is becoming more and more internationalized [1]. In order to control environmental pollution, great demands and tremendous efforts for new technology to detect hazard gases such as CH_4,

CO_2, CO, HONO, H_2S, and HCl have been performed. This would be beneficial for the implementation of global environmental protection policies for the reduction of gas pollution and for a general environmental management [2].

Several optical techniques have been developed to detect these hazard gases in the atmosphere [3–10]. Cavity-enhanced spectroscopy (CEAS) or cavity ring-down spectroscopy (CRDS) has been demonstrated to enable measurements of multiple gases with a low detection limit of sub-ppb [4–6]. However, these two technologies require critical optical alignment and regular cleaning of mirrors of the external cavity which affects continuous monitoring of atmospheric species in the field. Quartz-enhanced photoacoustic spectroscopy (QEPAS) technique was also developed for environmental and biomedical measurements [7, 8]. Nevertheless, the high modulation frequencies used in QEPAS may represent a problem for multicomponent gas mixtures containing varying amounts of water vapor such as ambient air, due to the strong influence of water vapor on the molecular vibrational-translational (V-T) relaxation times. Other spectroscopic methods such as open path Fourier transform infrared spectrometry (FTIR) and differential optical absorption spectroscopy (DOAS) have been reported for atmospheric molecule detection [9, 10]. But the minimum detection limits (MDLs) of FTIR usually exceed the requirements for high sensitivity measurements of the atmospheric species. The main disadvantage of the DOAS system is that its spatial resolution is rather poor with a path length generally greater than 1 km.

The technique based on tunable diode laser absorption spectroscopy (TDLAS) is an effective method to measure gas mixing ratios and multiple parameters with high selectivity, high sensitivity, high precision, and high response time [11–18]. Especially, with the development of multi-pass absorption cells, the effective optical path length can be extended from a few meters to several hundred meters; the sensitivity is significantly improved [19–21]. In order to further improve the signal-to-noise ratio (SNR), the wavelength modulation spectroscopy (WMS) technology with second harmonic (2f) signals is usually employed in the TDLAS system to measure the gas concentration.

The first commercial TDLAS gas sensor was introduced on the market in 1995 using the trademark laser gas by Norsk Elektro Optikk Company. Over the past decades, TDLAS has been extensively investigated potentially as an effective method to measure multiple gas parameters and is widely used in various areas such as gas mixing ratio detection, vehicle emissions, gas exhaust temperature monitoring, carbon isotope measurements, and so on [22–37]. Now NEO Monitors is one of the world leading suppliers of the TDLAS-based gas analyzers and dust monitors. Its products are widely used in the field of industrial process control and emission monitoring; nearly 6000 sets of laser gas analyzers were installed in more than 40 countries and regions in the world currently. We are also developing instruments based on TDLAS technology to satisfy the needs of environmental monitoring and industrial process control in China. **Figure 1** shows several pictures of the gas sensors developed by our research team. In this chapter, we will briefly present several kinds of gas sensors developed by our research group for various field applications, which could expand from environment and public safety to medical science.

Figure 1. Several pictures of the TDLAS system developed by our research team.

2. Basic principles of TDLAS

Based on the Beer-Lambert law, the relationship between the incident intensity I_0 and the transmitted intensity I can be expressed as

$$I = I_0 \exp(-kL) \tag{1}$$

where k is the absorption coefficient and L denotes the path length (in cm). In the near-infrared region, the gas absorption coefficient is usually very small, i.e., $kL \leq 0.05$ [38]. Eq. (1) can thus be simplified as

$$I = I_0(1 - kL) = I_0[1 - \sigma(\nu)CL] \tag{2}$$

where $\sigma(v)$ is the absorption cross section (in [cm^2/molecule]) at frequency ν and C is the gas mixing ratio. The integrated absorbance A_I (in [cm^{-1}]) can be written as

$$A_I = \int A(v)dv = \int \ln(I_0(v)/I(v))dv = \mathrm{NL}\int \sigma(v)dv = \mathrm{NLS} \tag{3}$$

N is the number of absorbing molecules (in [molecules/cm^3]); S is the molecule absorption line strength (in [cm^2/(mol cm)]). Based on Eq. (3), the gas species mixing ratio can be retrieved from the integrated absorbance A_I measured at temperature T and pressure P [39]:

$$C(\mathrm{ppm}) = \frac{N}{N_T} \times 10^6 = \frac{A_I P_0 T}{N_L \mathrm{PT_0LS}} \times 10^6 \tag{4}$$

where N_L = 2.6868 × 10^{19} mol/cm^3 represents the Loschmidt number at T_0 = 273.15 K and P_0 = 760 Torr.

For gas mixing ratio detection, WMS is often adopted. The intensity of 2f signal can be expressed as [40]

$$I_{2f} \propto I_0 \sigma_0 CL \tag{5}$$

When the reference signal and nonlinear least square multiplication method are introduced to fit the 2f signals of the target gas [41], Eq. (5) can be rewritten as

$$C_{Mea} = a\frac{I_{Mea} C_{\mathrm{Ref}} L_{\mathrm{Ref}}}{I_{\mathrm{Ref}} L_{Mea}} \tag{6}$$

where a is fitting coefficient; C_{Mea} and C_{Ref} are the mixing ratios of the target gas to be measured and reference gas in the calibration cell, respectively; I_{Ref} and I_{Mea} denote the intensities of the two split laser beams; and L_{Ref} and L_{Mea} represent the calibration cell and the measurement optical path length, respectively. In general, the ratio of the 2f and 1f signals can be used to cancel any laser intensity differences. In this case, the mixing ratio from the following equation could be easily obtained:

$$C_{Mea} = \frac{\left(\frac{I_{2f}}{I_{1f}}\right)_{Mea} C_{\mathrm{Ref}} L_{\mathrm{Ref}}}{\left(\frac{I_{2f}}{I_{1f}} 1f\right)_{\mathrm{Ref}} L_{Mea}} \tag{7}$$

where $\left(\frac{I_{2f}}{I_{1f}}\right)_{\mathrm{Ref}}$ and $\left(\frac{I_{2f}}{I_{1f}}\right)_{Mea}$ represent the 2f/1f ratio value of the reference and target gas signals, respectively.

3. Methane (CH_4) monitoring

3.1. Introduction

With the increasing attention to environment, energy, and safety, natural gas has gradually replaced coal as the main energy source in China, and its use has been increasing year by year.

The extraction, transportation, and storage of natural gas have become an important part of social development. Equipment safety and high efficiency operation in gas transmission station are the keys to ensure the natural gas transportation. Once it is released, the serious safety accidents such as energy waste, environmental pollution, fire, and explosion will happen [42], which would directly threaten the safety of life and property of the countries and people [43].

The main component of natural gas is methane, accounting for 90%, and also contains a small amount of ethane, acetylene, butane, carbon dioxide, carbon monoxide, hydrogen sulfide, and so on. Traditional natural gas leakage detectors include flame ion detectors (FID), electronic detectors, electrochemical catalytic combustion detectors, and infrared absorption detectors [44]. However, these detectors are self-charging and have potential safety problems in the application of flammable, explosive, and other special environments. Moreover, these sensors are short in life, low in precision, poor in stability, and difficult in adjustment and often give the wrong results of measurements and misinformation. Recently, TDLAS technology has been widely used with the rapid development of narrow linewidth semiconductor laser technology [45]. The SRI International (Menlo Park, CA) company in America has developed a vehicular natural gas pipeline leakage detector, which improves the efficiency of pipeline leakage detection. However, they are all limited to the detection of methane and do not involve the detection of other gases in natural gas.

In view of the area of natural gas, gathering station is large, and the pipeline system of natural gas is gathered; point and portable measurement is not suitable in this situation. We designed an open, continuous detection and alarm system which has the characteristics of fast response speed and high detection precision based on TDLAS technology. Moreover, this system also detects ethylene, acetylene, and other gases, which improves the measurement precision and reduces the probability of false alarm.

3.2. Absorption line selection

The near-infrared absorption band matches with the low loss window of optical fiber and is convenient for long-distance transmission and multipoint distributed detection by using fiber and fiber devices. Therefore, the absorption lines of selected CH_4, C_2H_2, and C_2H_4 are 1653.72, 1531.59, and 1621.36 nm, respectively. There are three adjacent absorption lines at 1653.72 nm for CH_4, which are close to each other and cannot be separated in the atmospheric pressure by consulting the HITRAN 2008 database. In the experiment, they are processed as one absorption line. The C_2H_4 absorption lines are not included in HITRAN database. A large amount of absorption lines of C_2H_4 from 1600 to 1650 nm can be found from the PNNL25C (Northwest Pacific National Laboratory) database which have been already experimentally verified in the literature [46]. The parameters of three gases absorption lines are shown in **Table 1**.

3.3. DFB-based experimental platform

The system is designed mainly aimed at the gas gathering station, and the schematic diagram of the system is shown in **Figure 2**. Three butterfly-packaged distributed feedback (DFB) lasers are selected to detect CH_4, C_2H_2, and C_2H_4 with the center output wavelengths of 1653, 1531, and 1621 nm, respectively. The light sources are controlled by the corresponding temperature,

Molecule	Wavenumber (nm)	Line strength at 300 K (cm^{-2} atm^{-1})	Δv (cm^{-1})
CH_4	1653.7282 1653.7256 1653.7225	0.0206 0.0206 0.0368	0.14
C_2H_2	1531.5878	0.2916	0.23
C_2H_4	1621.3600		

Table 1. The parameters of absorption lines.

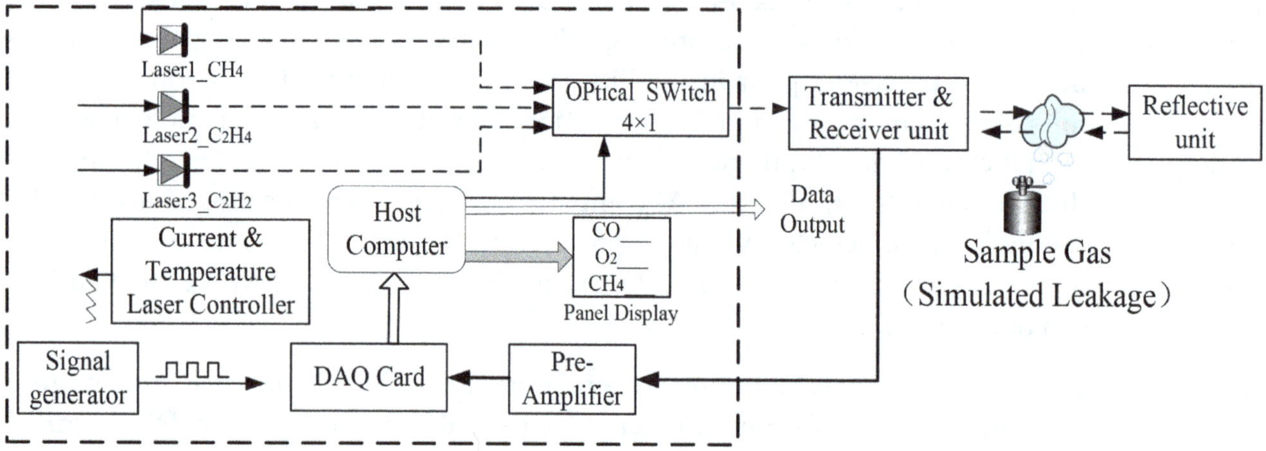

Figure 2. Schematic diagram of the experimental system.

current driver module, and signal generator module, respectively. Three modulation light beams are time-sharing output through a 3 × 1 optical switch which is controlled by a microprocessor and then the collimator and beam expander of the transmitter (THORLABS GBE10-C: ten times beam expander, 1050–1650 nm antireflective coating), passing through the measurement area to the corner cube mirror at the reflecting end. Then, returning to the receiving end along the parallel light path, the light beam containing the absorption signal is focused on the photosensitive surface of the photoelectric detector through an aspherical focusing lens and converted into electrical signals before entering the host control section. The amplified electrical signals are collected by the data acquisition card and transmitted to the microprocessor system after amplification by the preamplifier circuit. Finally, the online inversion of spectral data is carried out to obtain the gas concentration. Meanwhile, the early warning will be carried out according to the setting of the alarm limit. If the value exceeds the setting one; the system will send out light and sound alerting signal.

In order to decide the detection limit of the system, a calibration experiment was designed and shown in **Figure 3**. A calibrated absorption cell with a length of 1 m was placed on the laser path. In the calibration experiment, three gases CH_4, C_2H_2, and C_2H_4 with the mixing ratios of 1%, 500, and 500 ppm are mixed in the absorption cell, and the corresponding absorption signals are displayed in **Figure 4**. The absorption lines of CH_4 and C_2H_2 are independent, and there are no other spectral interferences, but there is a relatively weak absorption spectral line

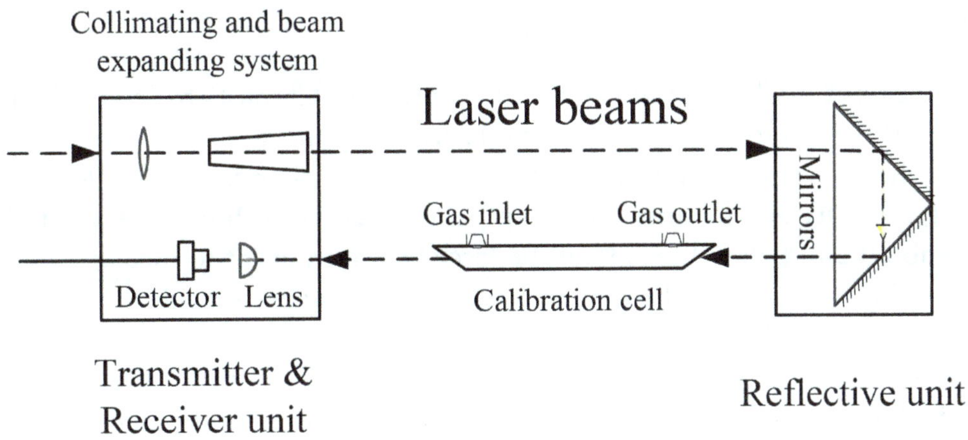

Figure 3. Schematic diagram of calibration principle.

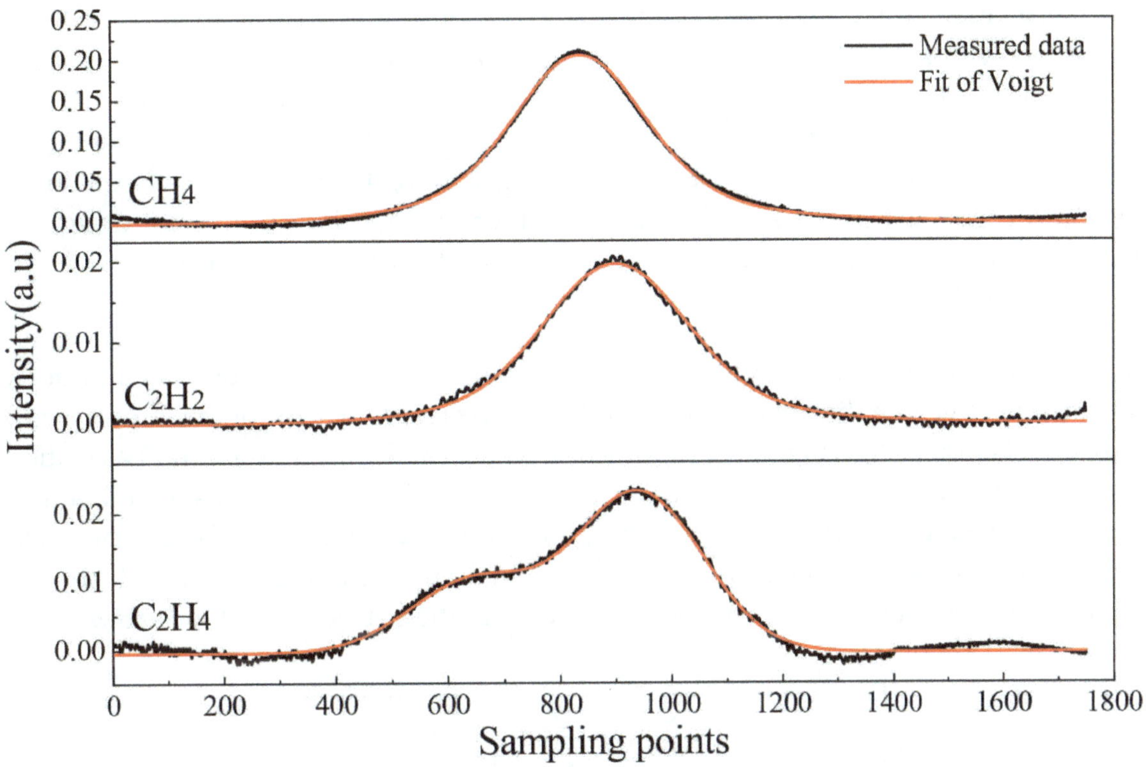

Figure 4. Direct absorption signal and fitting results.

on the left of the absorption line of C_2H_4. Therefore, the absorption lines of CH_4 and C_2H_2 are fitted using a single peak, and the absorption line of C_2H_4 is fitted by double peak in the fitting process. The absorbance A values of CH_4, C_2H_2, and C_2H_4 absorption spectral lines are 0.076, 0.012, and 0.014 cm^{-1}, respectively. The SNR of the absorption signals are 100, 12, and 10, respectively. According to the linear relationship between the direct absorbance and gas concentration, the obtained MDLs of CH_4, C_2H_2, and C_2H_4 were 100, 40, and 50 ppm-m, respectively, which completely satisfied the gas gathering station leakage test requirements [47].

3.4. Results and discussion

The system had been field-tested at the testing ground of China Petroleum Pipeline Bureau. The environment temperature was 35°C, air relative humidity was 45%, and wind speed was 1 m/s during the experiment. We used the gas which was mixed with 90% methane, 5% ethylene, and 5% acetylene to simulate gas pipeline leakage in the experiment. The leakage position was about 2 m below the side of the laser beam. In order to measure the three gases simultaneously, lasers were switched every 10 seconds using an optical switch. Three kinds of gases were detected circularly in the order of $CH_4 \rightarrow C_2H_2 \rightarrow C_2H_4$. The mixing ratios of the gases are exhausted 1 minute each time which was displayed in **Figure 5**. The reason for fluctuations is that the measured concentrations are the average of the paths along the line of sight. Due to the uncertainty of wind speed and gas diffusion in the measurement field, the concentration on the beam path fluctuates greatly. Meanwhile, this system is also equipped with an alarm limit for each gas, and the veracity of fire alarming system achieved 100%.

The system includes three DFB lasers which have an output power of about 20 mW higher than the other semiconductor lasers. Moreover, the optical fiber loss is less than 0.25 dB/km in this waveband. So the system can connect four pairs of transmitter and receiver units simultaneously. According to the requirements and distribution of gas pipeline, gas gathering device, housing, and other special places in the gas gathering station, the installation scheme including a host control machine and two pairs of transmitter and receiver units was designed and displayed in **Figure 6**. This system can be used to monitor the leakage of natural gas station in the range of 100 × 100 m.

The leakage detection system based on TDLAS can detect methane, ethylene, and acetylene rapidly and effectively in the open environment, and the response time of the three gases is less than 2 s. The accuracy of giving an alarm is 100%, which can be used in natural gas station and valve room gas leakage. Compared to other techniques, this technique has the advantages of safety in nature, no calibration, high accuracy, and little environmental effects. The MDLs for methane, acetylene, and ethylene gas are 100, 40, and 50 ppm-m, respectively, which meet the requirements for the detection of natural gas leakage in the petrochemical industry.

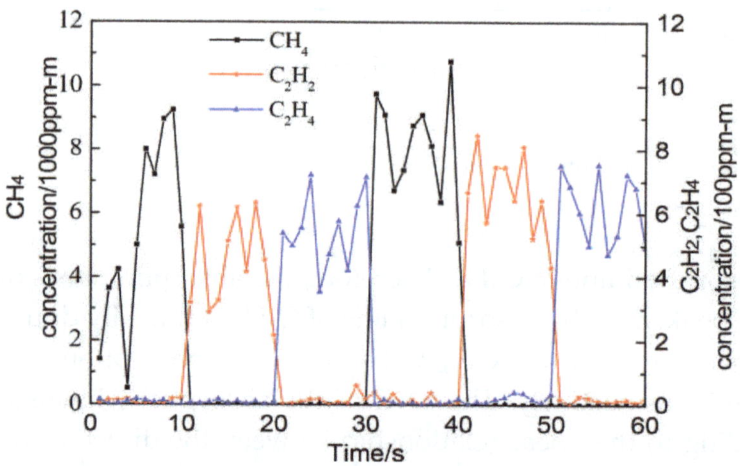

Figure 5. Concentration curves of experiment results.

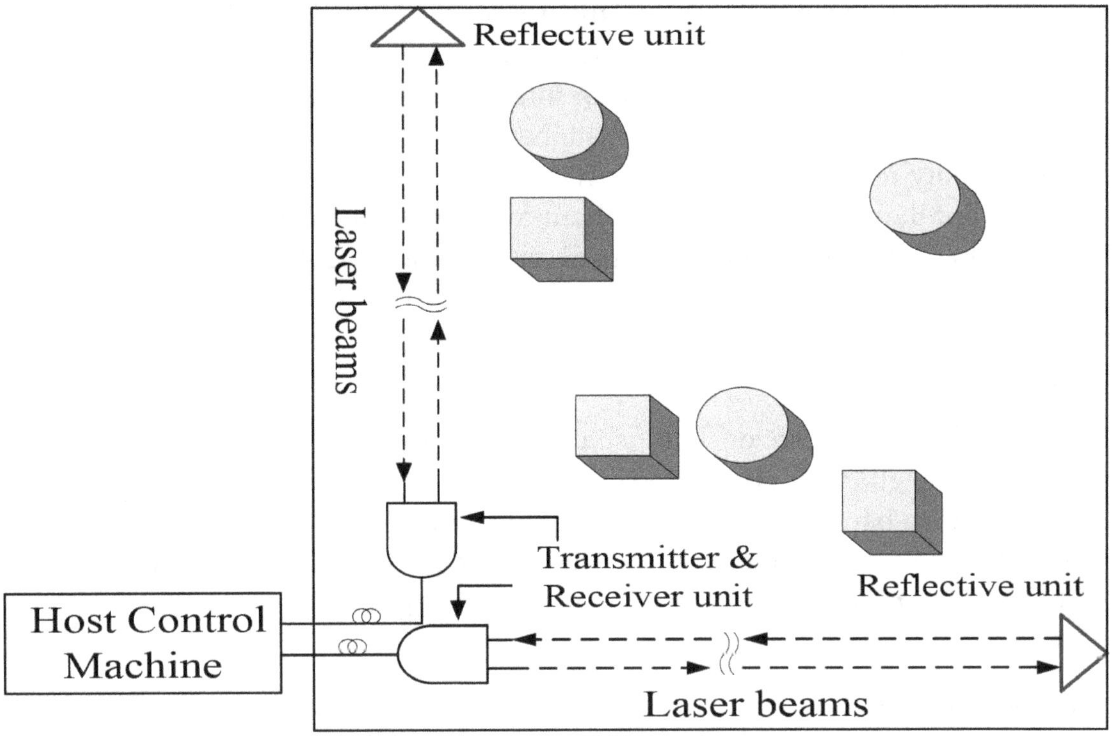

Figure 6. Installation scheme of natural gas gathering and transferring station.

4. Carbon monoxide (CO) monitoring

4.1. Introduction

CO is a kind of toxic, combustible, explosive gas and brings lots of hidden danger to the production and life of human beings. The research of coal spontaneous combustion suggests that a series of gases which could indicate the degree of oxidation and combustion of coal will be produced when coal seam is on fire. Using the relationship between the amount of indicator gases and the rate of change could predict coal seam fire at an early stage. Nowadays, CO is widely used as the main indicator gas for early warning of coal seam fire because the quantity of CO is closely related to the temperature of coal seam and the concentration change is obvious. In addition, the safety production under the mine has attracted much attention. In order to avoid accidents, gas monitoring has become a necessary means. The detection devices of the main gas constituents such as methane and carbon dioxide have been improved and widely used. With the improvement of security awareness, people have higher requirements on the accuracy of gas monitoring [48].

4.2. Absorption line selection

The absorption intensity of CO in the mid-infrared region is two orders of magnitude higher than that of overtone band in the near infrared. With the development of mid-infrared lasers, high sensitivity detection of CO has been obtained by some researchers [49]. But for the long-distance optical fiber transmission signals, the use of the mid-infrared laser is limited because the current optical fiber communication windows are mainly concentrated in the near infrared. The intensity of CO absorption line in the near infrared is weak, and the SNR is poor when low concentration is detected, which requires higher stability of the measurement system. At present, the stability research of high sensitivity detection of CO in the near infrared has not been reported. But there are some reports about measurement techniques such as the stability of DFB lasers [50], the application of signal processing in CO_2 and NO_2, and other gas measurements [51]. Therefore, it is of great practical significance to study the stability of whole measurement system and realize the high sensitivity detection of CO in the communication windows.

To select a unique gas absorption line usually adopts the following guide rules: (1) strong absorption line strength with good line profile and (2) free of interference from other gases. The second overtone band near 1.566 μm of CO was selected in this work to avoid interferences from other major ambient gases in the mixture. **Figure 7** shows the absorption spectrum of CO, CO_2, and H_2O near the wavelength range of 1.566 μm [52].

4.3. Experimental system design

The experimental system is shown in **Figure 8**. The system adopts balanced optical path detection method. The 2 * 1 beam combiner couples the collimated light and the measuring beam to the 1*3 beam splitter, after that the first beam through a multi-pass absorption cell filled with CO gas, marked as S (measuring light path); the second beam through a high

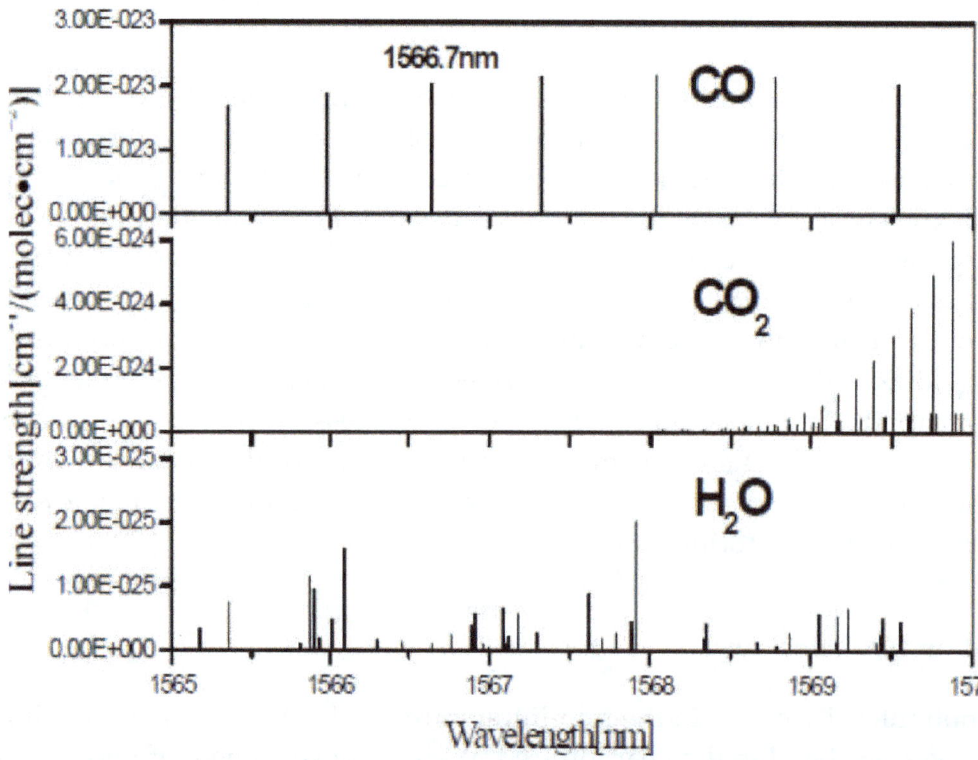

Figure 7. The absorption lines of CO, CO_2, and H_2O near the wavelength range of 1.566 μm (HITRAN 2008 database).

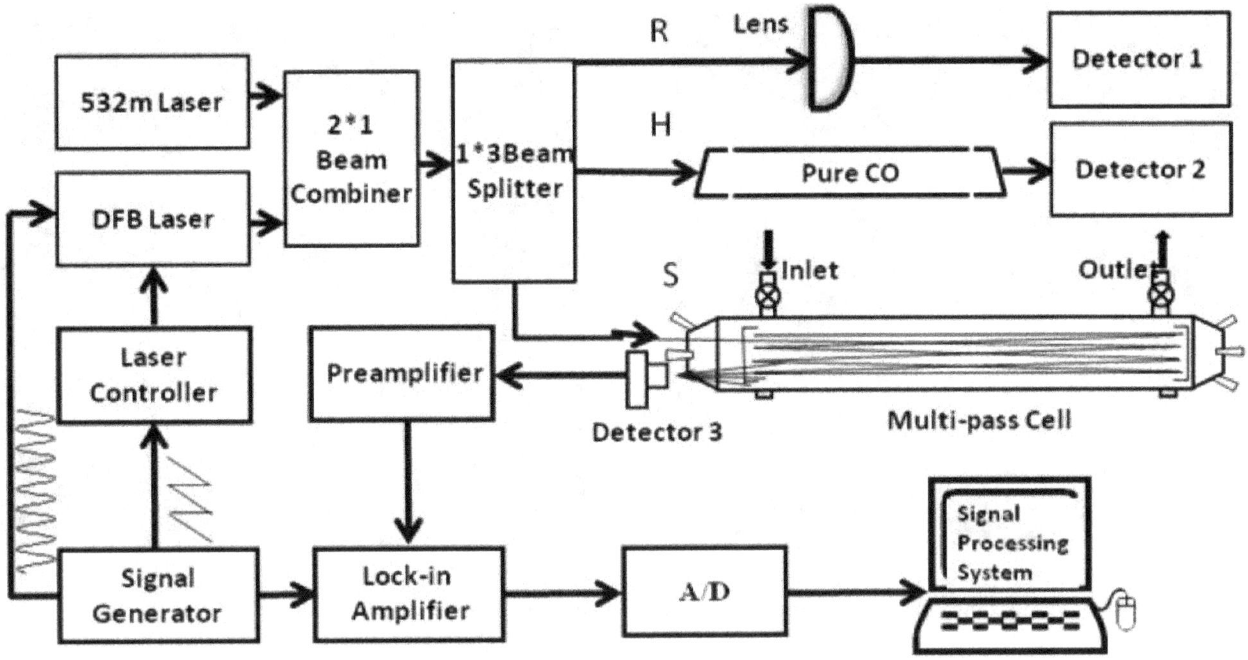

Figure 8. Schematic diagram of the TDLAS experimental system.

concentration reference cell with 100% CO, used to determine and control the position of absorption wavelength, marked as H; and the third beam is a reference light through the free space, used to monitor the changes of the laser background, marked as R. The three detection signals are sequential controlled by the switching circuit simultaneously. The wavelength is scanned with 100 Hz sawtooth wave and modulated with 10 kHz sine wave. More details about the electronics setup for the experiment could be found in [53]. Three modulation signals enter the lock-in amplifier through the switching circuit. In the lock-in amplifier, the detector output is mixed with the reference signal (10 kHz) to demodulate the 2f spectral signal. Then the 2f signal is simultaneously processed by a data acquisition card installed on a computer.

A new type of multi-pass absorption cell was developed and effectively improved the detection ability of the system. The new absorption cell has the advantages of simple structure, stable performance, effective use of the surface area, and solving the contradiction between the small volume and long-path length. The optical path length of 56.7 m was achieved in the volume of 1 L. At the same time, the optical path is adjustable; the spot array is uniform and in order, so that the optical path calculation is convenient; and the free spectral range is very narrow. The possible interference fringes in the cell are distributed in the high frequency region. By means of the digital averaging method, the influence of interference fringes on the second harmonic signals can be removed effectively and simply. The base length of the multi-pass absorption cell used in the system is 24.6 cm; the diameter of the mirror is 60 mm. According to the needs of TDLAS system for CO gas measurement, the mirror is coated with a dielectric film with a high reflectivity (typically 0.999) for wavelengths 532 and 1567 nm, wherein 532 nm is the collimated light during the alignment of the optical path. **Figure 9** shows the spot distribution of the mirrors at both ends of the absorption cell.

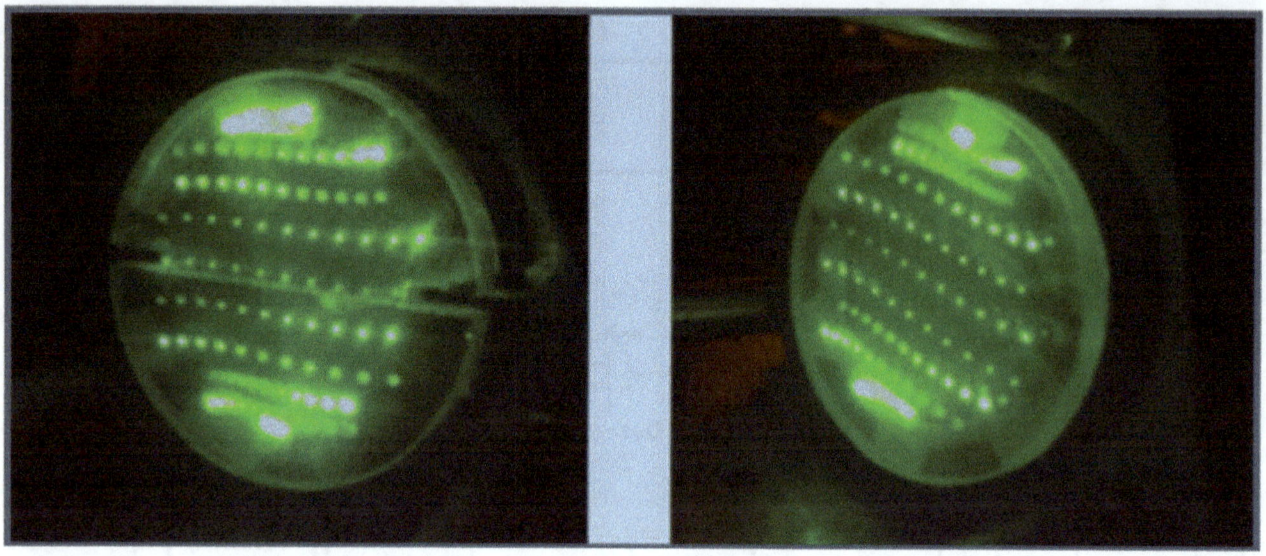

Figure 9. Light spot distribution of the mirrors at the both ends of the absorption cell.

4.4. Results and discussion

The CO standard gases of 10 and 200 ppm were measured in the laboratory by using the above described TDLAS system. The stability and detection limit of the system were analyzed. The linearity of the system was tested by measuring the CO standard gas at different concentrations. **Figures 10** and **11** display the measurement results of 10 and 200 ppm CO standard gases, respectively. After continuous measurements of 14 h, the average concentrations are 10.57 and 200.36 ppm, and the standard variance is 0.5 and 2.1 ppm, which can be found in **Table 2**. The standard variance reflects the stability of the system to a certain extent.

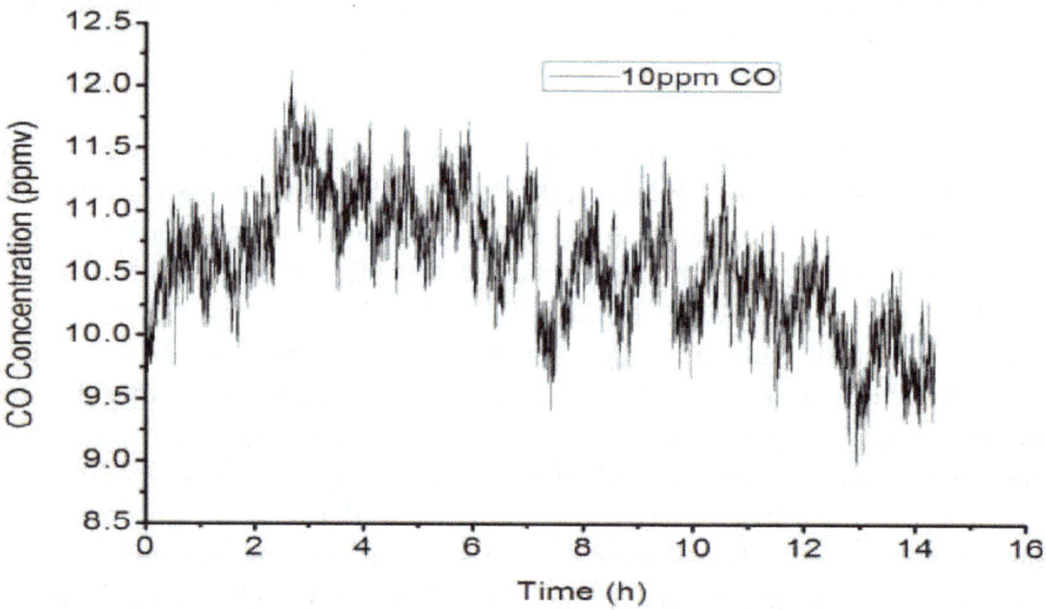

Figure 10. The measurement results of 10 ppm CO standard gas.

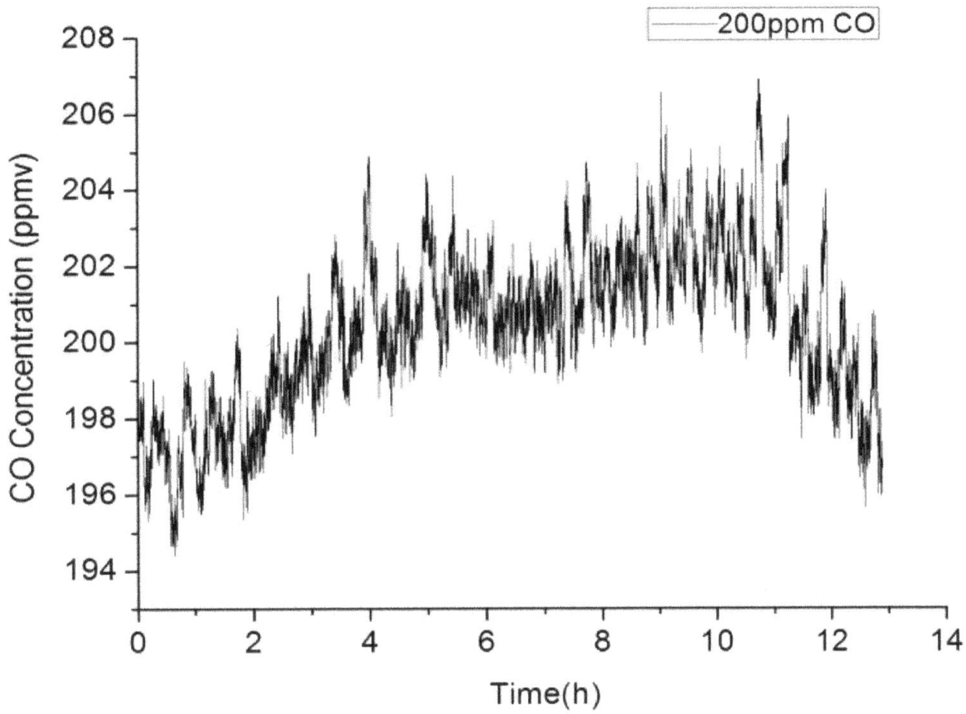

Figure 11. The measurement results of 200 ppm CO standard gas.

CO standard gas (ppm)	Mean value (ppm)	Standard deviation (ppm)	Fluctuation (%)
10	10.57	0.5	4.7
200	200.36	2.1	1

Table 2. Measured deviation of CO standard gas.

The measured concentration fluctuations of 10 and 200 ppm CO standard gas are 4.7 and 1% of the mean value, respectively. This illustrates that different concentration ranges should be divided when measuring low concentration gas with high sensitivity, such as 0–20, 20–50, 50–100 ppm, and so on, and different ranges have different stability indexes.

The results of Allan variance analysis of 10 ppm CO sample gas are shown in **Figure 12**. The corresponding integration time of the system is 30 s, the Allan variance is 0.067, and the predicted detection limit is 0.25 ppm. Moreover, if we continue to increase the integration time until the intersection with the slope of 1/2, the Allan variance decreases to 0.02, the corresponding detection limit is 0.14 ppm, but the long integration time will affect the sensitivity of the system [54]. Therefore, the integration time should be properly controlled when the requirement of detection limit is not very high. The measurement results of CO gas at different concentrations are shown in **Figure 13**, and the linear relationship between different concentrations and the peak values of second harmonic signal is displayed in **Figure 14**. The results illuminate that the measurement concentrations have a good linearity in the range of 10–250 ppm.

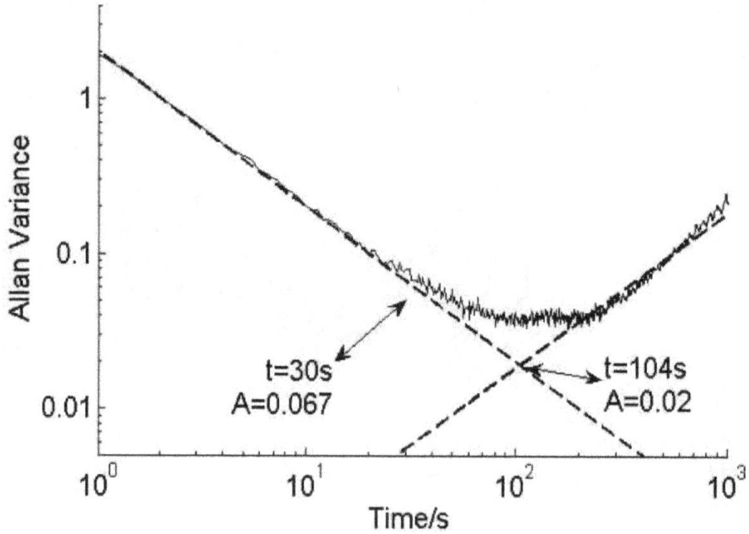

Figure 12. The Allan variance of 10 ppm CO.

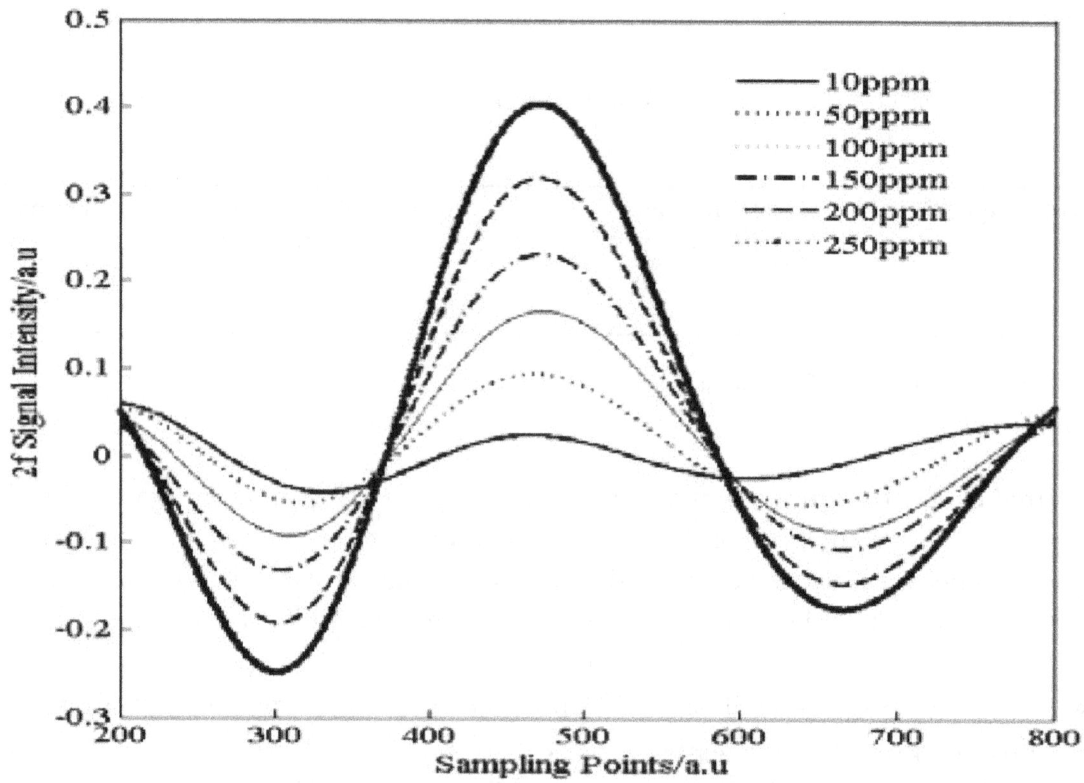

Figure 13. The 2f signals with different concentrations of CO.

The system of CO high sensitivity detection based on TDLAS technology combined with the new type of multi-pass absorption cell basically realizes the high sensitivity detection of CO in the near infrared. The system exhibits good stability and high linearity after long-term measurement experiments. According to the Allan variance analysis, the detection limit of the system is 0.25 ppm with an integration time of 30 s. The system meets the requirements for

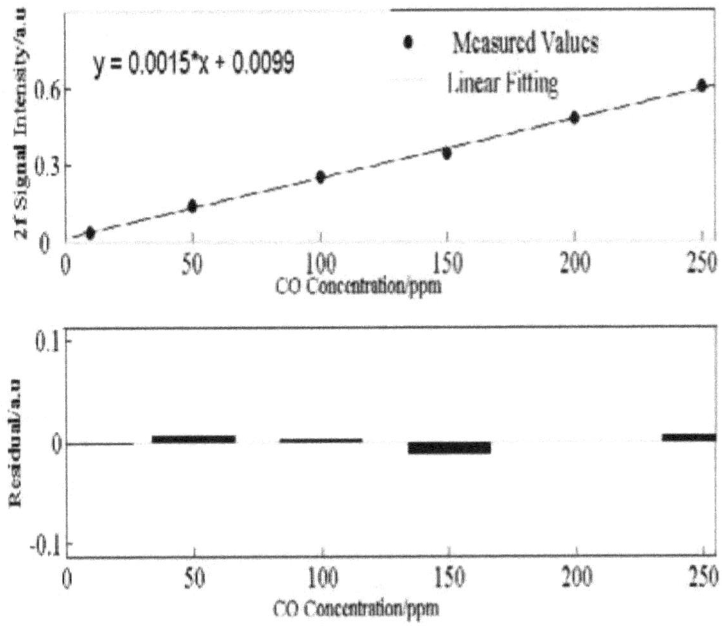

Figure 14. The linear relationship between different concentrations.

those situations which have a higher measurement requirement of CO such as alarming of coal spontaneous combustion and mine safety production. But it is only the results of experimental measurement under the laboratory conditions. For the mine environments with high temperature and humidity, the performances of the experimental relevant components need to be further tested.

5. Hydrogen sulfide (H_2S) monitoring

5.1. Introduction

Hydrogen sulfide (H_2S) is an important potential dangerous gas in oil drilling. It is colorless, highly toxic, and acidic; there is a special smell of rotten eggs; the olfactory threshold is 0.00041 ppm. Even low concentrations of H_2S can also damage people's sense of smell and have effects on the eye, respiratory system, and central nervous system. It is lethal to detect this kind of gas using a nose [55]. Because there is no smell when the concentration is high (high concentrations of hydrogen sulfide can paralyze olfactory nerves). Hence, sensitive H_2S detection is necessary in practical applications. In this part, a 1.578 μm distributed feedback (DFB) laser is used to detect H_2S of low concentration [56].

5.2. Wavelength modulation spectroscopy system

The WMS technique is used in the H_2S detection system, as shown in **Figure 15**. The used multi-pass absorption cell in this experiment is also homemade with a total optical path length of 56 m and a total volume of 0.8 L. A single-mode pigtailed DFB laser with a central wavelength of

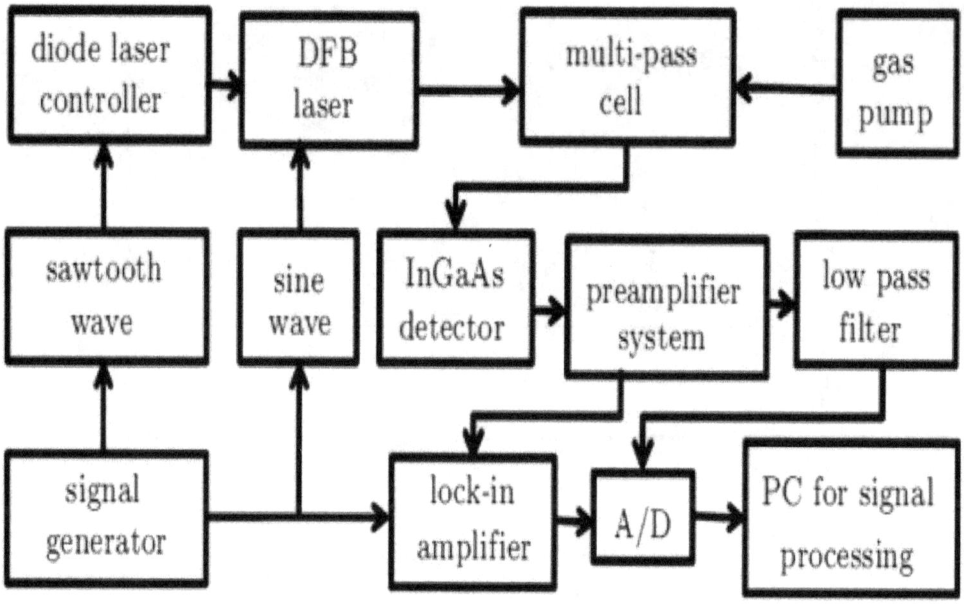

Figure 15. Sketch of the experimental setup for WMS system.

1.578 μm is employed in this system. The wavelength of the laser is controlled by a temperature and current controller, which can vary the laser wavelength with a magnitude of about 0.014 cm^{-1}/mA. The laser wavelength is scanned by a triangular wave of 30 Hz. A 20 kHz sine wave is used to modulate the laser output wavelength. The transmission signal was sent to the preamplifier system whose bias amplifier enhances the weak absorption signals. The parallel circuits amplify the signal and direct it to a lock-in amplifier for demodulation and to a low pass filter for obtaining the triangular wave after passing through the cell. Both signals are directed to a personal computer (PC) for signal processing via an A/D converter.

5.3. H_2S concentration measurements

5.3.1. Stability of the background signals

In order to improve the measurement accuracy and the detection limit, it is important to subtract the background spectrum in the spectral measurements. The background and the initial 2f signal are shown in **Figure 16**. Obviously, the symmetry of the demodulated signal was considerably improved after the background correction.

5.3.2. Linearity and response time

Linearity is an important parameter in measuring instruments. In this TDLAS system, certified H_2S gases with mixing ratios of 5, 10, 20, and 45 ppm were tested successively to check the linearity. **Figure 17** displays the corresponding background corrected signals at different mixing ratios. Furthermore, the mixing ratios and the peak-to-peak values are linearly fitted in **Figure 18**. The results illustrate that the system has a good linearity with a fitting coefficient of 0.998.

The fitting is given by y = 0.00143 + 7.52459x with a fitting coefficient of 0.998.

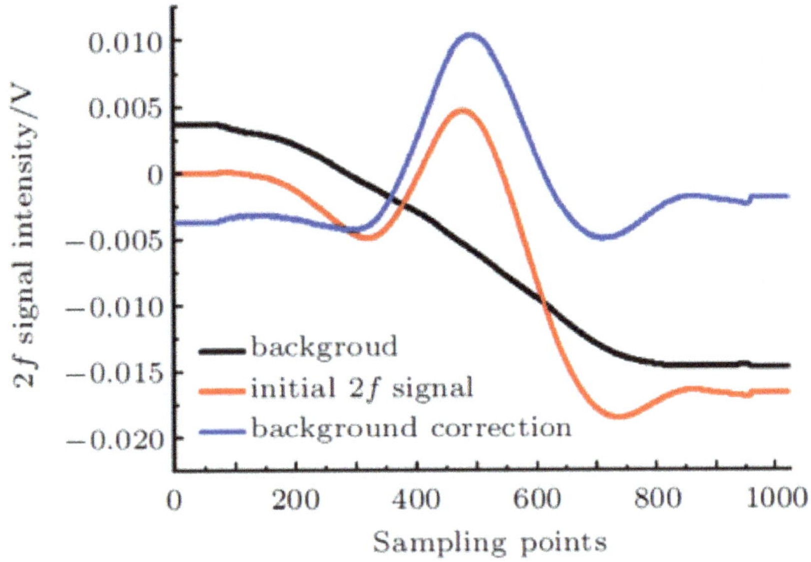

Figure 16. Background baseline.

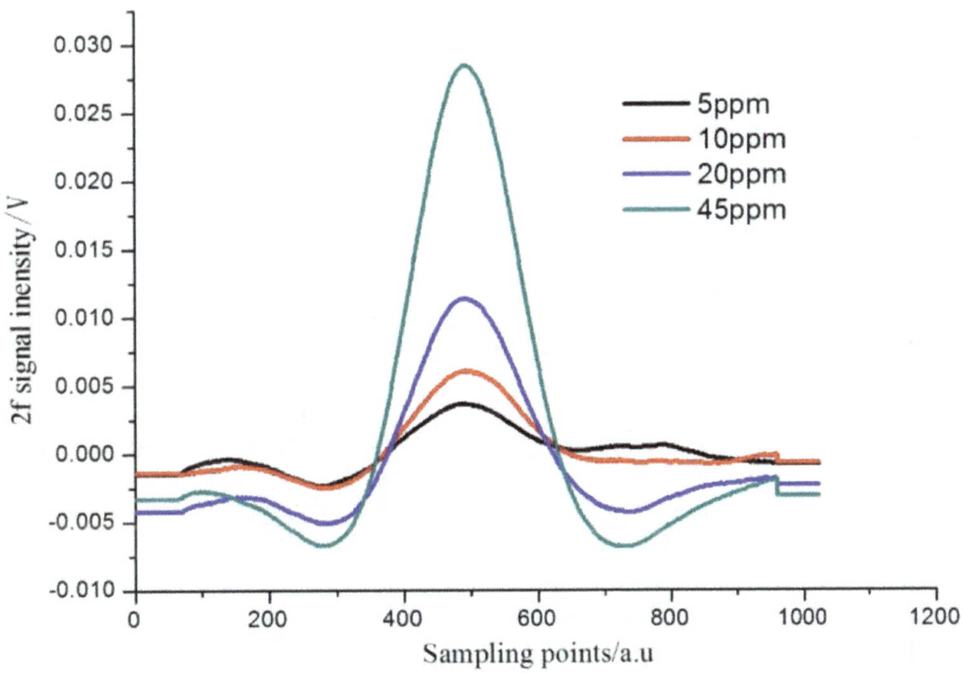

Figure 17. 2f signals with background correction for different H_2S concentrations.

The repeatability and the response time are also very important for H_2S detection. In the extraction system, a 1-L gasbag is used to get different H_2S standard gases into the multi-pass absorption cell successively. The response time depends on both the volume of the multi-pass absorption cell and the speed of releasing the airbag. The response time for filling 5 ppm H_2S standard gas mixture into the 0.8 L cell with an evacuation flow rate of 3 L/min is plotted in **Figure 19**. Obviously, the response is accurate and fast. In the system, the data sampling rate is 90 Hz, so the response time is about 4 s.

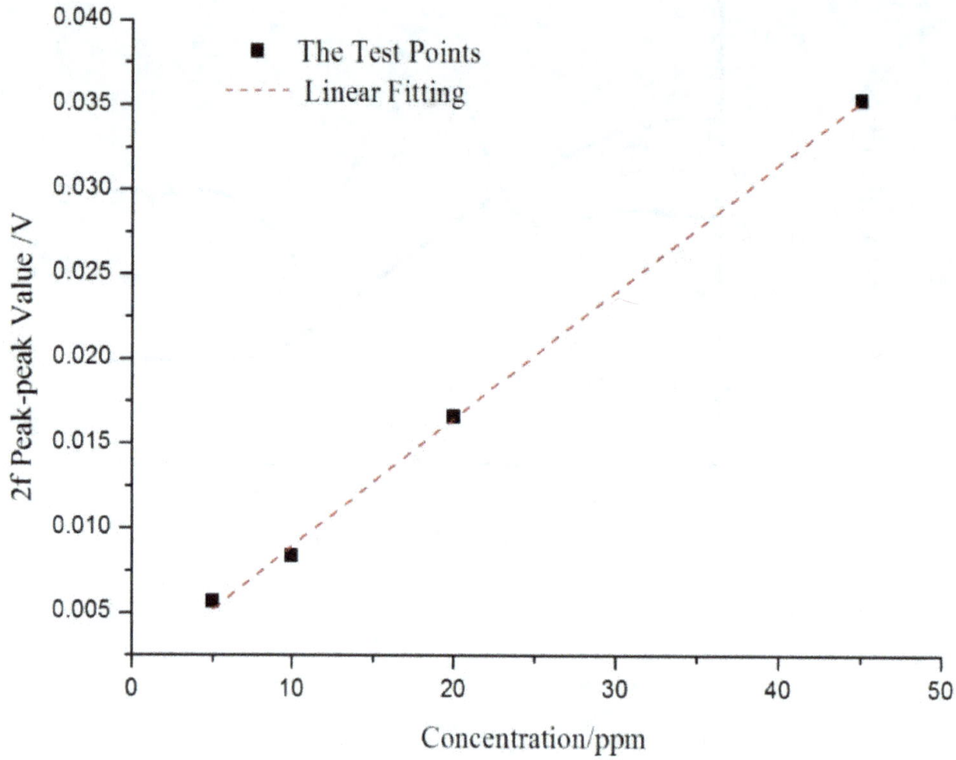

Figure 18. Linearity of the measurement system.

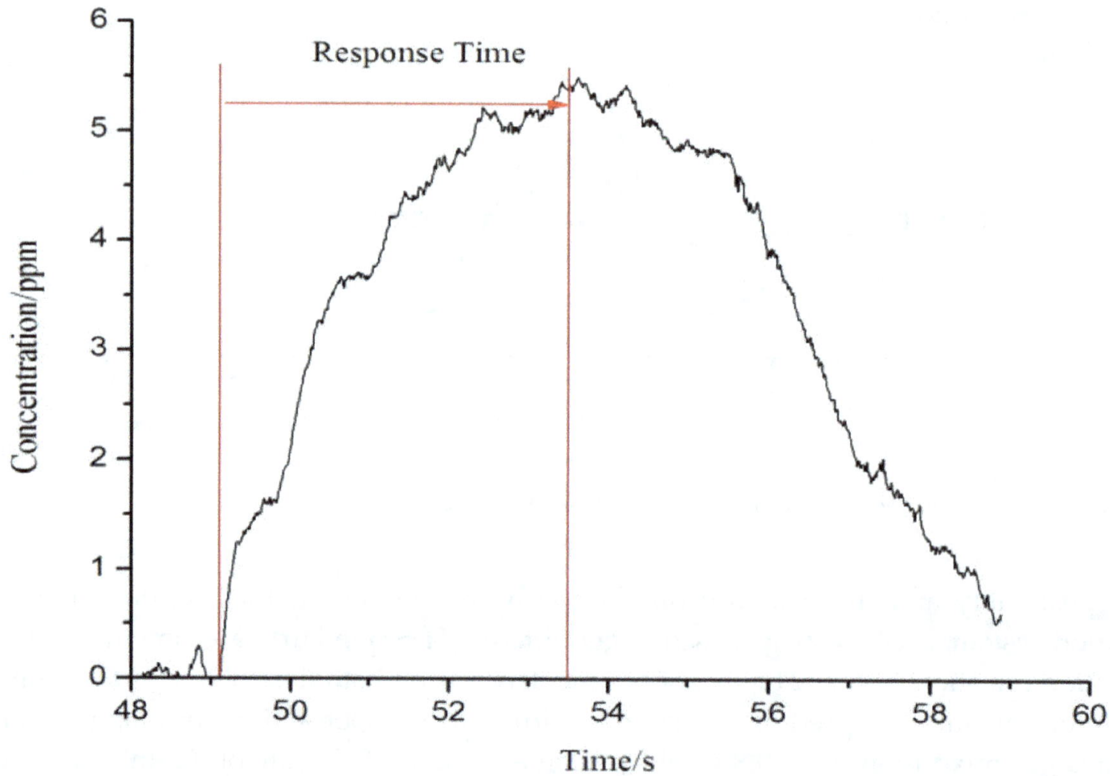

Figure 19. Response time of the measurement system.

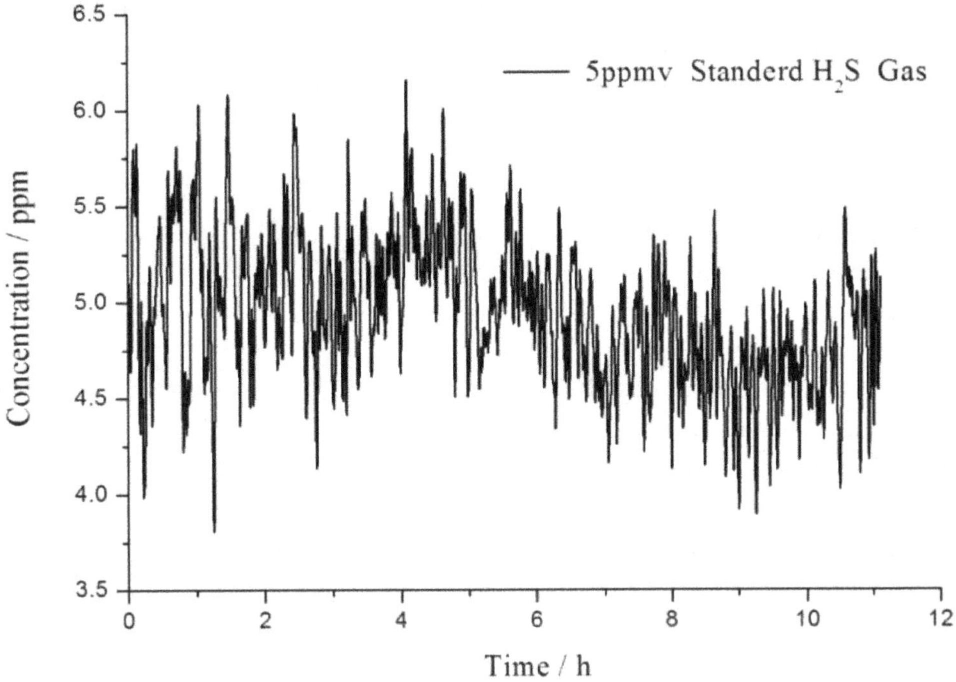

Figure 20. Measurement of 5 ppm H_2S standard gas.

5.3.3. System stability and detection limit

The Allan variance is usually used to analyze the temporal stability of the instrument performance. H_2S standard gas of 5 ppm was measured for a period time of 11 h and shown in **Figure 20** with the fluctuations of less than 1 ppm. Moreover, the Allan variance in **Figure 21** indicates a detection limit of 240 ppb with an integration time of 24 s for eliminating the white noise. When the time is increased to 60 s, the detection limit is reduced to 140 ppb for removing the 1/f noise.

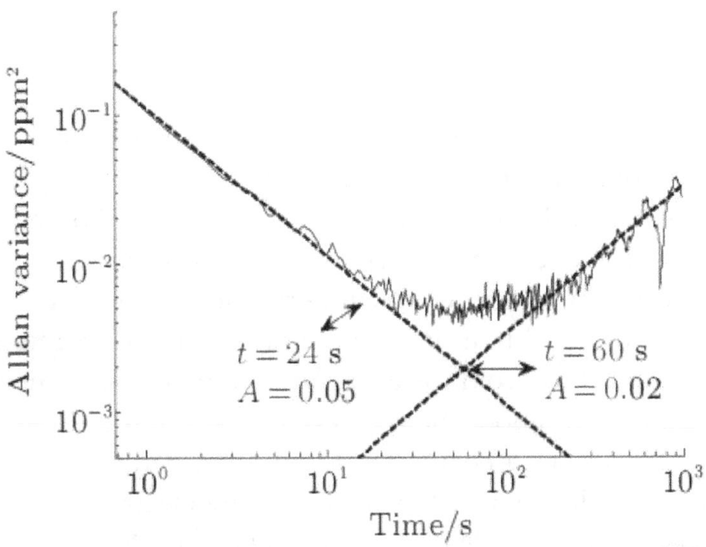

Figure 21. The Allan variance for 5 ppm H_2S.

The experimental results indicate that the system has good linearity, stability, and repeatability, combined with a quick response time and a low detection limit. The H_2S detection system based on TDLAS has the feasibility of online monitoring in many applications.

6. CO_2 isotope measurements

6.1. Introduction

Global warming is a serious problem that may lead to natural disasters, destroys the biological chain, and thus threats the existence and development of human beings. As one of the most important greenhouse gases, releasing of carbon dioxide must be controlled. Measuring and analyzing stable isotopes of atmospheric carbon dioxide are very useful to search sources and sinks of carbon dioxide in this area and seek the processes which are caused by human's activities. Moreover, human enzyme activities assessment, organ functions, and transport processes in the medical area could be achieved by noninvasive ^{13}C-breath analysis. For example, possible *Helicobacter pylori* infection of the stomach or the duodenum can be detected via ^{13}C-breath analysis. Thus ^{13}C-breath test can be easily performed and have a high patient acceptance [57].

The primary technology for determination isotopic ratio is isotope ratio mass spectrometry (IRMS) with a measurement precision from 0.01 to 0.1‰ by testing the mass of each isotope of samples. Although this method has high precision, the disadvantages of IRMS are obvious. For example, the instrument of IRMS is too large to move easily, and the sample must be pre-treated in the case of the influence of other substances whose numbers of molecules are same with those need to be tested. These drawbacks make it impossible to measure the isotopic ratio in situ or online. TDLAS is a popular way to measure concentrations of gases. According to direct absorption, concentration and isotope ratio can be easily calculated when temperature, pressure, optical path length, and absorption line strength of gases are certain.

6.2. Experimental setup

The experimental setup is depicted in **Figure 22**. The laser source is a room temperature operated DFB laser (nanoplus GmbH) with a center wavelength of 2.74 μm and a tuning range of 5 cm^{-1}. A visible He-Ne laser beam was used to do coalignment of the optical path since the mid-infrared light is not visible to human eyes. Positions of water vapor absorption lines from the HITRAN 2008 database provided an absolute frequency reference for frequency calibration. The laser beam was directed to a homemade multi-pass absorption cell with an optical path length of 107 m. In order to avoid the absorption line intensity fluctuation caused by the absorption cell temperature variation, the temperature of the multi-pass absorption cell was maintained at 30°C by the use of a heater band and a temperature controller. The emerging absorption signal from the cell was focused onto a thermoelectrically cooled (TEC) photovoltaic VIGO detector (PVI-4TE-3). The detector output was sampled with a fast data acquisition card and then transferred to a personal computer for further data processing.

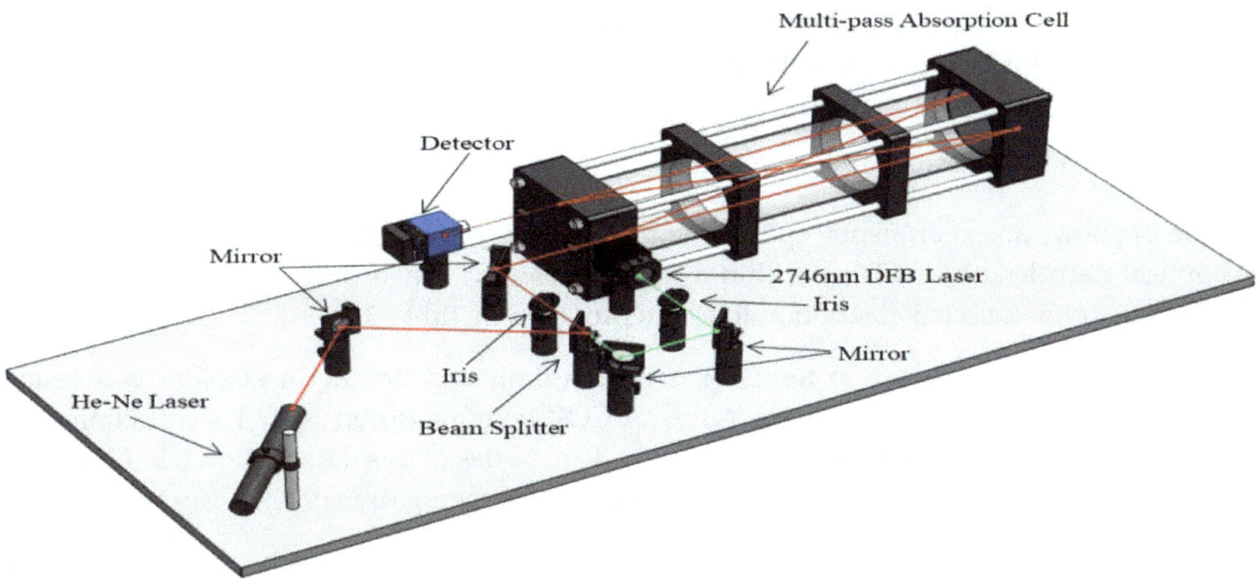

Figure 22. Three-dimensional view of the experimental setup.

6.3. Absorption line selection

For high-precision isotopic-ratio determination, it is necessary to select absorption lines which simultaneously fulfill the following conditions: (1) they should be located within the scanning range of the laser; (2) there should be no interferences from other atmospheric species, primarily water vapor; (3) the isotopologues of interest should have similar absorption strength to obtain an optimal SNR [58, 59].

Using the above requirements, it is rather straightforward to identify spectral regions that may contain suitable sets of isotopic absorption lines. Spectra simulation of 5% H_2O and 500 ppm CO_2 based on line positions and line strengths reported in the HITRAN 2008 database in the spectral range of the DFB laser scanned is displayed in **Figure 23**. The two absorption lines of

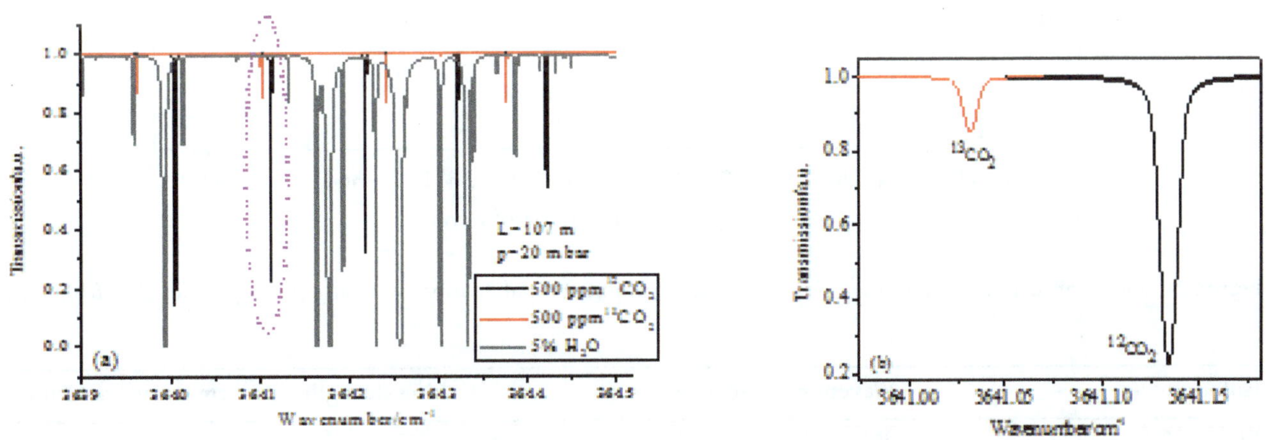

Figure 23. (a) Simulated absorption spectrum of 5% H_2O and 500 ppm CO_2 in the spectral range of 3639–3645 cm^{-1}. (b) Signal simulation of 500 ppm $^{12}CO_2$ and $^{13}CO_2$ based on HITRAN 2008 database with a path length of 107 m at a pressure of 20 mbar. $^{12}CO_2$, $^{13}CO_2$, and H_2O absorption lines are shown in black, red, and gray, respectively.

3641.0311 cm^{-1} for $^{13}CO_2$ and 3641.1338 cm^{-1} for $^{12}CO_2$ were selected for isotope analysis of CO_2 and free of interferences of water vapor absorption lines.

6.4. Results and discussion

Figure 24 shows an experimental spectrum of $^{12}CO_2$ and $^{13}CO_2$ in ambient air at 20 mbar with an optical path length of 107 m within a narrow scanned range of 0.1 cm^{-1}. Spectroscopic parameters of the selected absorption lines are provided in **Table 3**.

The instrument performance in terms of detection limit and long-term stability was tested using the Allan variance. The mixing ratios of CO_2 were measured with 1 s collection time from a standard gas cylinder with 197 ppm CO_2. Time series of this data is shown in **Figure 25**. From the associated Allan variance plot, an optimum averaging time of 130 s can be derived.

This instrument was used to measure the isotope ratios of CO_2 in the ambient air. Time series of CO_2 mixing ratio profiles and the derived $\delta^{13}C$ values with 1 s average time are shown in **Figure 26**. The measured mean value of CO_2 mixing ratios and $\delta^{13}C$ is 454 ppm and −98.75‰, respectively. The 1σ standard deviation of $\delta^{13}C$ is 1.8‰. According to the Allan variance, the optimum integration time is 130 s; the corresponding measurement precision can reach to

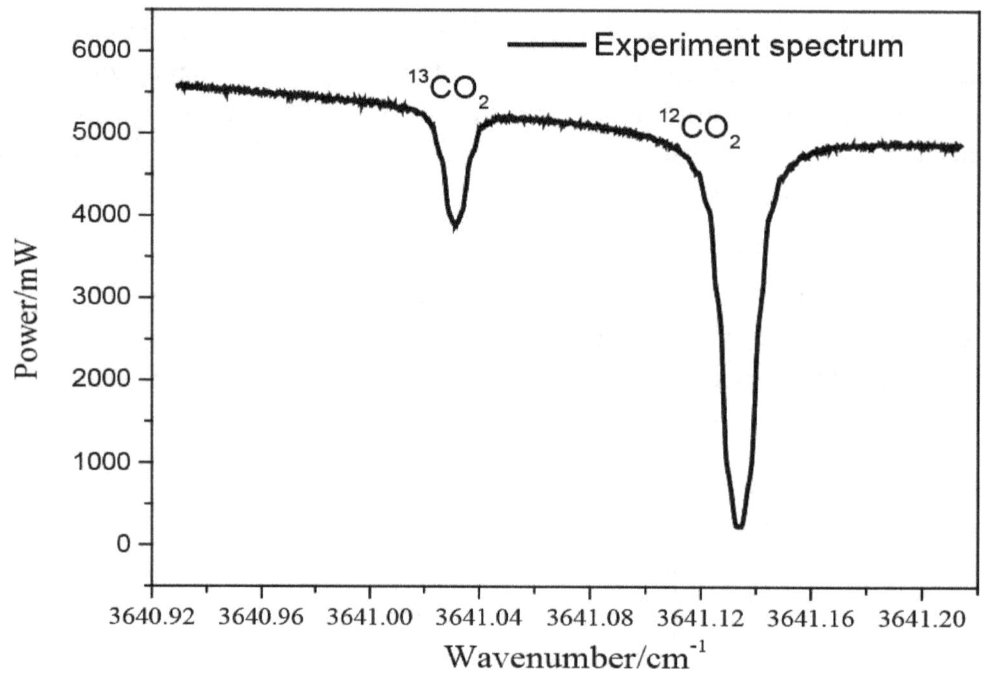

Figure 24. Direct absorption signals of $^{12}CO_2$ and $^{13}CO_2$ in ambient air at 20 mbar with an optical path length of 107 m.

Isotopologue	Wavenumber (cm^{-1})	Line strength (10^{-21} cm^{-1} cm^2/molecule)
$^{16}O^{12}C^{16}O$	3641.1338	5.637
$^{16}O^{13}C^{16}O$	3641.0311	0.641

Table 3. Spectroscopic parameters of the selected absorption lines for this work.

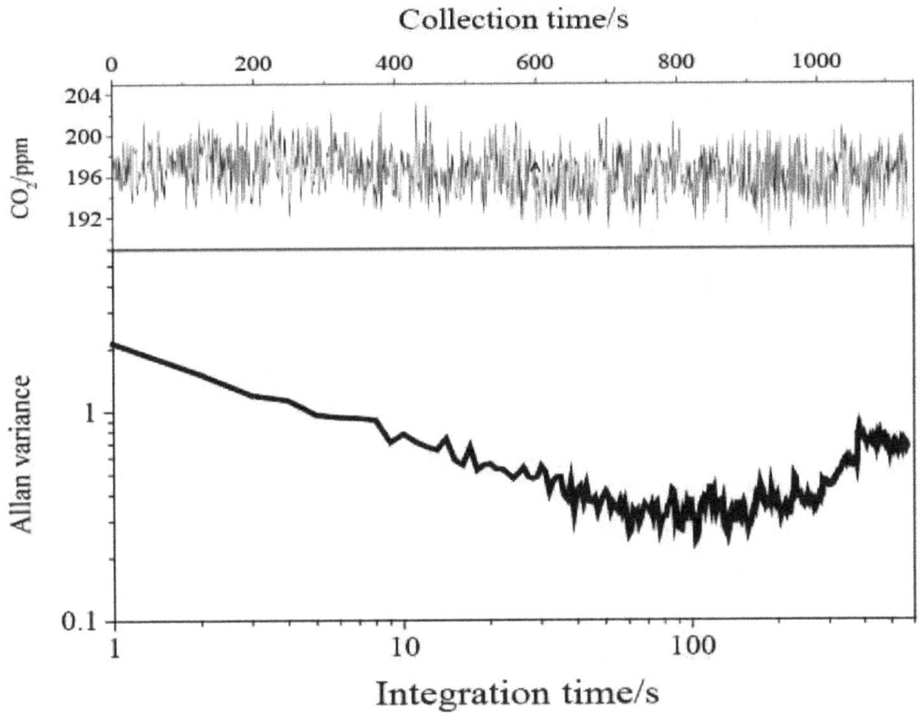

Figure 25. Time series and Allan plot of CO_2 from a standard gas cylinder.

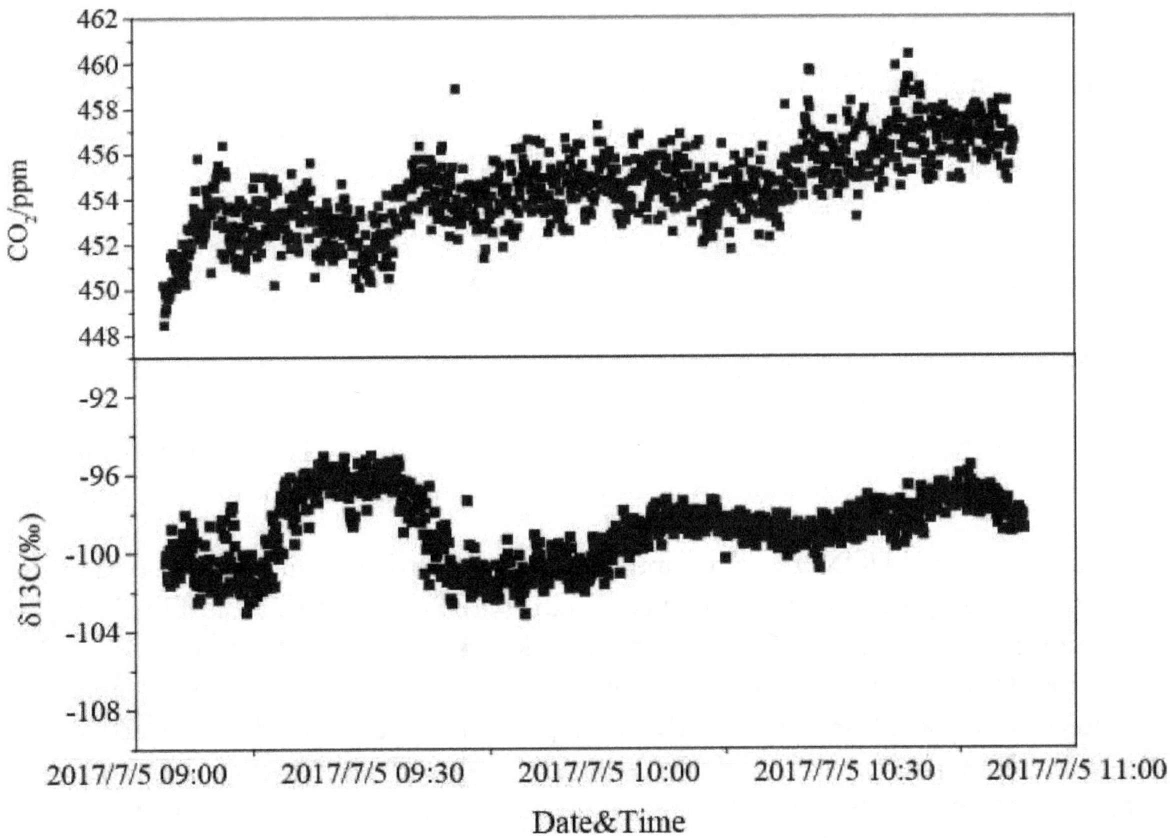

Figure 26. Time series of CO_2 mixing ratios and $\delta^{13}C$ measured by the DFB spectrometer.

0.2‰. For our CO_2 isotopologue measurement system based on TDLAS, high measurement precision has been obtained; the next step is to further improve the long-term stability of the system and perform calibration to get the correct isotope ratios and after that apply it to the medical area.

7. HONO measurements

7.1. Introduction

Gaseous nitrous acid (HONO) is a highly reactive short-lived species playing a significant role in tropospheric photochemistry. The photolysis of HONO in the wavelength range of 300–400 nm is an important source of the primary hydroxyl free radical (OH) in the lower atmosphere, up to 80% of the integrated source strength [60, 61]:

$$\mathrm{HONO} + h\nu\ (300\ \mathrm{nm} < \lambda < 400\ \mathrm{nm}) \rightarrow \mathrm{OH} + \mathrm{NO} \tag{8}$$

The OH radical governs the oxidation and removal of most pollutants from the atmosphere and is also a key species in photochemical cycles responsible for ozone formation leading to the so-called "photochemical smog" pollution. Therefore, HONO directly affects the oxidative capacity of the troposphere and indirectly contributes to production of secondary pollutants via the oxidation. Knowledge of atmospheric HONO concentration is very important for precise estimation of the OH radical budget and hence precise prediction of the impact on climate and air quality [62, 63]. In the lower atmosphere, the following formation pathways of gaseous HONO are commonly considered: (1) homogenous reaction [64, 65], (2) direct emission (i.e., by traffic) [66], and (3) heterogeneous conversion of NO_2 to HONO on the ground and other surfaces [67–71]. Homogeneous reaction and direct emissions have been identified, but these two sources are not sufficient to explain the observed atmospheric concentrations of HONO. At present, it is generally considered that HONO is mainly produced from heterogeneous process, namely, the heterogeneous reactions of NO_2 on wet surfaces as well as on surface of reducing substances such as carbon black aerosol surface [72–74]. Despite a large amount of research, the sources and the formation mechanisms of HONO in the atmosphere are still not well understood and identified due to the lack of accurate local measurements [75].

Good understanding of HONO sources and sinks requires instruments capable of performing high sensitivity, high precision, high specificity, high spatial resolution, and fast in situ measurements. Among various analytical instruments developed for field HONO monitoring [76–78], spectroscopic detection techniques capable of performing in situ measurements without any sample preparation have been increasingly developed since last decade as an attractive alternative for quantitative assessments of HONO in the atmosphere. Methods such as DOAS, incoherent broadband cavity-enhanced absorption spectroscopy (IBBCEAS), and the long-path absorption photometer (LOPAP) used in the ultraviolet region usually can get several hundred ppt-level detection limits, but the integration time of several minutes is long and cannot satisfy the requirement of fast measurements [6, 79]. In the mid-infrared region, the continuous-wave quantum cascade lasers (cw-QCL) combined with a multi-pass absorption cell based on TDLAS technology were applied to the measurement of atmospheric HONO with a sub-ppb detection limit [80–85]. The advantages of using cw-QCLs in TDLAS over lead salt lasers are better mode stability, higher laser output power, and room temperature operation without the need for cryogenic cooling, which facilitates long-term field measurements.

7.2. QCL-based instrumental platform

The developed QCL instrumental approach is depicted in **Figure 27**. It was based on a room temperature operation cw distributed feedback (DFB) quantum cascade laser (DQ7-M776H, Maxion Technologies, Inc.). It emitted single-mode laser power of up to 35 mW. The wavelength tuning of ~2 cm^{-1} around 1254 cm^{-1} might be achieved by ramping laser injection current and/or temperature tuning. The pre-collimated laser beam from the QCL was first coupled to a beam splitter (with 90% transmission and 10% reflection). In order to make the optical alignment easy, a visible He-Ne laser beam was adjusted to be coaxial with the invisible infrared beam from the QCL. The transmitted light was directed to a multi-pass cell with a base length of 0.8 m and a folded path length of 158 m. The emerging absorption signal from the multi-pass cell was focused onto a thermoelectrically cooled (TEC) photovoltaic VIGO detector (detector 1: PVI-4TE-10.6). The reflected beam was directed to a homemade Fabry-Perot etalon with a free spectral range of 0.03 cm^{-1}. The optical fringe signal was recorded with another VIGO detector (detector 2: PVMI-10.6) and used for relative wavelength metrology. The pressure in the multi-pass cell was

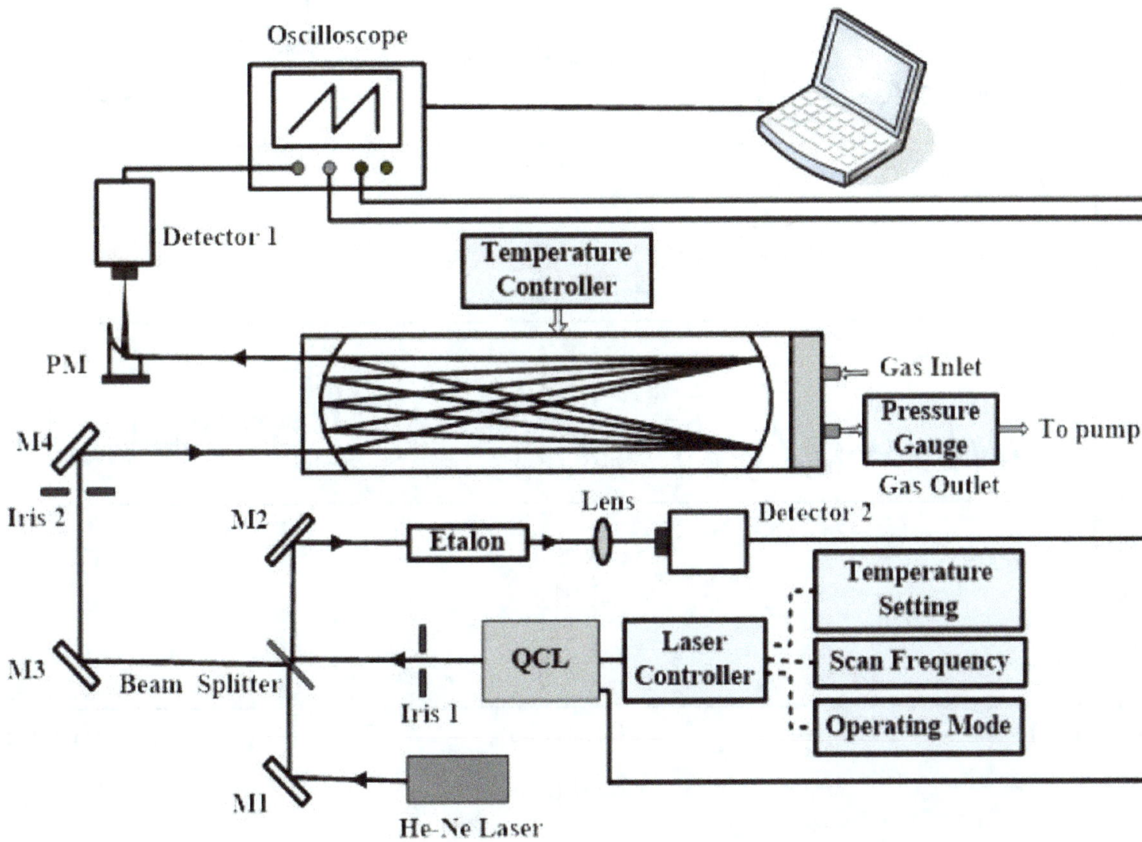

Figure 27. Schematic diagram of the experimental setup. Lens: f = 50 mm. PM (parabolic mirror): f = 25 mm. M: Mirror.

measured with a pressure transducer (Pfeiffer Vacuum, CMR 361). Temperature of the multi-pass cell was maintained at 30°C (within ±0.1°C) in order to avoid deposit of aqueous nitrous acid on the optical cell wall (especially on the cell mirrors) and to avoid any artifact production due to heterogeneous reaction inside the cell. The two detector outputs were sampled with a fast data acquisition digital oscilloscope (LeCroy Wavesurfer 104Xs-A). The data was then transferred to a personal computer for further data processing.

7.3. Results and discussion

7.3.1. Continuous monitoring of atmospheric HONO

The developed QCL instrument was employed for monitoring daytime and nighttime variation of HONO in an urban environment near a road with moderate traffic. Continuous monitoring of HONO mixing ratio variation was performed during a campaign of several days. **Figure 28** shows time series of the mixing ratios of 15 min averages of HONO and NO_2 and the corresponding ratios of HONO/NO_2 along with the solar radiation for the field measurements from 14 to 18 January 2013 (16–18 of them are snowy days). NO_2 was measured by a NO_x analyzer (Environmental SA). The solar radiation was recorded by a weather station (Davis Vantage Pro2, Montanay). The measured HONO mixing ratios ranged from 1.40 ppb to

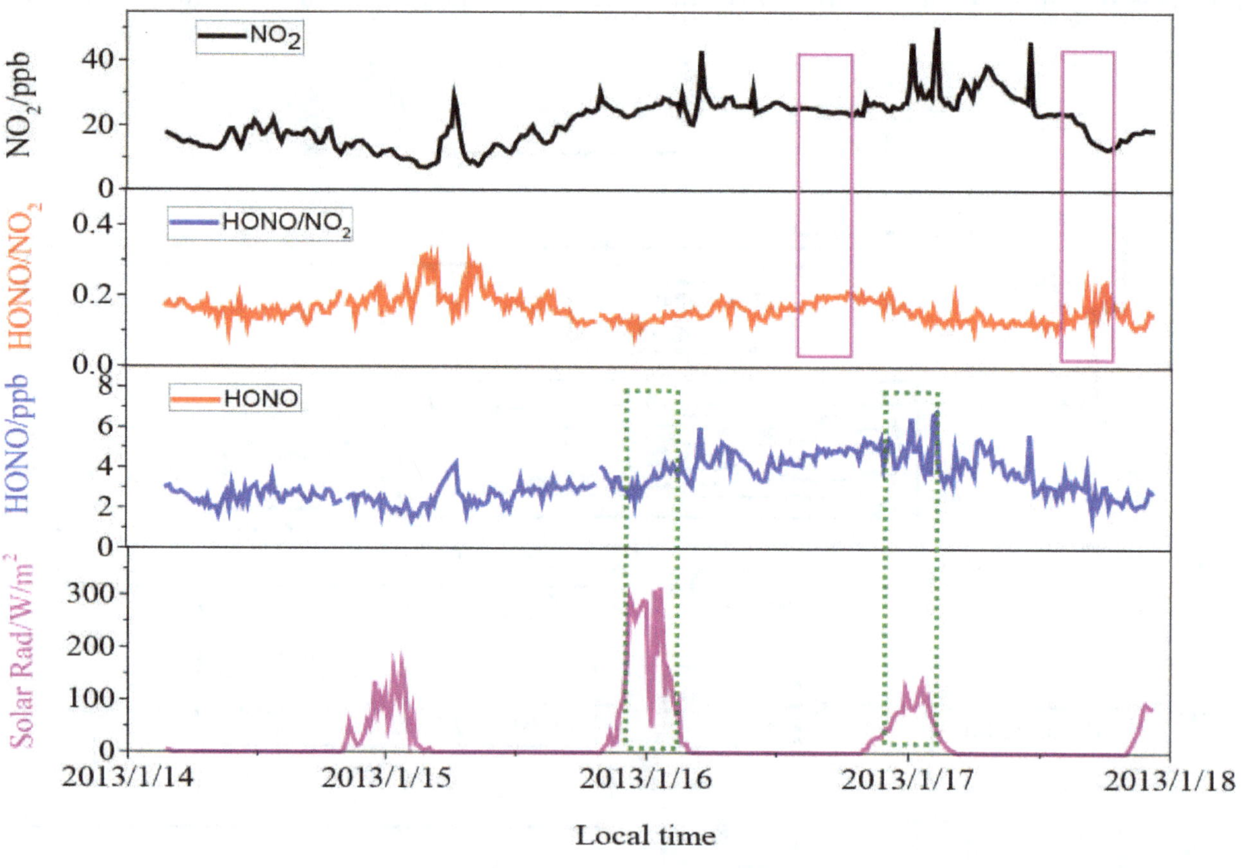

Figure 28. Time series of HONO, NO_2, the solar radiation, and HONO/NO_2 during the field measurements from 14 to 18 January 2013.

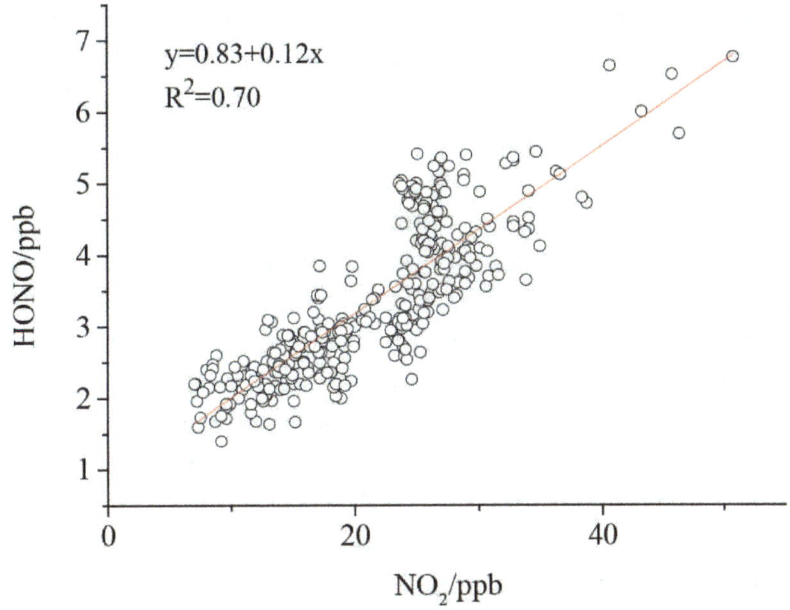

Figure 29. Correlation between HONO and NO_2 during the measurement period.

6.76 ppb, with a mean value of 3.33 ± 1.03 ppb, whereas the mean and maximum mixing ratios of NO_2 were 21.32 ± 7.36 ppb and 50.70 ppb, respectively.

7.3.2. Possible sources of HONO

NO_2 is known to be an important precursor for the formation of HONO or to have a common source. As mentioned in the introduction, the mixing ratios of HONO and NO_2 were found to be highly correlated in many field observations [77, 86]. The regression analysis (shown in **Figure 29**) of the combined data sets indicates good correlation between HONO and NO_2 mixing ratios, displaying an intercept of 0.83, slope of 0.12, and R^2 of 0.70. This slope can be interpreted as an upper limit for estimate of the HONO exhaust fraction of NO_2 emissions. The two parts marked with purple rectangles in **Figure 28** implied other sources of HONO formation, because of the increasing fraction of HONO/NO_2 with decreasing NO_2 mixing ratios. The higher mixing ratios of HONO in the morning are considered as products of heterogeneous reactions of NO_2 on wet surfaces during nighttime. The two green rectangles in **Figure 28** show a record of HONO mixing ratio variation with solar radiation on snow days. A photochemically enhanced HONO production from snowpack under solar radiation can be seen [73]. Finding the missing sources and the formation mechanism of HONO in the atmosphere is still the actual topic for tropospheric HONO chemistry.

8. Summary and outlook

In conclusion, we overviewed our recent developments of several gas sensors based on TDLAS technology for in situ monitoring of hazard gases, including CH_4, CO_2, CO, HONO, H_2S, and $^{13}CO_2/^{12}CO_2$. Good understanding of the sources and sinks of these hazard gases requires instruments capable of performing high sensitivity, high precision, high specificity, high spatial resolution, and fast in situ measurements. TDLAS is an effective method to measure these gases' mixing ratios and multiple parameters with these advantages. The methane detection system based on TDLAS can simultaneously detect CH_4, C_2H_2, and C_2H_4 rapidly and effectively in open environment, and the response time is less than 2 s. The MDLs of these three gases can meet the requirements for the detection of natural gas leakage to petrochemical industry. The accuracy of making an alarm is 100%, which can be used in natural gas station and valve room gas leakage detection. The detection limit of CO detection system based on TDLAS technology is 0.25 ppm with an integration time of 30 s, which basically realizes the high sensitivity detection of CO in the near infrared and satisfies the requirements for those situation that have a higher measurement requirement of CO such as alarming of coal spontaneous combustion and mine safety production. The experimental results of H_2S show that the system based on TDLAS has a good linearity and stability with a quick response time of 24 s and a low detection limit of 240 ppb. This indicates that the system has the feasibility of real-time online monitoring in many applications. The measurement system of CO_2 isotopologues has realized the high measurement precision of 0.2‰ for $\delta^{13}C$; the next step is to carry out calibration to get the correct isotope ratios and achieve long-term stability measurements. Good understanding of the important roles of HONO in the key chemical processes of hydroxyl radicals and the sources of HONO requires correct detection of the HONO mixing ratios. A QCL-based instrumental system was designed to measure the atmospheric HONO. The regression analysis indicates good correlation between HONO and NO_2. But increasing HONO mixing ratios with decreasing NO_2 also indicates other sources of HONO formation. Finding the missing sources and the formation mechanism of HONO in the atmosphere is still a great challenge for tropospheric HONO chemistry.

Although parts of these gas analysis experiments are just results under laboratory conditions, we are improving the stability and SNR of these systems with the aim of putting them into practical application. To date we have developed all-fiber gas sensor to detect CH_4, O_2, C_2H_2, and C_2H_4, portable CH_4 sensors, CO_2 analyzer, CO analyzer, and so on. Some of them have been put into the application. The development of these gas sensors would be beneficial for the implementation of environmental protection policies and expand their application in energy, public safety, and medical science. The TDLAS technology also shows high potential for monitoring all kinds of hazardous gases in the atmosphere from surface layer to troposphere combined with a wide spectral application range from the near infrared to mid-infrared.

Acknowledgements

This work is partly supported by National Natural Science Foundation of China (Grant No. 41775128, 41405034, 11204319), the External Cooperation Program of the Chinese Academy of Sciences (Grant No. GJHZ1726), the Special Fund for Basic Research on Scientific Instruments of the Chinese Academy of Science (Grant No. YZ201315), and the Chinese Academy of Science President's International Fellowship Initiative (PIFI, 2015VMA007). Finally, the Scientific Research Foundation for the Returned Overseas Chinese Scholars, State Education Ministry, is gratefully acknowledged.

Author details

Xiaojuan Cui[1,2], Fengzhong Dong[1,2,3*], Zhirong Zhang[1,2], Hua Xia[1,2], Tao Pang[1], Pengshuai Sun[1], Bian Wu[1], Shuo Liu[1,3], Luo Han[1,3], Zhe Li[1,3] and Runqing Yu[1,3]

*Address all correspondence to: fzdong@aiofm.ac.cn

1 Anhui Provincial Key Laboratory of Photonic Devices and Materials, Anhui Institute of Optics and Fine Mechanics, Chinese Academy of Sciences, Hefei Anhui, China

2 Key Laboratory of Environmental Optics and Technology, Anhui Institute of Optics and Fine Mechanics, Chinese Academy of Sciences, Hefei, China

3 University of Science and Technology of China, Hefei, China

References

[1] Zhang Z, Pang T, Yang Y, Xia H, Cui X, Sun P, Wu B, Wang Y, Sigrist MW, Dong F. Development of a tunable diode laser absorption sensor for online monitoring of industrial gas total emissions based on optical scintillation cross-correlation technique. Optics Express. 2016;**24**(10):A943-A955. DOI: 10.1364/OE.24.00A943

[2] Dong F, Liu W, Chu Y, Li J, Zhang Z, Wang Y, Pang T, Wu B, Tu G, Xia H, Yang Y, Shen C, Wang Y, Ni Z, Liu J. Real-time in situ measurements of industrial hazardous gas concentrations and their emission gross. InTech Publisher. 2011:66-90. DOI: 10.5772/27161

[3] Nelson DD, McManus B, Urbanski S, Herndon S, Zahniser MS. High precision measurements of atmospheric nitrous oxide and methane using thermoelectrically cooled mid-infrared quantum cascade lasers and detectors. Spectrochimica Acta A. 2004;**60**(14):3325-3335. DOI: 10.1016/j.saa.2004.01.033

[4] Wojtas J, Bielecki Z, Stacewicz T, Mikolajczyk J, Rutecka B, Medrzycki R. Nitrogen oxides optoelectronic sensors operating in infrared range of wavelengths. Acta Physica Polonica. 2013;**124**(124):592-594. DOI: 10.12693/APhysPolA.124.592

[5] Vardag SN, Hammer S, O'Doherty S, Spain TG, Wastine B, Jordan A, Levin I. Comparisons of continuous atmospheric CH_4, CO_2 and N_2O measurements-results of InGOS travelling instrument campaign at Mace head. Atmospheric Chemistry and Physics. 2014;**14**(7):8403-8418. DOI: 10.5194/acpd-14-10429-2014

[6] Wu T, Zha Q, Chen W, Xu Z, Wang T, He X. Development and deployment of a cavity enhanced UV-LED spectrometer for measurements of atmospheric HONO and NO_2 in Hong Kong. Atmospheric Environment. 2014;**95**(1):544-551. DOI: 10.1016/j.atmosenv.2014.07.016

[7] Lima JP, Vargas H. Mikl'os a, Angelmahr M, Hess P. Photoacoustic detection of NO_2 and N_2O using quantum cascade lasers. Applied Physics B: Lasers and Optics. 2006;**85**(2):279-284. DOI: 10.1007/s00340-006-2357-0

[8] Ma Y, Lewicki R, Razeghi M, Tittel FK. QEPAS based ppb-level detection of CO and N_2O using a high power CW DFB-QCL. Optics Express. 2013;**21**(1):1008-1019. DOI: 10.1364/OE.21.001008

[9] Bjorneberg DL, Leytem AB, Westermann DT, Griffiths PR, Shao L, Pollard MJ. Measurement of atmospheric ammonia, methane, and nitrous oxide at a concentrated dairy production facility in southern Idaho using open-path FTIR spectrometry. Transactions of the ASABE. 2009;**52**(5):1749-1756. DOI: 10.13031/2013.29137

[10] Kleffmann J, Lorzer JC, Wiesen P, Kern C, Trick S, Volkamer R, Rodenas M, Wirtz K. Intercomparison of the DOAS and LOPAP techniques for the detection of nitrous acid (HONO). Atmospheric Environment. 2006;**40**(20):3640-3652. DOI: 10.1016/j.atmosenv.2006.03.027

[11] Cui X, Dong F, Sigrist MW, Zhang Z, Wu B, Xia H, Pang T, Sun P, Fertein E, Chen W. Investigation of effective line intensities of trans-HONO near 1255 cm^{-1} using continuous-wave quantum cascade laser spectrometers. Journal of Quantitative Spectroscopy and Radiation Transfer. 2016;**182**:277-285. DOI: 10.1016/j.jqsrt.2016.06.014

[12] Rocco A, Natale GD, Natale PD, Gagliardi G, Gianfrani L. A diode-laser-based spectrometer for in-situ measurements of volcanic gases. Applied Physics B: Lasers and Optics. 2004;**78**(2):235-240. DOI: 10.1007/s00340-003-1339-8

[13] D'Amato F, Mazzinghi P, Castagnoli F. Methane analyzer based on TDL's for measurements in the lower stratosphere: Design and laboratory tests. Applied Physics B: Lasers and Optics. 2002;**75**(2–3):195-202. DOI: 10.1007/s00340-002-0981-x

[14] Sauer GC, Pisano TJ, Fitz RD. Tunable diode laser absorption spectrometer measurements of ambient nitrogen dioxide, nitric acid, formaldehyde, and hydrogen peroxide in Parlier, California. Atmospheric Environment. 2003;**37**(12):1583-1591. DOI: 10.1016/S1352-2310(03)00004-9

[15] Nelson DD, Shorter JH, Mcmanus JB, Zahniser MS. Sub-part-per-billion detection of nitric oxide in air using a thermo electrically cooled mid-infrared quantum cascade laser spectrometer. Applied Physics B: Lasers and Optics. 2002;**75**(2–3):343-350. DOI: 10.1007/s00340-002-0979-4

[16] Hancock G, Kasyutich VL, Ritchie GAD. Wavelength-modulation spectroscopy using a frequency-doubled current-modulated diode laser. Applied Physics B: Lasers and Optics. 2002;**74**(6):569-575. DOI: 10.1007/s003400200863

[17] Werle PW, Mazzinghi P, Amato FD, Rosa MD, Maurer K, Slemr F. Signal processing and calibration procedures for in situ diode-laser absorption spectroscopy. Spectrochimica Acta Part A. 2004;**60**(8–9):1685-1705. DOI: 10.1016/j.saa.2003.10.013

[18] Liu X, Jeffries JB, Hanson RK, Hinckley KM, Woodmansee MA. Development of a tunable diode laser sensor for measurements of gas turbine exhaust temperature. Applied Physics B: Lasers and Optics. 2006;**82**(3):469-478. DOI: 10.1007/s00340-005-2078-9

[19] Roller C, Fried A, Walega J, Weibring P, Tittel F. Advances in hardware, system diagnostics software, and acquisition procedures for high performance airborne tunable diode laser measurements of formaldehyde. Applied Physics B: Lasers and Optics. 2006;**82**(2): 247-264. DOI: 10.1007/s00340-005-1998-8

[20] Varga A, Bozoki Z, Szakall M, Szabo G. Photoacoustic system for on-line process monitoring of hydrogen sulfide (H_2S) concentration in natural gas streams. Applied Physics B: Lasers and Optics. 2006;**85**(2–3):315-321. DOI: 10.1007/s00340-006-2388-6

[21] Werle P. Accuracy and precision of laser spectrometers for trace gas sensing in the presence of optical fringes and atmospheric turbulence. Applied Physics B: Lasers and Optics. 2011;**102**(2):313-329. DOI: 10.1007/s00340-010-4165-9

[22] Cassidy DT, Reid J. Baseline correction method for second-harmonic detection with tunable diode lasers. Applied Optics. 1982;**183**(86):141-154. DOI: 10.1016/0003-2670(86)80082-4

[23] Tanaka K, Tonokura K. Sensitive measurements of stable carbon isotopes of CO_2 with wavelength modulation spectroscopy near 2 μm. Applied Physics B: Lasers and Optics. 2011;**105**(2):463-469. DOI: 10.1007/s00340-011-4666-1

[24] Sappev AD, Masterson P, Huelson E, Howell J, Estes M, Hofvander H, Jobson A. Results of closed-loop coal-fired boiler operation using a TDLAS sensor and smart process control software. Combustion Science and Technology. 2011;**183**(11):1282-1295. DOI: 10.1080/00102202.2011.590560

[25] Zhang Z, Xia H, Dong F, Pang T, Wu B, Sun P, Wang G, Wang Y. Simultaneous detection of multiple gas concentrations with multi-frequency wavelength modulation spectroscopy. Europhysics Letters. 2013;**104**(4), 44002. DOI: 10.1209/0295-5075/104/44002

[26] Witzel O, Klein A, Meffert C, Wagner S, Kaiser S, Schulz C, Ebert V. VCSEL-based, high-speed, in situ TDLAS for in-cylinder water vapor measurements in IC engines. Optics Express. 2013;**21**(17):19951-19965. DOI: 10.1364/OE.21.019951

[27] Liu C, Xu L, Chen J, Cao Z, Lin Y, Cai W. Development of a fan-beam TDLAS-based tomographic sensor for rapid imaging of temperature and gas concentration. Optics Express. 2015;**23**(17):22494-22511. DOI: 10.1364/OE.23.022494

[28] Cassidy DT, Reid J. Atmospheric pressure monitoring of trace gases using tunable diode lasers. Applied Optics. 1982;**21**(7):1185-1190. DOI: 10.1364/AO.21.001185

[29] Sepulveda-Jauregui A, Martinez-Cruz K, Strohm A, Katey MWA, Thalasso F. A new method for field measurement of dissolved methane in water using infrared tunable diode laser absorption spectroscopy. Limnology and Oceanography-Methods. 2012;**10**(7):560-567. DOI: 10.4319/lom.2012.10.560

[30] Svanberg S. Gas in scattering media absorption spectroscopy-from basic studies to biomedical applications. Laser & Photonics Reviews. 2013;**7**(5):779-796. DOI: 10.4319/lom.2012.10.560

[31] Deguchi Y, Kamimoto T, Wang ZZ, Yan JJ, Liu JP, Watanabe H, Kurose R. Applications of laser diagnostics to thermal power plants and engines. Applied Thermal Engineering. 2014;**73**(2):1453-1464. DOI: 10.1016/j.applthermaleng.2014.05.063

[32] Pogány A, Klein A, Ebert V. Measurement of water vapor line strengths in the 1.4-2.7μm range by tunable diode laser absorption spectroscopy. Journal of Quantiative Spectroscopy and Radiative Transfer. 2015;**165**:108-122. DOI: 10.1016/j.jqsrt.2015.06.023

[33] Choi DW, Jeon MG, Cho GR, Kamimoto T, Deguchi Y, Doh DH. Performance improvements in temperature reconstructions of 2-D tunable diode laser absorption spectroscopy (TDLAS). Journal of Thermal Science. 2016;**25**(1):84-89. DOI: 10.1007/s11630-016-0837-z

[34] Werle P. Tunable diode laser absorption spectroscopy: Recent findings and novel approaches. Infrared Physics and Technology. 1996;**37**(1):59-66. DOI: 10.1016/1350-4495(95)00113-1

[35] Riise H, Carlisle CB, Carr LW, Cooper DE, Martinelli RU, Menna RJ. Design of an open path near infrared diode laser sensor: Application to oxygen, water and carbon monoxide. Applied Optics. 1994;**33**(30):7059-7066. DOI: 10.1364/AO.33.007059

[36] Zhang Z, Dong F, Xia H, Tu G, Wu B, Wang Y. High sensitive monitoring of carbon monoxide in industry flue gases using tunable diode lasers. Proceedings of SPIE. 2009; **7382**:73823V-1-7. DOI: 10.1117/12.836230

[37] Viciani S, Amato FD, Mazzinghi P, Castagnoli F, Toci G, Werle P. A cryogenically operated laser diode spectrometer for airborne measurement of stratospheric trace gases. Applied Physics B: Lasers and Optics. 2008;**90**(3–4):581-592. DOI: 10.1007/s00340-007-2885-2

[38] Reid J, Labrie D. Second-harmonic detection with tunable diode lasers–comparison of experiment and theory. Applied Physics B. 1981;**26**(3):203-210. DOI: 10.1007/BF00692448

[39] Cui X, Lengignon C, Wu T, Zhao W, Wysocki G, Fertein E, Coeur C, Cassez A, Croize L, Chen W, Wang Y, Zhang W, Gao X, Liu W, Zhang Y, Dong F. Photonic sensing of the atmosphere by absorption spectroscopy. Journal of Quantitative Spectroscopy and Radiation Transfer. 2012;**113**(11):1300-1316. DOI: 10.1016/j.jqsrt.2011.11.008

[40] Wang S, Li J, Che R, Wang TA. Methane gas sensor with optic Fiber based on frequency harmonic detection technique. Journal of Applied Optics. 2004;**25**(2):44-47

[41] Kan R, Liu W, Zhang Y, Liu J, Wang M, Chen D, Chen J, Cui Y. A high sensitivity spectrometer with tunable diode laser for ambient methane monitoring. Chinese Optics Letters. 2007;**5**(1):54-57

[42] Xu Y, Qian X, Liu Z. Quantitative risk analysis on the leakage of compressed natural gas pipeline. Chinese Safety Science. 2008;**18**(1):146-149

[43] Liu X, Wang L, Jin W, et al. The development of optical remote measurement for hazardous gas leakage. Infrared Technology. 2009;**31**(10):563-572

[44] Liu M, Xiang Q, Dai Z. Studies on the leakage calculation models and optimal measurement conditions of natural gas transmission pipe line systems. Chemical Engineering of Oil and Gas. 2002;**31**(1):47-51

[45] Durry G, Li JS, Vinogradov I, Titov A, Joly L, Cousin J, Decarpenterie T, Amarouche N, Liu X, Parvitte B, Korablev O, Gerasimov M, Zéninari V. Near infrared diode laser spectroscopy of C_2H_2, H_2O, CO_2 and their isotopologues and the application to TDLAS, a tunable diode laser spectrometer for the martian PHOBOS-GRUNT space mission. Applied Physics B: Lasers and Optics. 2010;**99**(1):339-351. DOI: 10.1007/s00340-010-3924-y

[46] Nikiforova OY, Kapitanov V, Ponomarev YN. Influence of ethylene spectral lines on methane concentration measurements with a diode laser methane sensor in the 1.65 μm region. Applied Physics B: Lasers and Optics. 2008;**90**(2):263-268. DOI: 10.1007/s00340-007-2919-9

[47] Sun P, Zhang Z, Li J, Xia H, Han L, Li Z, Tan D, Ma Y, Dong F. Research on natural gas leakage monitoring technology based on the open-path measurement technique of methane. Optics Optoelectronics Technology. 2016;**14**(5):63-67

[48] Xia H, Wu B, Zhang Z, Pang T, Dong F, Wang Y. Stability study on high sensitive CO monitoring in near-infrared. Acta Physica Sinica. 2013;**62**(21):214208-1-7. DOI: 10.7498/aps.62.214208

[49] Curl RF, Tittel FK. Tunable infrared laser spectroscopy. Annual Reports Progress Chemical Section C. 2002;**98**(4050):217-270. DOI: 10.1039/b111194a

[50] Li JY, Du ZH, Ma YW, Xu KX. Dynamic thermal modeling and parameter identification for monolithic laser diode module. Chinese Physics B. 2013;**22**(3):317-322. DOI: 10.1088/1674-1056/22/3/034203

[51] Che L, Ding YJ, Peng ZM, Li XH. Calibration-free wavelength modulation spectroscopy for gas concentration measurements under low-absorbance conditions. Chinese Physics B. 2012;**21**(12):463-470. DOI: 10.1088/1674-1056/21/12/127803

[52] Rothman LS, Gamache RR, Goldman A, Brown LR, Toth RA, Pickett HM, Poynter RL, Flaud JM, Camy-Peyret C, Barbe A, Husson N, Rinsland CP, MAH S. The HITRAN database: 1986 edition. Applied Optics. 1987;**26**(19):4058-4097. DOI: 10.1364/AO.26.004058

[53] Tu G, Dong F, Wang Y, Culshaw B, Zhang Z, Pang T, Xia H, Wu B. Analysis of random noise and long-term drift for tunable diode laser absorption spectroscopy system at atmospheric pressure. IEEE Sensors Journal. 2015;**15**(6):3535-3541. DOI: 10.1109/JSEN.2015.2393861

[54] Werle P, Mucke R, Slemr F. The limits of signal averaging in atmospheric trace-gas monitoring by tunable diode-laser absorption spectroscopy (TDLAS). Applied Physics B: Lasers and Optics. 1993;**57**(2):131-139. DOI: 10.1007/BF00425997

[55] Chen WD, Kostere A, Tittel FK, Gao XM, Zhao WX. H_2S trace concentration measurements using off-axis integrated cavity output spectroscopy in the near-infrared. Applied Physics B: Lasers and Optics. 2008;**90**(2):311-315. DOI: 10.1007/s00340-007-2858-5

[56] Xia H, Dong F, Wu B, Zhang Z, Pang T, Sun P, Cui X, Han L, Wang Y. Sensitive absorption measurements of hydrogen sulfide at 1.578 μm using wavelength modulation spectroscopy. Chinese Physics B. 2015;**24**(3):180-184. DOI: 10.1088/1674-1056/24/3/034204

[57] Hofstetter D, Francesco JD, Hvozdara L, Herzig HP, Beck M. CO_2 isotope sensor using a broadband infrared source, a spectrally narrow 4.4 μm quantum cascade detector, and a Fourier spectrometer. Applied Physics B: Lasers and Optics. 2011;**103**(4):967-970. DOI: 10.1007/s00340-011-4532-1

[58] Tuzson B, Zeeman MJ, Zahniser MS, Emmenegger L. Quantum cascade laser based spectrometer for in situ stable carbon dioxide isotope measurements. Infrared Physics Technology. 2008;**51**(3):198-206. DOI: 10.1016/j.infrared.2007.05.006

[59] Tanaka K, Kojima R, Takahashi K, Tonokura K. Continuous measurements of stable carbon isotopes in CO_2 with a near-IR laser absorption spectrometer. Infrared Physics Technology. 2013;**60**(5):281-287. DOI: 10.1016/j.infrared.2013.05.011

[60] Stemmer K, Ammann M, Donders C, Kleffmannm J, George C. Photosensitized reduction of nitrogen dioxide on humic acid as a source of nitrous acid. Nature. 2006;**440**(7081):195-198. DOI: 10.1038/nature0460

[61] Li X, Rohrer F, Hofzumahaus A, Brauers T, Häseler R, Bohn B, Broch B, Fuchs H, Gomm S, Holland F, Jäger J, Kaiser J, Keutsch FN, Lohse I, Tillmann R, Wegener R, Wolfe GM, Mentel TF, Kiendler-Scharr A, Wahner A. Missing gas-phase source of HONO inferred from zeppelin measurements in the troposphere. Science. 2014;**344**(6181):292-296. DOI: 10.1126/science.aaa3777

[62] Alicke B, Geyer A, Hofzumahaus A, Holland F, Konrad S, Patz HW, Schafer J, Stutz J, Volz-Thomas A, Platt U. OH formation by HONO photolysis during the BERLIOZ experiment. Journal of Geophysical Research. 2003;**108**(D4):8247-8264. DOI: 10.1029/2001JD000579

[63] Acker K, Moller D, Wieprecht W, Meixner FX, Bohn B, Gilge S, Plass-Dulmer C, Berresheim H. Strong daytime production of OH from HNO_2 at a rural mountain site. Journal of Geophysics Letters. 2006;**33**(2). DOI: L02809, 10.1029/2005GL024643

[64] Pagsberg P, Bjergbakke E, Ratajczak E, Sillesen A. Kinetics of the gas phase reaction OH + NO (+M) → HONO (+M) and the determination of the UV absorption cross sections of HONO. Chemical Physics Letters. 1997;**272**(5–6):383-390. DOI: 10.1016/S0009-2614(97)00576-9

[65] Djehiche M, Tomas A, Fittschen C, Coddeville P. First direct detection of HONO in the reaction of Methylnitrite (CH_3ONO) with OH radicals. Environmental Science & Technology. 2010;**45**(2):608-614. DOI: 10.1021/es103076e

[66] Febo A, Perrino C. Measurement of high concentration of nitrous acid inside automobiles. Atmospheric Environment. 1995;**29**(29):345-351. DOI: 10.1016/1352-2310(94)00260-R

[67] Heikes BG, Thompson AM. Effects of heterogeneous processes on NO_3, HONO, and HNO_3 in the troposphere. Journal of Geophysical Research. 1983;**88**(C15):10883-10895. DOI: 10.1029/JC088iC15p10883

[68] Wong KW, Oh HJ, Lefer BL, Rappengluck B, Stutz J. Vertical profiles of nitrous acid in the nocturnal urban atmosphere of Houston, TX. Atmospheric Chemistry and Physics. 2011;**11**(8):3595-3609. DOI: 10.5194/acp-11-3595-2011

[69] Costabile F, Amoroso A, Wang F. Sub-mm particle size distributions in a suburban Mediterranean area. Aerosol populations and their possible relationship with HONO mixing ratios. Atmospheric Environment. 2010;**44**(39):5258-5268. DOI: 10.1016/j.atmosenv.2010.08.018

[70] Kleffmann J, Wiesen P. Heterogeneous conversion of NO_2 and NO on HNO_3 treated soot surfaces: atmospheric implications. Atmospheric Chemistry and Physics. 2005;**5**(1):77-83. DOI: 10.5194/acp-5-77-2005

[71] Stemmler K, Ammann M, Donders C, Kleffmann J, George C. Photosensitized reduction of nitroged dioxide on humic acid as a source of nitrous acid. Nature. 2006;**440**(7081):195-198. DOI: 10.1038/nature04603

[72] Ziemba LD, Dibb JE, Griffin RJ, Anderson CH, Whitlo SI, Lefer BL, Rappengluck B, Flynn J. Heterogeneous conversion of nitric acid to nitrous acid on the surface of primary organic aerosol in an urban atmosphere. Atmospheric Environment. 2010;**44**(33):4081-4089. DOI: 10.1016/j.atmosenv.2008.12.024

[73] Acker K, Mfller D, Auel R, Wieprecht W, Kalaβ D. Concentrations of nitrous acid, nitric acid, nitrite and nitrate in the gas and aerosol phase at a site in the emission zone during ESCOMPTE 2001 experiment. Atmospheric Research. 2005;**74**(1):507-524. DOI: 10.1016/j.atmosres.2004.04.009

[74] Monge ME, D'Anna B, Mazri L, Fendler AG, Ammann M, Donaldson DJ, George C. Light changes the atmospheric reactivity of soot. Proceedings of the National Academy of Sciences. 2010;**107**(15):6605-6609. DOI: 10.1073/pnas.0908341107

[75] Su H, Cheng Y, Oswald R, Behrendt T, Trebs I, Meixner FX, Andreae MO, Cheng P, Zhang Y, Pöschl U. Soil nitrite as a source of atmospheric HONO and OH radicals. Science. 2011; **333**(6039):1616-1618. DOI: 10.1126/science.1207687

[76] Wu T, Chen W, Fertein E, Cazier F, Dewaele D, Gao X. Development of an open-path incoherent broadband cavity enhanced spectroscopy based instrument for simultaneous measurement of HONO and NO_2 in ambient air. Applied Physics B: Lasers and Optics. 2012;**106**(2):501-509. DOI: 10.1007/s00340-011-4818-3

[77] Qin M, Xie P, Su H, Gu J, Peng F, Li S, Zeng L, Liu J, Liu W, Zhang Y. An observational study of the HONO–NO_2 coupling at an urban site in Guangzhou City, South China. Atmospheric Environment. 2009;**43**(36):5731-5742. DOI: 10.1016/j.atmosenv.2009.08.017

[78] Wang S, Zhou R, Zhao H, Wang Z, Chen L, Zhou B. Long-term observation of atmospheric nitrous acid (HONO) and its implication to local NO_2 levels in Shanghai, China. Atmospheric Environment. 2013;**77**(7):718-724. DOI: 10.1016/j.atmosenv.2013.05.071

[79] Gherman T, Venables DS, Vaughan S, Orphal J, Ruth AA. Incoherent broadband cavity enhanced absorption spectroscopy in the near-ultraviolet: Application to HONO and NO_2. Environmental Science & Technology. 2008;**42**(3):890-895. DOI: 10.1021/es0716913

[80] Becker K, Kleffmann J. KurtenbachR, Wiesen P. Line strength measurements of trans-HONO near 1255 cm^{-1} by tunable diode laser spectrometry. Geophysical Research Letters. 1995;**22**(18):2485-2488. DOI: 10.1029/95GL02471

[81] Schiller CL, Locquiao S, Johnson TJ, Harris GW. Atmospheric measurements of HONO by tunable diode laser absorption spectroscopy. Journal of Atomspheric Chemistry. 2001; **40**(3):275-293. DOI: 10.1023/A:1012264601306

[82] Li YQ, Schwab JJ, Demerjian KL. Fast time response measurement of gaseous nitrous acid using a tunable diode laser absorption spectrometer: HONO emission source from vehicle exhausts. Geophysical Research Letters. 2008;**35**(4):121-124. DOI: 10.1029/2007GL031218

[83] Lee BH, Wood EC, Zahniser MS, McManus JB, Nelson DD, Herndon SC, Santoni GW, Wofsy TC, Munger JW. Simultaneous measurements of atmospheric HONO and NO_2 via absorption spectroscopy using tunable mid-infrared continuous-wave quantum cascade lasers. Applied Physics B: Lasers and Optics. 2011;**102**(2):417-423. DOI: 10.1007/s00340-010-4266-5

[84] Craig IM, Taubman MS, Lea AS, Phillips MC, Josberger EE, Raschke MB. Infrared near-field spectroscopy of trace explosives using an external cavity quantum cascade laser. Optics Express. 2013;**21**(25):30401-30414. DOI: 10.1364/OE.21.030401

[85] Suter J, Bernacki B, Phillips M. Spectral and angular dependence of mid-infrared diffuse scattering from explosives residues for standoff detection using external cavity quantum cascade lasers. Applied Physics B: Lasers and Optics. 2012;**108**(4):965-974. DOI: 10.1007/s00340-012-5134-2

[86] Su H, Cheng Y, Cheng P, Zhang Y, Dong S, Zeng L, Wang X, Slanina J, Shao M, Wiedensohler A. Observation of nighttime nitrous acid (HONO) formation at a non-urban site during PRIDE-PRD2004 in China. Atmospheric Environment. 2008;**42**(25): 6219-6232. DOI: 10.1093/emboj/17.2.455

4

Vehicle Dynamics and Green Electronic Differential for eKart

Golebiowski Wlodzimierz, Kubiak Przemysław and
Szymon Madziara

Abstract

Today, electric vehicles are becoming increasingly popular in our lives. In motorsport, however, they are not as widely used. Formula E, which was created to boost electric motorsport, is not enough to popularize it. Every driver who wants to advance to F1 (highest rank racing series) has to start from karting between the ages of 5 and 8. But, today, gokarts are only powered by combustion engines. In order to provide young drivers with the possibility of racing small green electric vehicles, the so-called eKarts, combustion engines have to be replaced with electric motors. eKarts should offer similar performance to combustion engine go-karts. Therefore, to determine the required power of the electric motors and the capacity of the batteries in eKarts for different age categories, the technical parameters of the different age categories of combustion engine racing go-karts were analyzed. In this chapter, the present Li-ion battery technology makes it possible to construct eKarts for children and junior categories. With the current technology, it is not possible to create an eKart for senior categories (15 and over) in line with the current regulations for go-karts. Chance such green electronics solution with torque vectoring will provide better efficiency of energy consumption and lover impact on natural environment in reduced emission of both noise and greenhouse gases.

Keywords: eKart, electric motorsport, Formula E, go-karts, green electronic differential

1. Introduction

The history of the electric car started at the beginning of the automobile era, that is, in the nineteenth century. At that time, electric vehicles competed with combustion and steam engine ones not only on the streets but also on racetracks. The world land speed record of 105.88 km/h was established by the Belgian Camille Jenats in 1899 with an electric vehicle and was unbeaten for another 3 years [1, 2]. The Paris-Bordeaux-Paris race was attended by the French electric constructor Charles Jeantud. His car offered great performance, but its 950-kg batteries had to be replaced 15 times during the race [3]. The problem with storing electricity was the reason why electric vehicles have disappeared from the streets and from motorsport for a long time.

Rising awareness of ecology and the search for new fields of experiments in motorsport have led to the development of hybrid technology, which was the first sign of a return to electricity. In the F1 racing series, fuel efficiency regulations have been intensified and the share of hybrid energy recovery systems has increased [5]. In 2006, regulations were announced, and in 2009, a kinetic energy recovery system was introduced in the F1 racing series. The system called kinetic energy recovery system (KERS) allowed the release of energy up to 400 kJ on one lap [6, 8] and no more than 60 kW. In long-distance races, the importance of the hybrid system has been confirmed during the most prestigious race. In the LeMans race on June 16–17, 2012, two Audi hybrid cars were classified in the first two places. This was the first long-distance race for hybrid vehicles; apart from the Audi LMP1 e-tron, Toyota also entered its TS030 hybrid vehicle [7].

Another step toward ecological motorsport was the creation of the electric Formula E series [4]. In Formula E, the race takes about 50 min, and the problem of storing a sufficient amount of energy has been resolved by the obligatory pit stop and car replacement. Formula E takes advantage of easy energy control and makes the sport more entertaining. In racing mode, maximum power is limited to 170 kW. However, the three drivers who win a fan online vote can receive an additional 100 kJ, increasing maximum power to 200 kW for couple of seconds [9].

Formula E has received credit from racing drivers. Former F1 driver and current Formula E driver, Nelson Piquet Jr., said that "if you care about the fame and flashes the F1 is the best racing series, but equally exciting for racing drivers is competing in Formula E. Formula E was created to boost electric motorsport, but still very few series are purely electric."

To get to F1, each driver has to start from karting between the ages of 5 and 8, but present-day go-karts are only powered by combustion engines. In order to introduce the possibility of racing small green electric racing vehicles for young drivers, the so-called eKarts, combustion engines have to be replaced with electric motors. It was assumed that for eKarts to gain a foothold, they should provide similar parameters to those of combustion engine go-karts. Therefore, to determine the required power of the electric motors and the capacity of the batteries in eKarts for different age categories, the technical parameters of the different age categories of combustion engine racing go-karts have been analyzed.

The Rotax Max Challenge series has been chosen for analysis due to the fact that it is one of the most popular karting series in the world and also in Poland. **Table 1** shows the main features of selected Rotax categories [19].

An analysis of the power used by go-karts was made on the basis of data from racing and official trainings on the following certified karting tracks:

- Speedworld in Bruck, Austria—length 1060, 1120, 1140 m (depending on configuration), 8–10 m wide, clockwise direction.
- Goethe Stadium in Kecskemet, Hungary—length 935 m, 7 m wide, anticlockwise.
- Pann Ring in Ostffyasszonyfa, Hungary—length 1071 m, 8 m wide counterclockwise direction.

- Autodromo Vysoke Myto in Vysoke Myto, Czech Republic—length 1142 m, 8 m clockwise directional width.
- Tor Radom, in Radom, Poland—length 820 m, 8–12 m in clockwise direction (possibly also in the opposite direction).
- Bydgoszcz in the city of Bydgoszcz, Poland—length 1017 m, 8–10 m wide counterclockwise direction
- 7 Laghi Kart—International Circuit in Castelletto di Branduzzo, Italy—length 1017 m, 8–10 m in clockwise direction (**Figure 1**—Picture of 7 Laghi Kart—International Circuit)

In each of the age categories, data from a dozen race laps were analyzed, and for each six races were selected, three fastest and three slowest lap times for three different racetracks. An Off Camber Data tool for Unipro devices was used for the analysis. **Figure 2** shows the Off Camber Data analysis panel.

	MicroMax	MiniMax	JuniorMax	SeniorMax	DD2
Driver age (years)	8–10	10–13	13–16	15+	15+
Min. mass of vehicle with driver (kg)	110	130	147	165	173
Chassis type, wheelbase (mm)	950	1050	1050	1050	1050
Power max. (kW/rpm)	6/7500	11/8500	17/8500	22/8500	25/8500
Torque (Nm/rpm)	9/6000	13/8000	19/8500	21/9000	22/10500
Mass of drive train (kg)	21.6	23.0	23.0	23.1	28.8
Mass of fuel (kg)	~4.0	~5.0	~6.0	~7.0	~7.0
Race distance (km)	14	16	18	20	20

Table 1. Main parameters of the Rotax series in Poland.

Figure 1. Picture of 7 Laghi Kart—International Circuit © Google Maps.

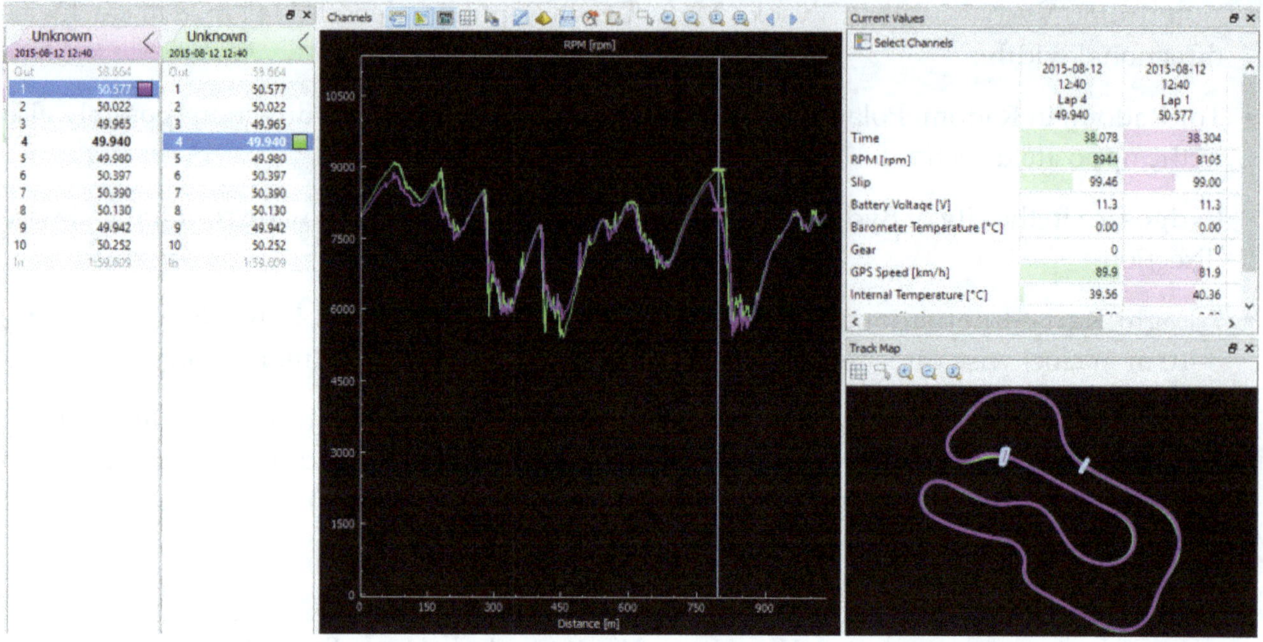

Figure 2. Analytical panel Off Camber Data application.

2. Model of go-kart energy usage

In order to determine energy requirements, the model assumes that the energy stored in batteries E_B must provide the possibility to finish a maximum duration race (based on the Rotax Max Challenge 2016 Cup) with the go-kart using maximum power, which was experimentally derived on different racetracks in Poland and Europe:

$$E_B = W_{Rmax} \times n \tag{1}$$

where E_B—energy stored in batteries (kWh), W_{Rmax}—work done by running go-kart during the race with maximum energy consumption (kWh), n—coefficient of securing sufficient level of energy (%).

In simplified terms, it was assumed that W_{Rmax} is the work done by the combustion engine of a go-kart of average power P_{Omax} throughout the maximum duration of the race t_{Rmax}.

$$W_{Rmax} = P_{Omax} \times t_{Rmax} \tag{2}$$

where t_{Rmax}—the maximum duration of the race based on an analysis of the Rotax Max Challenge 2016 Polish Championships (s), P_{Omax}—average power used by go-kart during a lap with the highest power usage (kW).

P_{Omax} was obtained by selecting the highest value from the calculated average power P_Os, for six different cases of go-kart racing in a given age category. We chose the shortest and longest run times on three racetracks with different characteristics.

The model assumes that the energy used on one lap P_O is equal to the sum of instantaneous power calculations in tenths of a second of the lap.

$$P_O = \frac{\sum_{i=0}^{N_O} P_{ch}}{N_O} \tag{3}$$

where P_O—average power value used during lap (kW), P_{ch}—instantaneous power (kW), N_O—number of measurements (measurement in 0.1-s intervals).

In the presented approach, it was assumed that P_{ch}—instantaneous power is calculated based on instantaneous engine rpm and the engine power/rpm specification curve presented in **Figure 4**. Due to the permanent coupling of the engine with the drive axle, in the course of the analysis, braking periods during the lap were identified and excluded from the calculation of average power. Based on observations and analysis, it was assumed that braking is a decrease in engine speed of at least 300 rpm with at least three measuring periods, that is, over 0.3 s and is associated with a significant decrease in vehicle speed, that is, at least 2 km in 0.1 s.

2.1. Analysis of race durations in the Rotax max challenge Poland championships in 2016

Following Eq. (2), it was necessary to analyze the duration of races t_{Rmax} to calculate the work performed by the power train. This analysis is based on the duration of go-kart races in the 12 rounds of the Rotax Max Challenge Poland Championships in 2016.

The analysis allowed us to determine maximum race times for the drivers who had completed full races in each of the age categories. **Figure 3** shows the time intervals in which the drivers finished the races of the given round of the Rotax Max Challenge Poland Championships in 2016. The data for wet races of the first round of the Rotax Max Challenge Poland Championships in 2016 are shown separately.

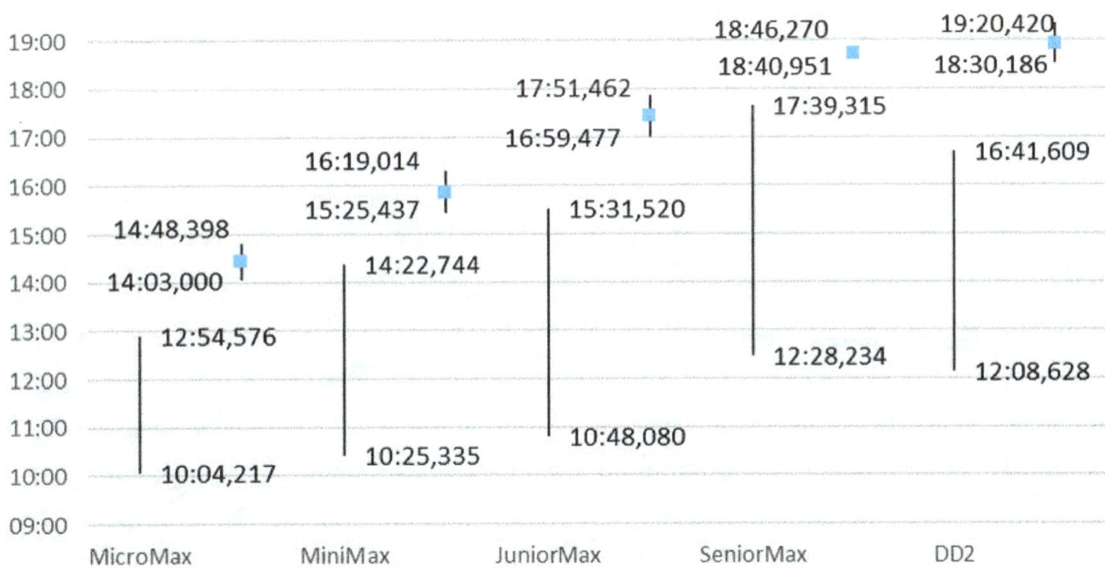

Figure 3. Race durations in Rotax Max Challenge Poland Championships 2016 for given categories.

2.2. Go-kart used power analysis based on Rotax max challenge series

The MicroMax Rotax Max Challenge category according to **Table 1** is a category for children aged 8–10. In this category of go-karts, maximum engine power is 6 kW with engine power/rpm curve shown in **Figure 4**.

The chassis of this category is the so-called "small frame" with a wheelbase of 950 mm, with brakes only on the rear axle. Maximum race distance for this category is 14 km, which translates to a maximum race time of 12 min and 54.576 s.

The data for the analysis were collected during official races and trainings on the following tracks:

- Speedworld
- Go-kart Stadion
- Pannónia Ring

Two lap times from the session on August 12, 2015, on Speedworld track with the configuration of 1120 m shown in **Figure 5** were selected:

i. 49,940 (s)—the fastest one

ii. 50,577 (s)—the slowest one

The power usage diagram for lap (i) and (ii) is shown in **Figure 6**.

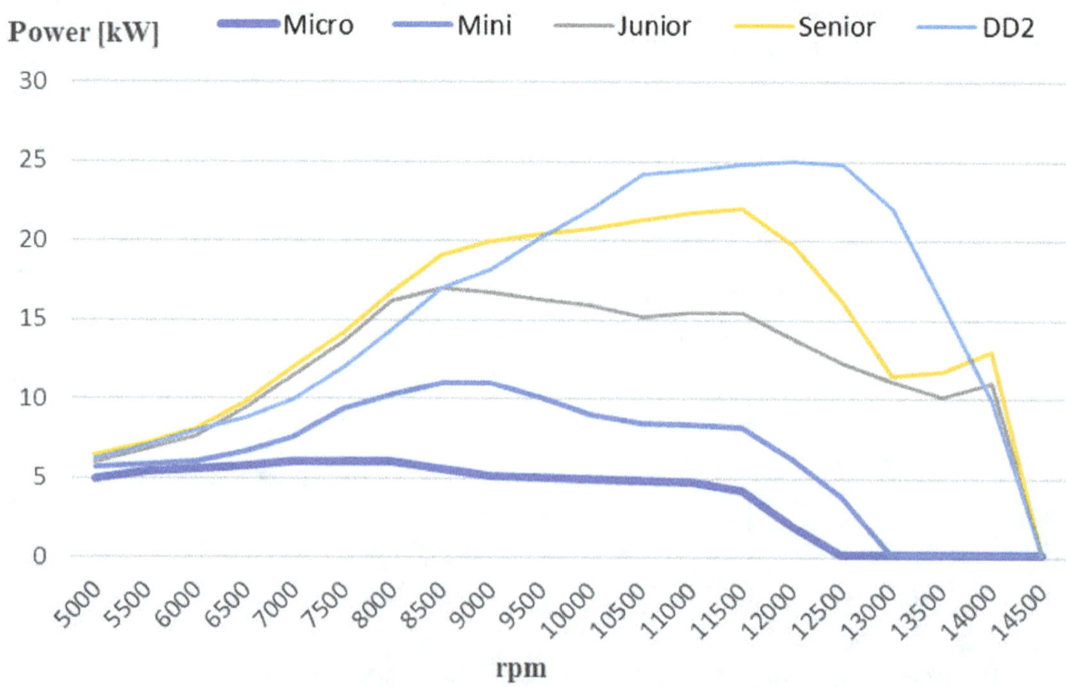

Figure 4. Rotax engine power/rpm curve for different categories [10].

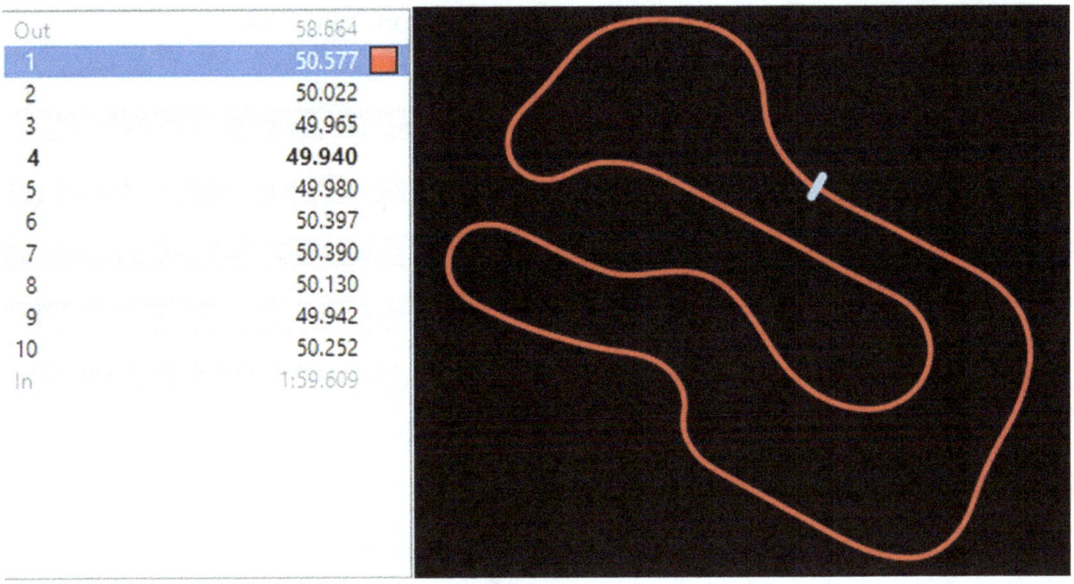

Figure 5. Race data from MicroMax category during Speedworld session on August 12, 2015, and track configuration Off Camber Data tool.

Based on **Figure 6**, the average levels of power used for laps were calculated:

i. 5.58 (kW)

ii. 5.64 (kW)

The maximum power was determined for both laps at the same level, that is, 6 kW.

Further laps were selected from the following sessions on the following tracks:

- Gokart Stadion session on June 21, 2015, data from two laps:

iii. 41,232 (s)—the fastest lap

iv. 41,760 (s)—the slowest lap

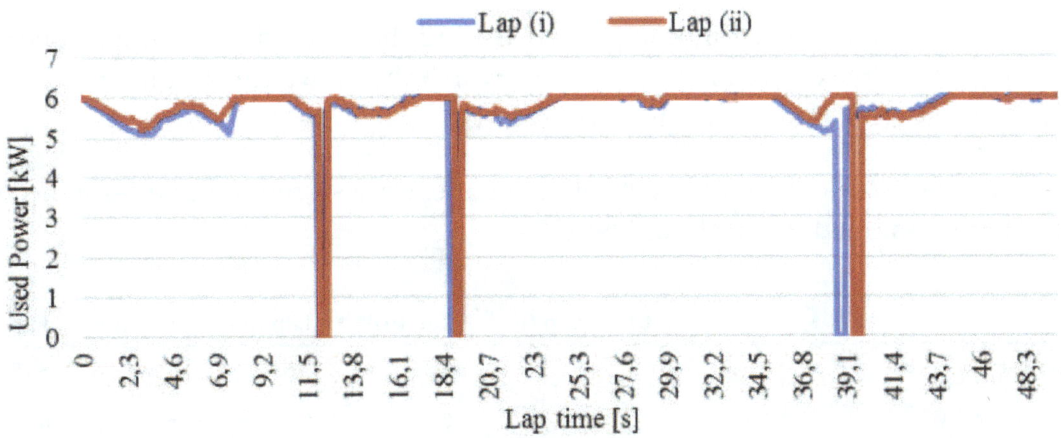

Figure 6. The power usage diagram for lap (i) and (ii) on Speedworld track, session on August 12, 2015.

- Pannónia Ring session on September 20, 2015, data from two laps:

v. 50,626 (s)—the fastest lap

vi. 52,015 (s)—the slowest lap

The average power value used during laps and the maximum power achieved in the MicroMax category were summarized in **Table 2.**

Maximum power of electric motors for MicroMax category was defined on the level of P_{max} = 6 kW. Average power value used during the lap with the highest power requirements P_{Omax} = 5.83 kW.

Knowing that the maximum race time is t_{Rmax} = 12: 54,576, the work done by the MicroMax go-kart drive systems according to Eq. (2) W_{Rmax}:

$$W_{Rmax} = P_{Omax} \times t_{Wmax} = 9.29 \times 0.239651 = 2.23 \text{ kWh}$$

Similar analyses and calculations were conducted for each Rotax Max Challenge category, and the results are presented in condensed form.

Knowing the value of average power used during a lap for each of the categories P_{Omax} and knowing maximum race time t_{Rmax} based on **Figure 3**, and Section 2.1. Analysis of race durations in the Rotax Max Challenge Poland Championships in 2016, we calculated the work done by a running go-kart during the race with maximum energy consumption W_{Rmax}. The results are presented in **Table 4**.

2.3. eKart: initial definition of main parameters

Based on the analysis of go-kart power usage presented in Section 2.2, the average power calculated based on Eq. (3) is assumed to be the minimum power that an eKart should generate. Maximum power for each category of eKarts is the maximum power achieved in a given age category for safety reasons. The engine power ranges for each eKart age category are shown in **Figure 7**.

Knowing the work done by a running go-kart during the race with maximum energy consumption for every Rotax Max Challenge category in **Table 4**, we can assume to conserve the same energy for each of the age categories of eKarts.

However, in order to calculate the practical value of energy stored in batteries in accordance with Eq. (1), it is assumed that n should be 120%.

	Lap (i)	Lap (ii)	Lap (iii)	Lap (iv)	Lap (v)	Lap (vi)
Average power used (kW)	5.58	5.64	5.81	5.83	5.64	5.78
Maximum power (kW)	6	6	6	6	6	6

Table 2. Average power used and maximum power of MicroMax category in six different laps of three sessions.

	Lap (i)	Lap (ii)	Lap (iii)	Lap (iv)	Lap (v)	Lap (vi)
MiniMax						
Session	Speedworld August 14, 2015		Motodrom Vysoke Myto July 12, 2015		Pannónia Ring May 28, 2016	
Lap time (s)	46,520	47,224	49,315	50,165	49,252	50,027
Fastest/slowest lap	Fastest	Slowest	Fastest	Slowest	Fastest	Slowest
Average power used (kW)	9.15	9.12	8.91	8.95	9.29	9.21
Maximum power (kW)	11	11	11	11	11	11
JuniorMax						
Session	Pannónia Ring May 29, 2016		Tor Radom May 22, 2016		Tor Bydgoszcz May 1, 2016	
Lap time (s)	47,152	48,356	32,480	33.34	46,331	46,821
Fastest/slowest lap	Fastest	Slowest	Fastest	Slowest	Fastest	Slowest
Average power used (kW)	13.70	13.41	13.64	14.00	13.45	13.75
Maximum power (kW)	17	17	17	17	16.99	17
SeniorMax						
Session	Speedworld May 21, 2015		Pannónia Ring September 19, 2015		–	
Lap time (s)	44,086	45,597	45,786	46,532	–	–
Fastest/slowest lap	Fastest	Slowest	Fastest	Slowest	–	–
Average power used (kW)	17.63	18.44	17.22	16.71	–	–
Maximum power (kW)	22	21.99	22	21.99	–	–
DD2						
Session	Speedworld August 29, 2015		7 Laghi Kart May 16, 2016		Pannónia Ring May 29, 2016	
Lap time (s)	56,173	57,048	49,568	51,918	45,455	46,527
Fastest/slowest lap	Fastest	Slowest	Fastest	Slowest	Fastest	Slowest
Average power used (kW)	19.05	18.80	20.16	19.59	20.79	20.16
Maximum power (kW)	24.99	25	25	25	24.99	25

Table 3. Summarized sessions and laps information with average power used and maximum power on the lap for rest of Rotax max challenge categories, that is, MiniMax, JuniorMax, SeniorMax and DD2 [17].

	8–10	10–13	13–16	15+	15+ gears
P_{Omax} (kW)	5.83	9.29	14.00	18.44	20.79
t_{Rmax} (mins:s)	12:54,576	14:22,744	15:31,520	17:39,315	16:41,609
t_{Rmax} (h)	0.21516	0.23965	0.25876	0.2942	0.2783
W_{Rmax} (kWh)	1.25	2.23	3.62	5.42	5.79

Table 4. Work done by running go-kart during the race with maximum energy consumption for every Rotax Max Challenge category.

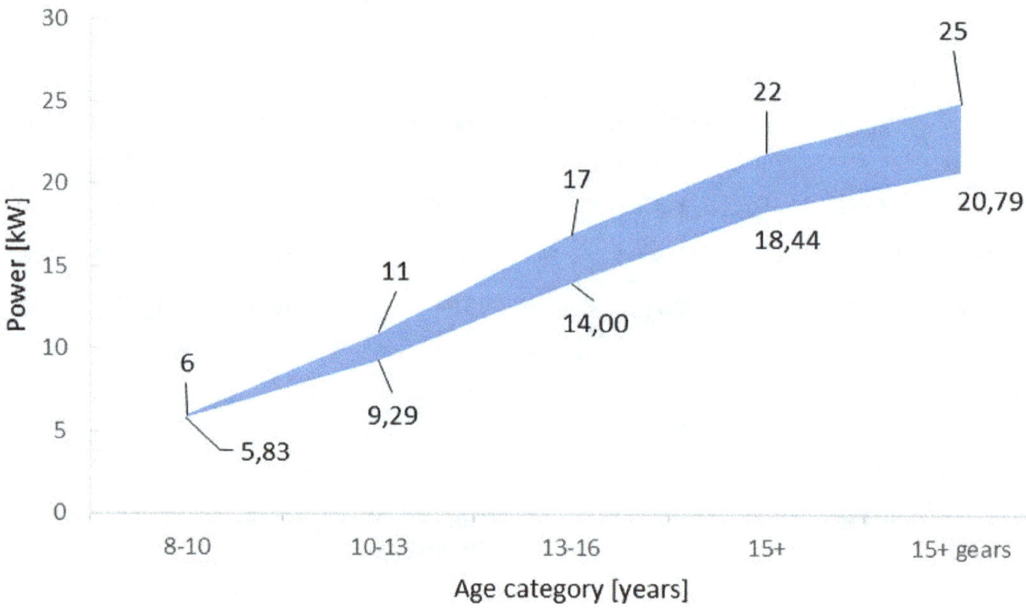

Figure 7. Power range of eKart drive system for different age categories.

$$E_A = W_{Rmax} \times 120\% \tag{4}$$

This assumption was stated because of the necessity to provide energy for additional formation laps and potentially higher consumption during rain races. On average, a rain race is longer than a race in normal conditions by 18% (**Figure 3**).

It was assumed that eKarts will be equipped with Li-ion batteries due to their best energy-to-weight ratio [11]. Degradation of Li-ion depends on the conditions, but it is up to 10% less capacity already at 300 charging-discharging cycles and 20% at about 1000 cycles [12, 13]. Therefore, it was proposed that the batteries should have 20% larger capacity than the expected E_A energy needed for an eKart race. Such a solution in addition will provide optimum, fast loading and braking capability from the first lap. The above can be described as:

$$Q_B = 120\,\% \times E_A \tag{5}$$

where Q_B—battery capacity (kWh).

	MicroMax	MiniMax	JuniorMax	SeniorMax	DD2
Average power used by eKart on lap, P_{Omax} (kW)	5.83	9.29	14.00	18.44	20.79
Expected race time, t_{Rmax} (min:s)	12:54,576	14:22,744	15:31,520	17:39,315	16:41,609
Expected race time, t_{Rmax} (h)	0.21516	0.23965	0.25876	0.2942	0.2783
Work expected to be done during the race, W_{Rmax} (kWh)	1.25	2.23	3.62	5.42	5.79
Energy stored in batteries, E_B (kWh)	1.50	2.68	4.34	6.50	6.95
Capacity of batteries, Q_B (kWh)	1.75	3.12	5.07	7.59	8.11

Table 5. Capacity of batteries for different age categories of eKarts.

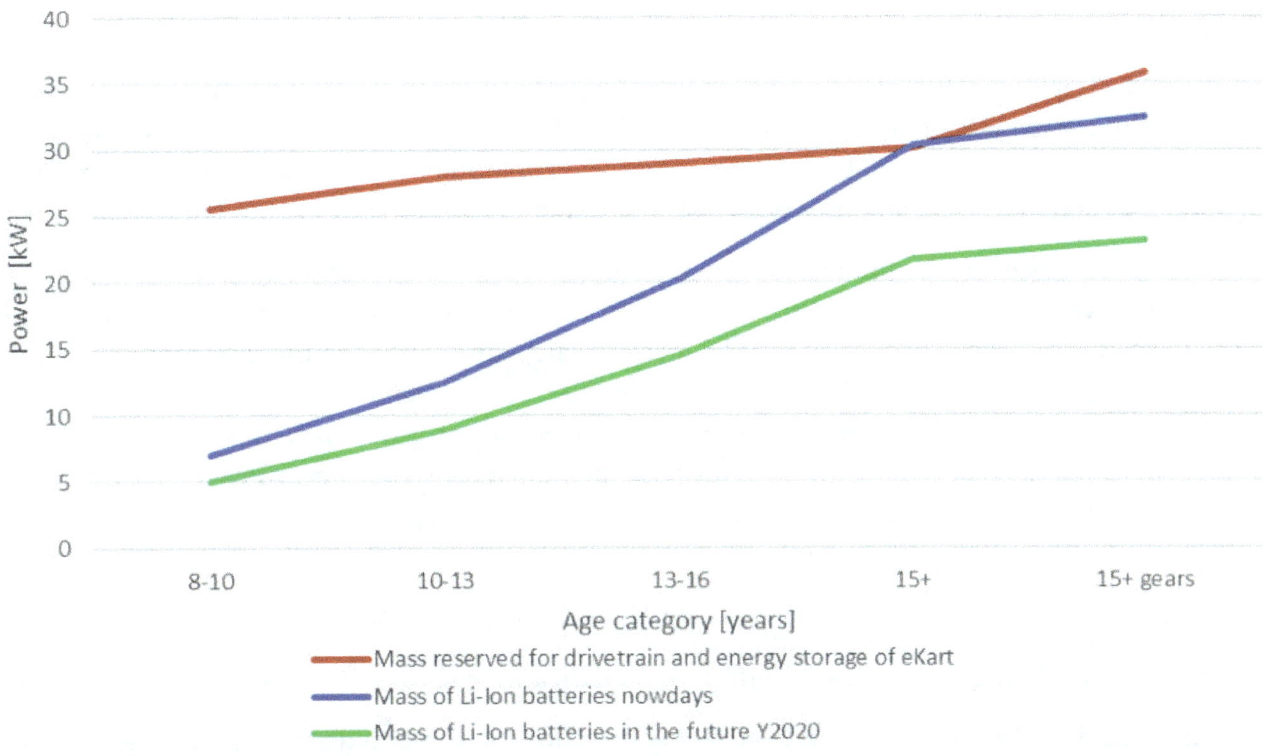

Figure 8. Maximum weight of drive train unit of eKart with batteries compared to potential mass of the batteries in Li-ion technology nowadays and in 3-year time for given age categories.

Therefore, we can rewrite **Table 4** for the eKart categories as following in **Table 5**.

An additional guideline for the selection of detailed electrical system parameters was the current weight of the go-kart in each age category. It was assumed that the weight of the eKart drive train system and the batteries should not exceed the weight of the current kart drive train system and weight of the fuel for the race shown in **Table 1**.

Based on battery capacity Q_B (6), we analyzed the options for designing eKarts for each category, within the limits of the maximum mass of the drivetrain system and the battery. We took into consideration the energy density of today's most efficient Li-ion battery technology (250 Wh/kg) and the technology expected in 3 years time (350 Wh/kg). The analysis is shown in **Figure 8**.

3. The concept of electric go-kart

For working out creation of the concept, it was assumed that the eKart design should be a simple four-wheeled vehicle with rear-wheel drive, using advantages of an electric drive. The key advantages of the electric drive are the compact size, power-to-weight ratio, high torque of the drive, and the ability to provide almost identical output parameters (speed, power, and torque) [18].

The ability to provide identical driving parameters like ICE go-kart has also become the basis for vehicle chassis design. It was assumed that for the simplicity of the eKart, a tubular flat frame without suspension element will form the chassis.

In the eKarta project, it was decided to use the compact dimensions of the electric motor. It was assumed that each wheel of the rear axle would be driven by a separate engine. This solution will allow to vary the speed of the inner wheel and the outer rear axle while cornering. This will help to achieve better stability with a similar balance of eKart versus the ICE go-kart [15].

The use of an electronic differential is an advantage of the electric motor to provide the same or better driving performance. In the case when the ICE go-kart is moving in a turn with rigidly join left and right wheel by rear axle, forces a large understeer as the inner wheel moves at a higher speed than that vehicle trajectory and the outer wheel moves at a lower speed than the trajectory. Thanks to the torque vectoring correction of independent rear axle motors, eKart handling will be neutral. Disconnection of rear axle wheels and powering them by separate motors make possible to adjust its settings so that, depending on the track, the effect can be configured under or above the steering.

According to the assumptions, the following solution was proposed.

eKart chassis is ICE go-kart chassis with two modifications shown in **Figure 9**.

The first modification marked 1 in **Figure 9** is the additional rear axle support to the left side of the chassis. Support is symmetrical longitudinally to the existing double supports on the right side, in a way that the rear axle can be divided into two components.

The second modification marked 2 in **Figure 9** is the division of the rear axle into two independent elements. Because the rear axle in ICE go-karts connects its fulcrum points (shown in **Figure 10**, marked 2), it is also the structural component of the rear frame of the vehicle. In the ICE go-karts, the rear axle has different stiffness parameters so allow to adjust the rear traction of the vehicle. The traction of the rear axle of the eKart will be adjusted by strut bar between the inner axle support points (not present ICE go-karts) and the spacer between the outer support points (solution in go-karts marked 2 in **Figure 10**).

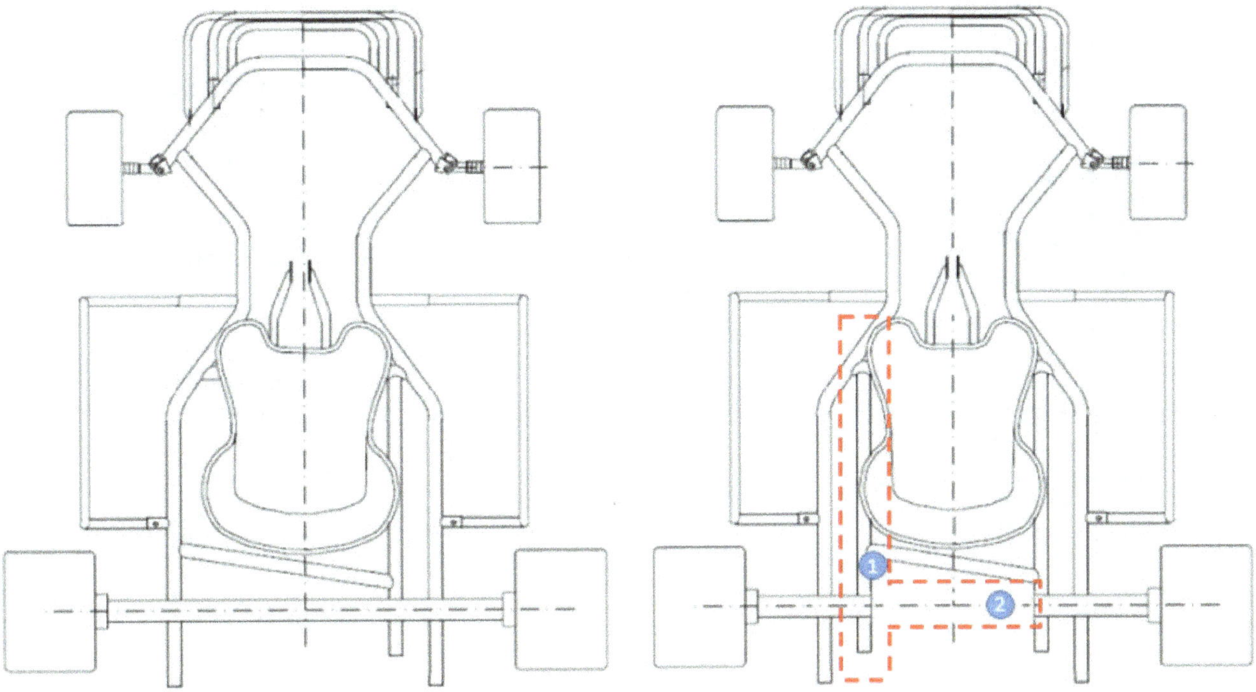

Figure 9. Sketch of modification of the ICE go-kart chassis for the eKart construction.

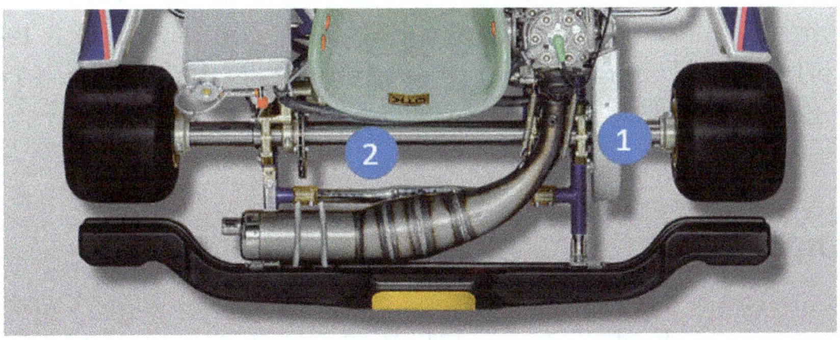

Figure 10. Rear suspension elements of Kosmic MERCURY MY15 [16].

Since in the ICE go-kart support of the rear axle on the right-hand side are at the same time structural parts of the engine mounting, modification marked 1 in **Figure 9** will allow the installation of separate motors for both wheels of the rear axles. At the design stage, it will be determined whether the motors are located ahead of or behind the rear axle of the vehicle; as illustrated in **Figure 11**, both variants are possible.

eKart will be equipped with electric motors of different power depending on the age category. There can be also different kinds of motor cooling, air cooled and liquid cooled, depending on its power.

Potential position of batteries is also shown in **Figure 11**. It was indicated to ensure the best balance of eKarts. The position of the batteries must at the same time ensure maximum protection due to poisoning or burns in case of damage. The position of the batteries has been proposed in places least exposed to other eKarts and track elements (barriers) during the

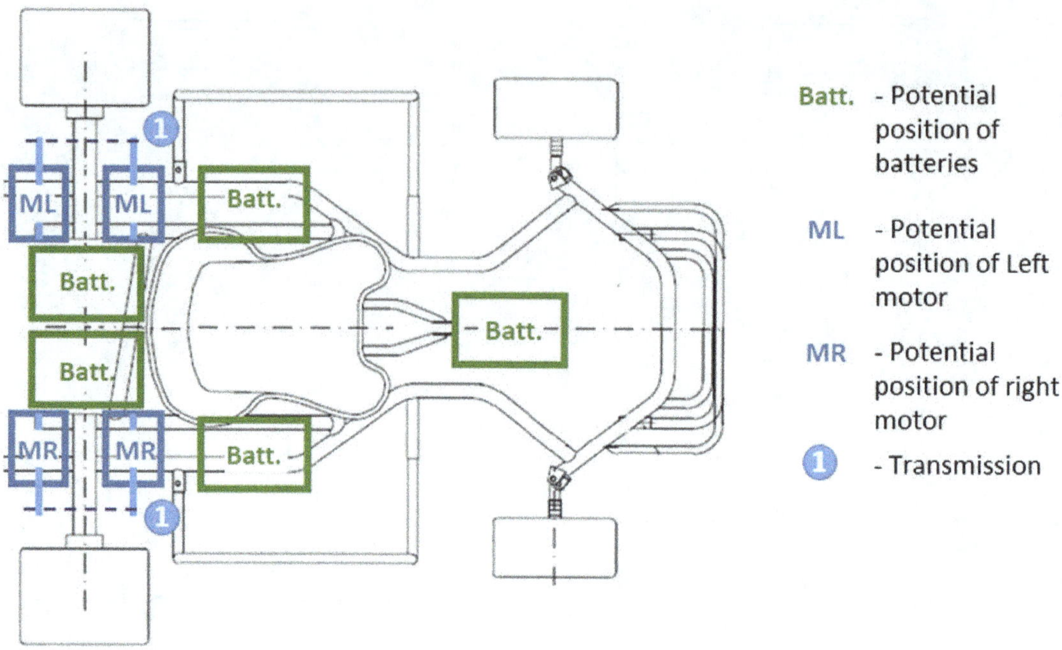

Figure 11. Potential position of the main components of eKart.

crash. The proposed battery position at the rear of the vehicle is located between the rear axles and is protected by a rear bumper, and lateral side seats are secured with side chassis, aerodynamic elements, and a bumper. The front battery compartment is located at the fuel tank of ICE go-kart. The advantage of rear or side mounts is the distance to the motor, while the advantage of the front position is a better eKart static balance.

4. Construction of eKart

For proving the concept of eKart, prototype of eKart was designed and built according to the concept [13]. After design phase where different technologies and component were analyzed, eKart was built with following parameters:

- Power transmission — Synchronous toothed belt
- Maximum power of electric motors — 2×5 kW
- Maximum torque of electric motors — 2×14 Nm
- Protection class — IP54
- Operating voltage — 39–58 V
- Battery capacity — 108 Ah
- Vehicle mass — 97.3 kg
- Dimensions — 1870 × 1340 mm

Figure 12. Assembly of eKart.

- Wheelbase 1050 mm
- Battery mass 23.2 kg

The construction of the eKart was realized in cooperation with car enthusiast student club from Faculty of Mechanical Engineering of the Technical University of Lodz. In **Figure 12**, assembly of eKart is presented.

5. Tests of eKart

The purpose of the test was to verify the concept and the design assumptions in experimental research. Test run was conducted on the Tor Łódź circuit in Stryków, Poland 596 m short track, 8–12 m wide, clockwise direction (**Figure 13**).

Test runs were carried out on November 19, 2016, air temperature 11°C and atmospheric pressure ~ 992 hPa. The surface at the start of the tests was wet, while the test surface was mostly wet.

Figure 13. Short track of Tor Łódź © circuit.

The analysis was based on data from RaceCapture Pro2 as shown in **Figure 14.**

The maximum speed on the test section was 52.64 mph, that is, 84.75 km/h (**Figure 9**). The maximum measured acceleration measured on a straight run was 13,787 s from 0 to 100 km/h. Acceleration was extrapolated from an increase in speed from 40.41 km/h to 84.75 km/h in 6.12 s, on a straight line allowing maximum acceleration. In **Figure 15** the extrapolated acceleration of the eKarts with comparison to three categories of the Rotax Max Challenge Series.

Comparing the maximum speed and acceleration of the eKart with ICE go-kart of the Rotax Max Challenge categories shown in **Table 6**; not considering the movement resistance and air resistance, it can be assumed that, by increasing the gear ratio by 22% in eKart, eKart would have similar parameters to the MiniMax category of the Rotax Max Challenge series.

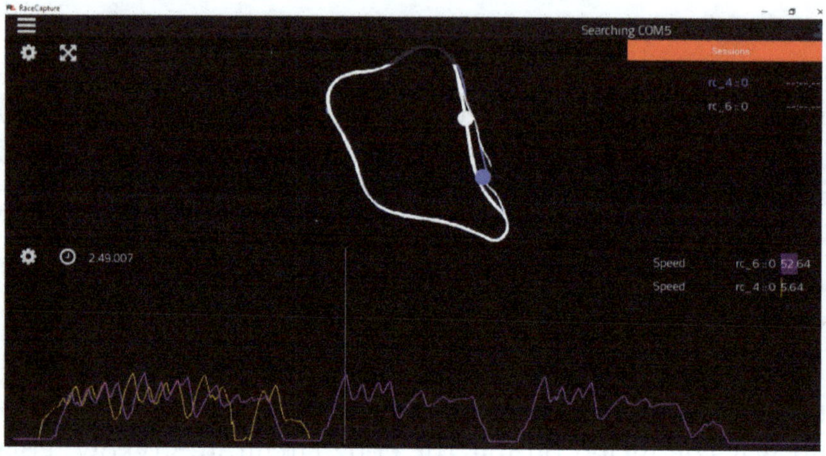

Figure 14. Data from eKart road tests on Tor Łódź © circuit.

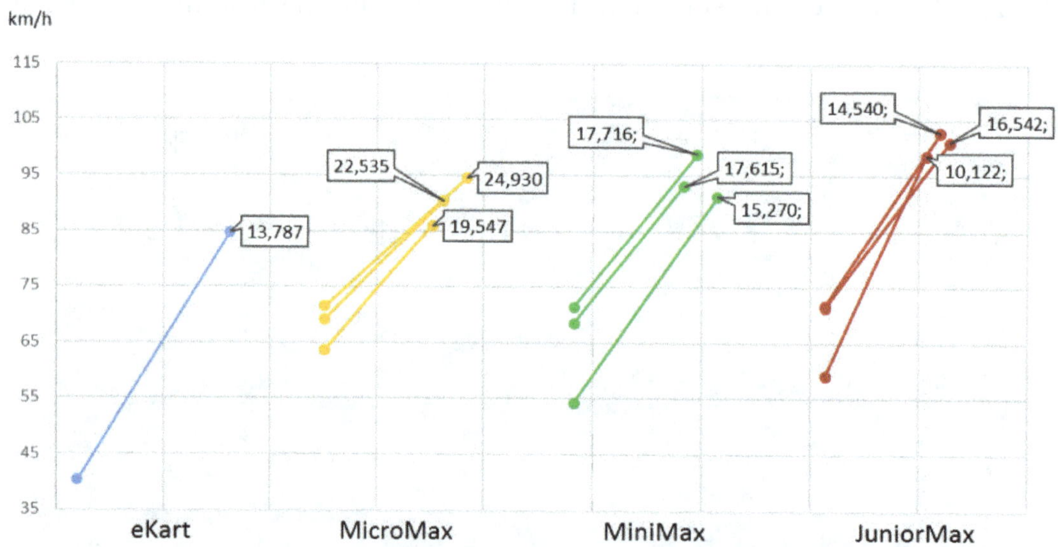

Figure 15. Comparison of extrapolated accelerations of eKart and the category Rotax Max.

	Acceleration (0–100 km/h)	Max velocity (km/h)	Acceleration diff. eKarta (%)	Max velocity diff. eKarta (%)
eKart	13.79	84.75	—	—
MicroMax	22.34	91.23	162	108
MiniMax	16.87	103.07	122	122
JuniorMax	13.74	110.67	100	131

Table 6. Comparison of speed and acceleration of eKart with ICE go-karts of Rotax Max Challenge categories.

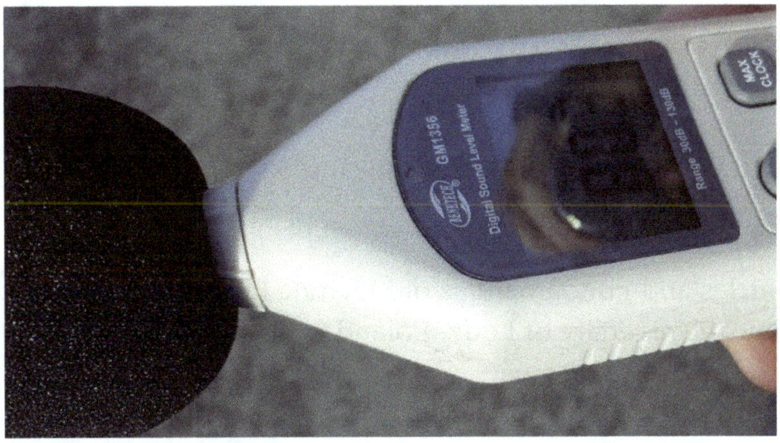

Figure 16. eKart noise intensity measurement.

During the tests, eKart sound loudness was measured. At a distance of 5 m from the track axis, the Benetech GM1356 was used, with the range of 30–130 dB and tolerance of ± 1.5 dB. In three tests, the noise level was measured at 80.3 dB; measurement is shown in **Figure 16**.

The measured value of the eKart noise is about six times lower than the noise level generated by the comparable class of ICE go-kart from Rotax Max Challenge series which is 90 dB [14].

6. Conclusions

With existing Li-ion battery technology, it is possible to construct eKarts for children and junior categories. With the current technology, it is not possible to create eKarts for senior categories in line with the current regulations for go-karts as the mass of the battery exceeds the weight of the entire drive train system of the go-kart. In order to provide a solution for senior categories, it would be necessary to change the regulations, for example, race time and vehicle weight, or wait for emerging battery technology to provide an energy density of at least 350 Wh/kg, which should take place within 3 years.

In this chapter, author proves that current technology is enabled to create eKart which is competitive and more efficient to ICE go-kart MiniMax category from Rotax Max Challenge series.

eKart also has better functional parameters; for example, noise generated by eKart is about six times lower than that for ICE go-kart MiniMax category from Rotax Max Challenge series. Due to electronic differential, which can actively manage torque and power on each of the rear wheels, called torque vectoring, eKart with 10 kW has parameters comparable with 12 kW ICE go-kart. That means that positive influence on natural environment is not only caused by applied electric motor but also due to algorithm in so-called green electronic differential. Due to electronic differential, which can actively manage torque and power on each of the rear wheels, called torque vectoring, eKart with 10 kW has parameters comparable with 12 kW ICE go-kart. That means that positive influence on natural environment is not only caused by applied electric motor but also due to algorithm in so-called green electronic differential.

Author details

Golebiowski Wlodzimierz*, Kubiak Przemysław and Szymon Madziara

*Address all correspondence to: golebiowskiw@wp.pl

Department of Vehicles and Fundamentals of Machine Design, Faculty of Mechanical Engineering, Technical University of Lodz, Poland

References

[1] Renault SAS. The Electric Revolution. youtube.com; 2011

[2] Jastrzębska G. Odnawialne źródła energii i pojazdy proekologiczne. Warszawa: Wydawnictwa Naukowo-Techniczne; 2009

[3] Arkadiusz Węglarz Michał Pleśniak. Samochód Elektryczny. Warszawa: Instytut na Rzecz Ekorozwoju; 2011. Krajowa Agencja Poszanowania Energii S.A

[4] Marks P. Dawn of motorsport's electric dream. www.newscientist.com 21.09.2011

[5] Marks P. Silent running: the price of a greener F1 car. www.newscientist.com 23.09.2011

[6] FIA. Formula One Technical Regulations. www.fia.com/regulations; 2009

[7] Audi R18 e-tron Quattro Flywheel KERS Hybrid System 2012, September 17, 2012; http://www.formula1-dictionary.net

[8] Jacob J, Colin JA, Montemayor H, Sepac D, Trinh HD, Voorderhake SF, Zidkova P, JJH P, Borisaljevic A, Lomonova EA. InMotion hybrid racecar: F1 performance with LeMans endurance. The International Journal for Computation and Mathematics in Electrical and Electronic Engineering. 2015

[9] FIA. http://www.fiaformulae.com; 2017

[10] BRM. http://www.rotax-kart.com/en/Products/MAX-Engines/; 2017

[11] Van Noorden R. A better battery—Chemists are reinventing rechargeable cells to drive down costs and boost capacity. Nature. 6 March 2014;**507**

[12] Xu B, Oudalov A, Ulbig A, Andersson G, Kirschen DS. Modeling of lithium-ion battery degradation for cell life assessment. IEEE Transactions on Smart Grid. June 2016

[13] Lawder MT, Northrop PWC, Subramanian VR. Model-based SEI layer growth and capacity fade analysis for EV and PHEV batteries and drive cycles. Journal of the Electrochemical Society. 2014;(14)

[14] Technical drawing No. 1 chassis frame and chassis main parts, http://www.cikfia.com/regulations/technical.html, CIK-FIA

[15] Milliken WF, Milliken DL. Race Car Vehicle Dynamics: Problems, Answers and Experiments. Society of Automotive Engineers Inc; 2003. http://www.sae.org/

[16] http://kosmickart.com/telai_mercurys_eng.php

[17] Golebiowski W, Kubiak P. Analysis of vehicle dynamics parameters for electric go-kart (eKart) design. Mechanics and Mechanical Engineering International Journal. Lodz University of Technology. 2017;**21**(3):763-777. ISSN 1428-1511

[18] Golebiowski W, Szosland A, Kubiak P. Concept of the construction of electric go-kart eKart. The Archives of Automotive Engineering. issn 1234/754X; accepted for publication

[19] Surmacz A. Regulamin Krajowy Pucharu Rotax Max Challenge Poland 2015 ROK, http://polskikarting.pl/wp-content/uploads/2015/04/regulamin_krajowy_pucharu_rotax_2015_-_wersja-ostateczna.pdf

5

Advanced Technologies for Large-Sized OLED Display

Chang Wook Han, Hong-Seok Choi, Chanki Ha,
Hongjae Shin, Hyun Chul Choi and In Byeong Kang

Abstract

Five years have passed, since the first 55″ full high-definition (FHD) OLED TV fabricated on Gen 8.5 glass was successfully launched into the TV market. For the time being, the size of OLED TV became diverse from 55″ to 77″, and the resolution was doubled into ultrahigh definition (UHD). The brightness and color gamut were enhanced, while the lower power consumption was realized. Utmost picture quality and slim form factor of OLED TV as well as the improved performance have made OLED TV recognized as the best premium TV. In this chapter, we describe the recent progress in three key technologies, which enable such an enhancement of performance in OLED TV, i.e., oxide thin-film transistor (TFT) and white organic light-emitting diode (WOLED), compensation circuit, and method to compensate the nonuniformity of oxide TFTs, OLED devices, and luminance.

Keywords: OLED TV, oxide TFT, white OLED, compensation circuit

1. Introduction

Organic light-emitting diode (OLED) is a very promising display. OLED provides a major technological enhancement to displays and TVs, such as wide viewing angle, high contrast ratio, and extremely fast response times. Furthermore, OLEDs can be used to make thin-light displays, transparent displays, and rollable displays.

OLED is a type of display technology that makes it possible to achieve dark black levels from ultrathin screens while at the same time making TVs more efficient and eco-friendly. OLED technology actively eliminated parts containing hazardous substances such as Cd, Hg, InP, etc. In addition, self-luminous OLEDs removed the backlight, helping designers achieve a lightweight and slim design that used less parts and attained greater resource efficiency and recycle rates [1].

LG commercialized 15-inch OLED TVs in 2009. The technologies to realize the TVs were RGB deposition technology using a shadow mask and thin film transistor (TFT) using low-temperature polycrystalline silicon. However, it was difficult to produce larger-sized OLED TVs with the same technology applied to 15-inch OLED TVs. The shadow mask is applied to the mass production of small-sized OLED displays, forming RGB subpixels. However, this method is not suitable for larger OLED displays due to sagging in metal mask and defects and color mixing due to misalignment between mask and glass substrate. Also, the development of TFT for large substrate was another important obstacle. We need to further improve reliability and production yield of TFT.

Recently, there are several breakthroughs in realizing large OLED TVs. The first one is to use white OLED as the light source and implement the color via color layer. When WOLED receives an electric current, it mixes two or three wavelengths of light and produces a white light. For RGB color, the color layer was used to filter the white light. WOLED can provide solutions for manufacturing processed on large-sized substrate [2, 3]. This technique does not require sophisticated metal mask taking into consideration the misalignment margin of pixel designs. In addition, due to good process yield and high productivity, the eight-generation glass substrate can be deposited to produce large-sized OLED TVs. The second is an oxide TFT that was developed intensively to meet the large backplane requirements for OLED TVs [4, 5]. Oxide TFT is also capable of processing on eighth-generation glass substrates because of its similarity to amorphous silicon TFT and may be compatible with TFT lines for LCDs in terms of production lines. By combining WOLED and oxide TFT that can be made from large substrate, larger OLED TVs can be produced with higher productivity and lower cost.

We commercialized 55-inch full HD (FHD) OLED TVs in the early 2013 [6]. And then in 2016, we launched not only 65-inch and 55-inch ultrahigh definition (UHD) OLED TVs but also 77-inch UHD OLED TVs. In this chapter, we will describe advanced technologies including oxide TFTs, WOLEDs, and compensation circuit and UHD OLED TV.

2. Oxide TFT technology

2.1. Overview of TFT structure

The structure of TFT is classified as top and bottom gate depending on the relative position of gate and active layer. Additionally, when the gate is on the same side of source/drain electrodes, it is called coplanar structure. Two types of oxide TFT structures are shown in **Figure 1**. Double-gate TFTs with both top and bottom gates are illustrated by **Figure 1(a)**. Gate and SD2 electrode function as a bottom and top gate, respectively. As the advantageous side, this structure was utilized to increase current flow and gets better output characteristics [7]. However, large parasitic capacitances between gates and source/drain metals are disadvantages, which are a big hurdle developing UHD OLED TV. Coplanar TFT was designed to reduce parasitic capacitance avoiding overlap between gate terminal and source/drain

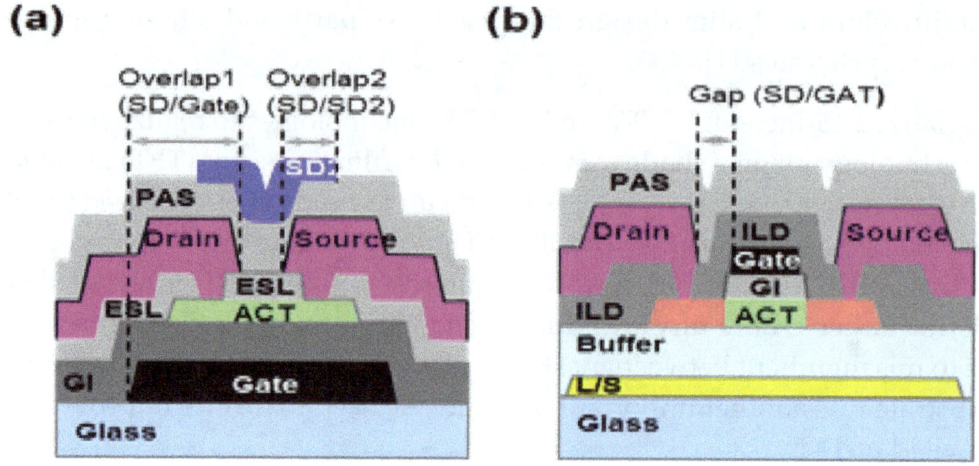

Figure 1. The cross-section of oxide TFT structures. (a) Etch-stopper structure with double gates and (b) self-aligned coplanar structure with top gate.

terminals as shown in **Figure 1(b)**. Light shield (LS) layer acts as blocking light coming to active layer, which is the main source to cause device degradation under negative bias temperature illumination stress (NBTIS). a-IGZO TFT with coplanar structure needs active metallization process to make ohmic contact between a-IGZO semiconductor and source/drain metals. This process was optimized to generate increased oxygen vacancies inside active film [8]. After metallization process, we obtained effective channel length, which is found to be shorter than nominal channel length. Effective channel length should be managed with controlled process.

2.2. Device characteristics

Driving TFTs for OLED TV operate in a current-driven mode, which control supplying current to OLED devices. Scan TFTs act as switches in active-matrix OLED (AM-OLED). It is important to obtain excellent device characteristics of these TFTs because device characteristics are directly connected to the performance and lifetime of OLED TV.

2.2.1. Electrical properties of a-IGZO TFT

Threshold voltage, subthreshold swing, field-effect mobility, and on/off current ratio can be extracted from I–V curve [9]. Typical values for each property are 0.5 V, 0.15 V/dec., 10㎠/Vs, and 10^7 at Vds = 10 V in the same order. Threshold voltage (Vth) is an important factor in the electrical characteristics, and it tells when the device turns on and off. Vth can be extracted from gate voltage when the drain current reaches 10 nA at the transfer curve under 10 V of drain voltage. Series resistance is also an important parameter in short-channel devices. The effective channel length can be obtained from the channel resistance method [10].

Figure 2(a) shows the transfer curves of 20pts TFTs located in six 55-inch panels which are fabricated on a Gen. 8.5 glass. The transfer curves were plotted by measuring the W/L = 26 μm/10 μm devices in the test element groups (TEG) through an inline probe station instrument. The inset picture shows the Vth distribution extracted from the real-time automatic Vth sensing instrument [11]. The variations of Vth extracted from transfer curve and automatic Vth sensing methods

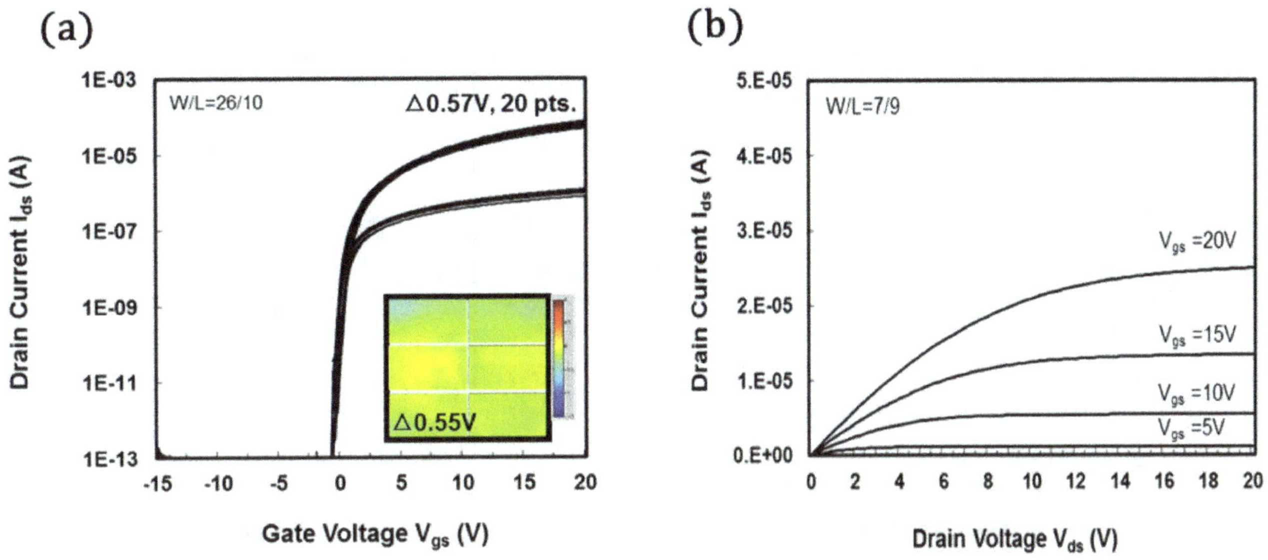

Figure 2. Transfer characteristic of a-IGZO TFTs on Gen. 8.5 glass (2200 × 2500mm).

were 0.57 V and 0.55 V, respectively. **Figure 2(b)** shows the output curves of the a-IGZO TFT. The output curve was plotted by sweeping the drain voltage from 0 V to 20 V, while gate voltages were applied from 5 to 20 V with 5 V steps.

Figure 3 is a graph plotting Rtot versus physical length of a-IGZO TFT with a width of 26 μm through the channel resistance method. ΔL, which is the difference between the physical channel length and the effective channel length of the coplanar a-IGZO TFT device, was calculated as 1.2 μm.

2.2.2. Reliability properties of a-IGZO TFT

Once OLED TV displays images, a-IGZO TFTs for OLED TV turn on and off repeatedly. In such operating environment, the device is stressed by continuous voltage, current, and temperature for a long time, which causes to degrade the electrical characteristics of a-IGZO TFT. The deteriorated a-IGZO TFT usually shows the change of Vth and drain current. This reduces the lifetime of

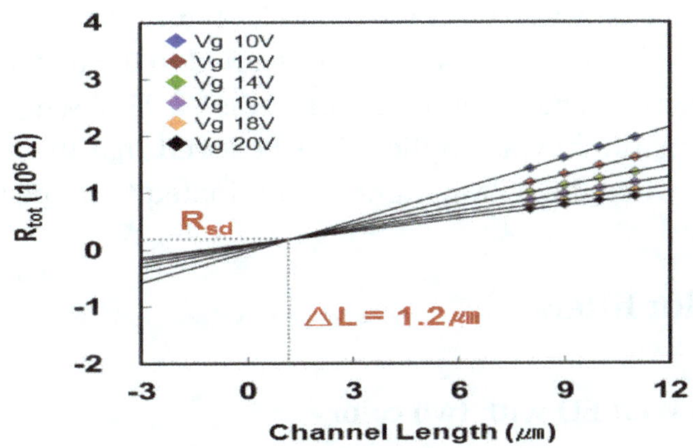

Figure 3. Illustration of the method used to extract effective channel length.

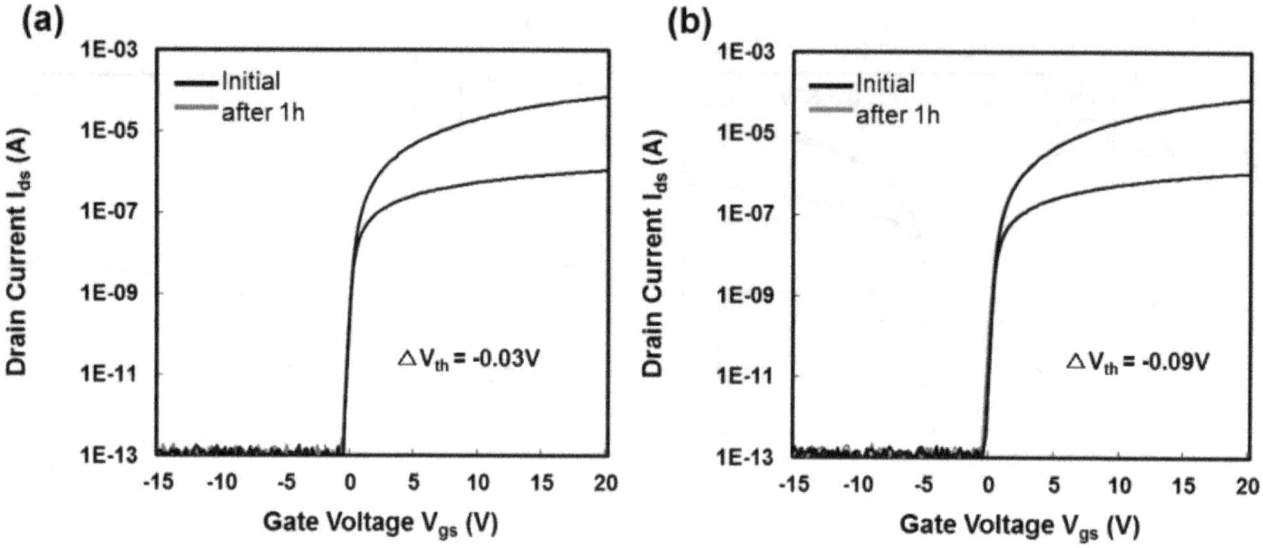

Figure 4. Shift of transfer curve after 1 h of negative gate bias at 60°C. Light emission from (a) top and (b) bottom.

the OLED TV or causes nonuniform luminance and image-sticking problems [12, 13, 14]. For this reason, a-IGZO TFT must be able to exhibit stable electrical characteristics for guaranteed lifetime. There are three typical methods to monitor device reliability such as NBTIS, positive bias temperature stress (PBTS), and current stress (CS) [15, 16, 17]. There are two well-known device degradation models of charge trapping in gate insulator [18] and defect creation in active layer models [19]. In addition, research has been going on the ambient effect model [20]. In order to improve the device reliability, it is necessary to optimize device fabrication process. Defect sites which are the source of charge trap should be reduced inside the oxide semiconductor thin film and at the interface between gate insulator and active layer. Studies have been conducted to improve device reliability. Oxide TFT was passivated to prevent external moisture and oxygen permeating into active layer [21, 22]. By applying light shield (LS) layer into TFT structure, it is possible to improve light-induced reliability degradation preventing active layer from the light [23].

Figure 4 shows the shift of transfer curve of a-IGZO TFT before and after NBTIS test. The NBTIS test was evaluated by applying Vgs = -30 V and Vds = 0 V at 60°C for 1 hour. The 4500 nit of visible light is also applied in the top and bottom directions of the device. **Figure 4** shows NBITS reliability characteristics of a-IGZO TFT under the top light (a) and bottom light (b). In both illumination tests, transfer curves shifted to the negative direction compared to the initial transfer curves. The amount of shifts was −0.03 V when top light was applied and − 0.09 V when bottom light was applied. It is believed that the amount of light is more injected into the device when the bottom light is illuminated than the top light.

3. WOLED and color filters

3.1. Two-stack tandem WOLED with two colors

WOLED employed to the first OLED TV launched in 2013 had a two-stack two-color tandem structure consisting of fluorescent blue and phosphorescent yellow-green (YG) stacks

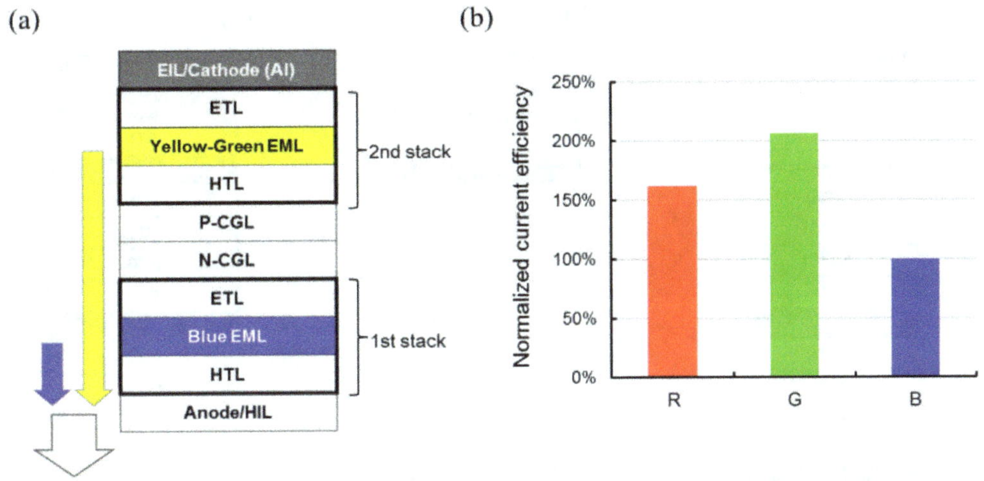

Figure 5. (a) Device structure of two-stack two-color tandem WOLED (b) relative current efficiency of red, green and blue subpixels normalized by the values needed to display full-window white pattern of 100 nit.

serially connected by n- and p-type charge generation layers, as shown in **Figure 5(a)** [2, 3, 24]. Specifications on luminance for the first OLED TV were 100 nit at full-window white pattern and 400 nit at peak-luminance pattern where 25% area is turned on. However, for the second models of OLED TV, 150 and 450 nit were demanded for the full-window and peak-luminance patterns, respectively, even at the lower power consumption.

Figure 5(b) shows relative current efficiencies (cd/A) of RGB subpixels after transmitting through color layer. These values are normalized by the efficiency for each subpixel, required to get a certain luminance. Since it is shown that blue subpixel is a restriction factor to the panel brightness, we need to drastically increase the efficiency of the blue device by more than 1.5 times. Two approaches could be considered in respect of external quantum efficiency and internal quantum efficiency. With regard to the external quantum efficiency, blue efficiency could be improved through a strong cavity effect using an electrode of thin metal. However, color shift varied by view angle could be worse. Regarding the internal quantum efficiency, phosphorescent blue [25, 26] or thermally activated delayed fluorescent (TADF) blue is still not around the corner.

3.2. Three-stack tandem WOLED with two colors

Our solution for the higher blue efficiency is three-stack tandem WOLED, namely, adding one more blue stack [27]. It is true that three-stack tandem WOLED needs higher applied voltage, which might become a factor to increase the power consumption. Nonetheless, three-stack WOLED can realize lower power consumption as well as higher luminance, which will be demonstrated later. As shown in **Figure 6**, three device architectures with different sequences of the unit devices are considered: (a) YG unit adjacent to the cathode (W1), (b) in the middle of the device (W2), and (c) adjacent to anode (W3). The best device architecture for three-stack two-color (3S2C) tandem WOLED is decided in terms of efficiency and color shift by viewing angle, with an assistance of optical simulation [28].

The viewing angle dependence of blue and YG efficiencies in each WOLED device can be predicted on the basis of the fact that angular dependence of blue and YG mono-devices is

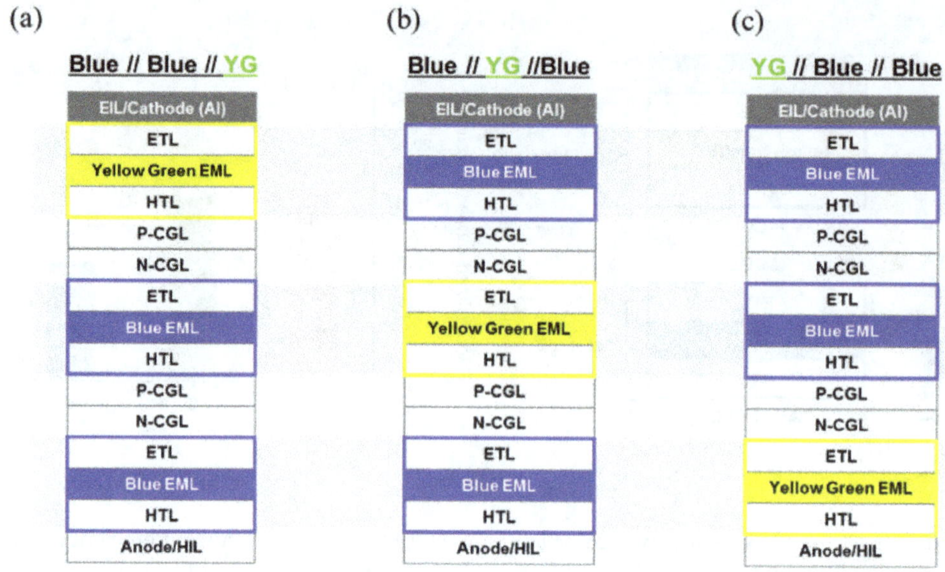

Figure 6. Three candidates for three-stack WOLED consisting of two fluorescent blue stacks and 1 phosphorescent YG stack.

correlated with the distance of EML from the cathode [27]. W1 whose both blue EMLs are farther from the cathode than YG EML shows that the blue intensity falls faster than the YG. W3 whose YG EML is farther from the cathode than the two blue EMLs shows that the YG intensity is reduced faster than the blue. In case of W2, as the blue intensities of two blue EMLs of which one is closer to and the other is farther from the cathode than YG EML, the blue efficiency is supposed to fall at the same trend with YG efficiency when the viewing angle is varied, resulting in little color shift. To comment on the viewing angle dependence of our two-stack WOLED depicted in **Figure 5(a)**, as blue EML is more distant from the cathode than YG EML, its color is supposed to shift toward YG at high incident angle.

We fabricated three-stack WOLED device where B, YG, and B units are sequentially formed, based on the optical simulation. **Table 1** summarized its actual performance compared with two-stack WOLED device. Since another blue device unit is supplemented, the voltage applied across the three-stack WOLED is enlarged by 4.5 V at the same current density of 10 mA/cm^2. However, the three-stack WOLED has the current efficiency (cd/A) enhanced by 8% and emits cool white light with high CCT of 8500 K, owing to its enhanced blue intensity by two blue units. Generally, TV panel displays white color with high CCT up to 9300~10,000 K. To realize such a cool white color using two-stack WOLED emitting white light of 6300 K CCT, the blue subpixel should be turned on with a high intensity, resulting in a high power consumption. The experimental electroluminance (EL) spectra varied by viewing angle, as depicted in **Figure 7(a)**, show that the intensities at blue and YG regions drop simultaneously at the same rate, resulting in very low color shift of 0.011 at 60°. The efficiency of blue subpixel for the three-stack WOLED is found to be enhanced by 75%, through adding one more blue stack, as shown in **Figure 7(b)**.

As mentioned above, the three-stack WOLED employing two stacks of blue device is advantageous to lowering the power consumption as well as boosting luminance of OLED panel.

	Two-stack WOLED	Three-stack WOLED
Voltage (V)	7.1	11.6
Efficiency (cd/A)	78.6	85.0
Color (Wx, Wy)	(0.317, 0.332)	(0.287, 0.310)
Color shift ($\Delta u'v'$)*	0.020	0.011

Table 1. The comparison of device characteristics between two-stack WOLED and three-stack WOLED at 10 mA/cm^2.

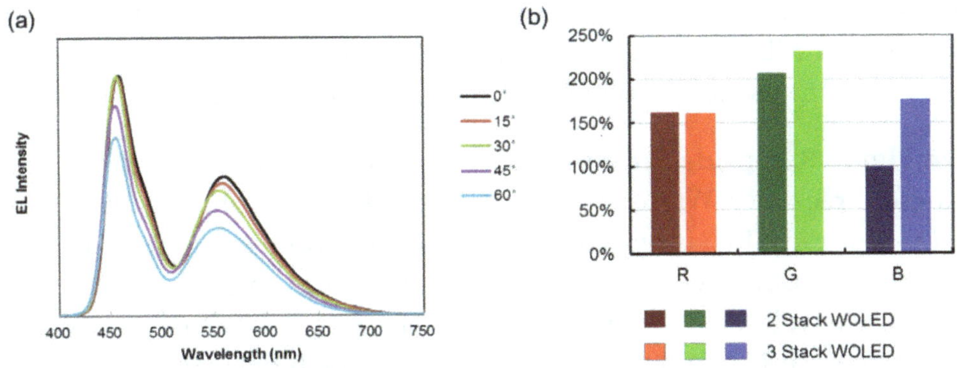

Figure 7. (a) Experimenal EL spectra of our real three-stack two-color tandem WOLED at various viewing angles showing that blue and YG intensities drop at the same ratio and (b) the comparison of efficiency of R, G, B subpixels between the two-stack tandem WOLED and three-stack tandem WOLED.

By enhancing the blue efficiency, the three-stack WOLED made our new OLED TV realize the luminance of 150 nit at full-window pattern which is higher by 50% compared to the first model using two-stack WOLED.

As regards to power consumption, we calculated time-average current values spent at each pixel of WRGB for OLED panels during playing a movie file ruled by the international standard CIE60287 at a fixed maximum luminance. The total current of OLED panel employing the three-stack WOLED is estimated to be at the 68% level of two-stack WOLED. Considering voltage rise of the three-stack WOLED, total voltage across WOLED and driving transistor in the backplane is increased by 10%. Namely, the final effect of the three-stack WOLED for saving the power consumption of OLED panel is calculated to be 24%.

To analyze the current of each subpixel in detail, OLED panel of the two-stack WOLED shows that a half of the total current is applied to the blue subpixel. In that case, by extending aperture ratio of the blue subpixel and lessening the current density, electrical stress on the blue subpixel could be reduced.

For the three-stack WOLED emitting cool white, red subpixel is more used to display warm colors, and the current to the red subpixel is slightly increased. However, the current decrease of blue subpixel as well as white subpixel is overwhelming. As a result, balancing of currents applied to four subpixels becomes better in the three-stack WOLED, which provides an advantage to OLED panel design. In addition, the lifetime for blue subpixel can be extended significantly.

3.3. Three-stack three-color tandem WOLED

The next urgent demand by our customers was to widen color gamut up to digital cinema initiatives (DCI) standard color space, after the luminance was enhanced by three-stack two-color (3S2C) tandem WOLED. The 3S2C WOLED has achieved 114% color gamut area compared to sRGB standard color but covered only 90% area in DCI standard color space [27]. In order to widen color gamut in DCI color space, deep green color with high CIE.y value and deep red color with high CIE.x value are necessary.

Since phosphorescent green EML did not show long enough lifetime for WOLED TV, we considered the insertion of an additional red EML in the three-stacked WOLED structure while maintaining the YG EML. By adding red EML, red color point was supposed to move toward higher CIE.x value. In case of green color point, we relied on optimizing green color layer.

As shown in **Figure 8**, considering contour map of the emittance [27, 28], three ways to add red EML in the three-stack WOLED could be taken into account. The two ways are as follows: a fluorescent red EML is added, adjacent to the fluorescent blue EML in the first- or third-stack unit. The other way is that a phosphorescent red EML is inserted, adjacent to the phosphorescent yellow-green in the second-stack layer. In the case of the three-stack three-color (3S3C) WOLED with fluorescent red EML, the blue EML in the third stack shares the exciton with the red EML so that the blue efficiency is decreased a little, but the overall efficiency (cd/A) is raised, thanks to the contribution of the red EML. As a result, the WOLED has warmer white. On the other hand, in the case of the WOLED with phosphorescent red EML, the YG EML in the second stack shares the exciton with the red EML. As the efficiency

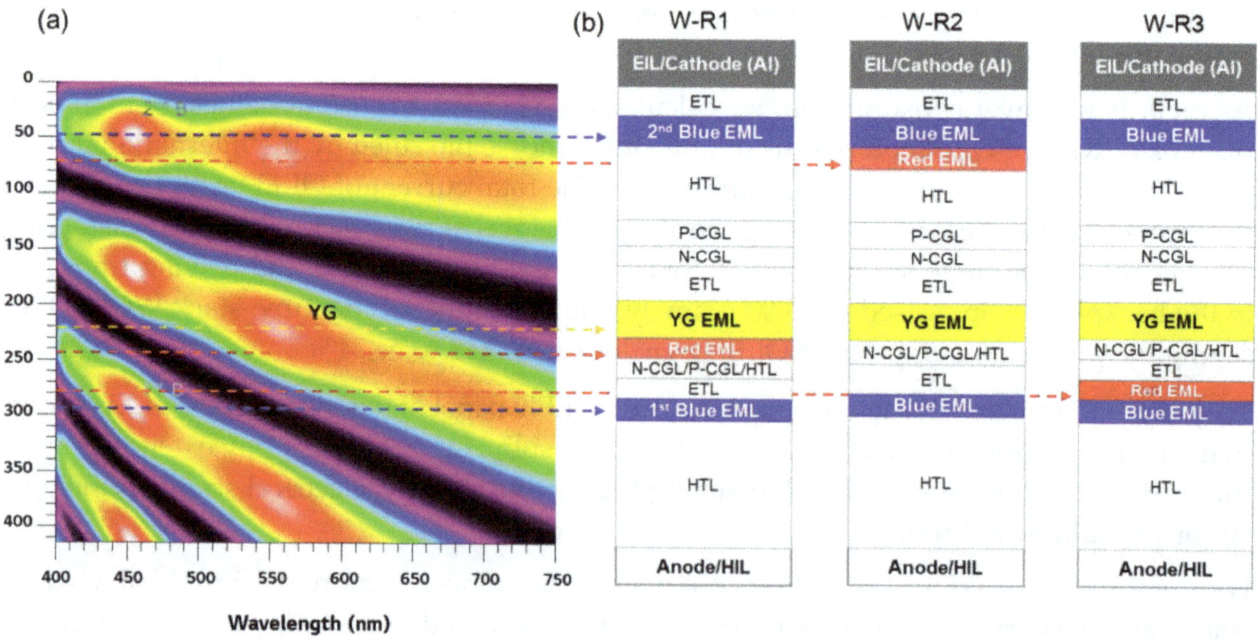

Figure 8. (a) Emittance contour map for three-stack three color tandem WOLED where red dotted lines are drawn at the position having relatively high emittance value near 620 nm. (b) Three device architectures for three-stack three color tandem WOLED; red EML in W-R1 is fabricaed before YG EML in the second stack, red EML in W-R2, before the second blue EML in the third stack, and red EML in W-R3, after the first blue EML in the first stack.

(cd/A) of the red EML is lower than that of the YG EML, this WOLED has the less efficiency than the 3S2C WOLED, but it emits the cool white light which is more suitable for the display application. Moreover, it is well known that fluorescent red materials do not show good lifetime performance. Thus, we should choose the WOLED involving the phosphorescent red (Ph-R) EML.

In the second case, or inserting phosphorescent R-EML, we fabricated a unit device of the phosphorescent red and YG (Ph-R/Ph-YG) EMLs whose structure is in the following, ITO/HTL/x% red dopant/12% YG dopant/ETL/EIL/Al, to examine the influence between red dopant ratio and YG dopant ratio. By controlling the concentration of R dopant at 12% YG dopant concentration, we obtained the best result at 2% R dopant concentration. **Figure 9(c)** illustrates EL spectra at the R doping ratio of 2 and 4% with the YG dopant fixed at 12% concentration. As R doping ratio is increased, the intensity in green region is reduced, and the intensity in red region is enhanced, while the external quantum efficiency (EQE) is slightly changed. These phenomena can be caused by the increase in the exciton energy transfer from YG EML to red EML owing to the increased red dopant concentration.

In addition to the red dopant concentration, it is also possible to control the relative intensities of red and green colors by changing the ratio of hole transport host and electron transport host in YG EML [3]. We have been adopting the mixed-host structure of a hole-type host and an electron-type host as a phosphorescent EML layer for high efficiency and long lifetime of WOLED. We fabricated Ph-R/Ph-YG unit devices along with the host ratio in YG EML of 0:10, 7:3, and 10:0 ratios of hole transport host and electron transport host at 12% YG dopant and 2% red dopant concentration.

Table 2 summarizes electrooptical performance of the Ph-R/Ph-YG unit devices, and **Figure 9(b)** shows EL spectra of the devices. When the hole-type host only (h-type host:e-type host =10:0) is used, the EL spectrum shows mainly YG peak with a slight red peak, which means that excitons were formed largely in the YG EML. As the ratio of the electron-type host in the YG EML is increased (h-type host:e-type host =7:3, 0:10), the YG peak

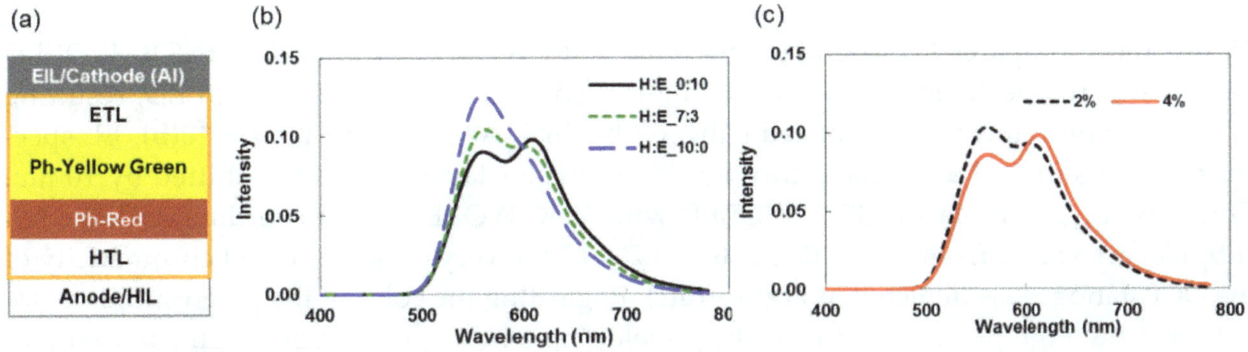

Figure 9. (a) Schematic diagram of phosphorescent unit device, (b) EL spectra of Ph-devices with different mixed host ratio at fixed YG and R dopant concentration, and (c) EL spectra of Ph-devices with the different R dopant concentration at fixed YG dopant concentration.

	Volt (V)	cd/A	EQE (%)	CIEx	CIEy
Host ratio (H:E)					
10:0	3.9	60.0	19.5	0.467	0.524
7:3	4.0	52.9	19.2	0.487	0.505
0:10	4.1	49.1	19.3	0.501	0.492
Red dopant Conc. (%)					
2	4.0	52.9	19.2	0.487	0.505
4	4.1	47.0	19.1	0.507	0.486

Table 2. Summary of device performance of Ph-devices (upper) with mixed host ratio at 10 mA/cm^2 (lower) with R dopant concentration at 10 mA/cm^2.

intensity is reduced, whereas the red peak intensity grows higher with approximately equal EQE. The increasing ratio of the electron-type host makes the facile electron transfer from YG EML to R EML. As a result, the electron-hole recombination zone will shift from the YG EML to the R EML/YG EML interface, which makes the rise of the R peak in the EL spectrum. We found the optimal spectrum at h-type host:e-type host = 7:3 ratio in YG EML. In case of Ph-R/Ph-YG unit device, we can control the red intensity with the host ratio as well as the dopant ratio.

There is an alternative way to increase the red intensity in WOLED, i.e., co-deposition of phosphorescent red and YG dopants at one EML in second unit device. Since the optimal doping ratio for the red dopant is in the range from 0.2 to 0.4%, the doping ratio should be precisely controlled. As thc doping concentration of the red dopant was raised a little above the range, it was found that the intensity in green region is reduced, while the intensity in red region is increased. However, when the red doping ratio was too low, the exact ratio control between the red dopant and green dopant was not easy in process. Therefore, we considered inserting a red EML separately.

Figure 10(a) compares EL spectra of 3S3C and 3S2C WOLEDs. It found that 3S3C WOLED has distinct red peak at 620 nm and the higher red intensity than 3S2C WOLED, resulting in an efficiency enhancement of red subpixel by 38%. As shown in **Figure 10(b)**, EL spectrum at red subpixel after going through the red color layer (CL) is red-shifted by 10 nm. Consequently, by replacing 3S2C WOLED with 3S3C WOLED, color coordinates of the red subpixel are varied from (0.666, 0.332) to (0.678, 0.321), very close to the red chromaticity of the DCI standard, as depicted in **Figure 10(c)**. Regarding the color of the green subpixel, we obtained the high purity green color by developing a new green CL for high color gamut. Thanks to the new green CL, the color coordinates of the green subpixel were shifted from (0.300, 0.645) to (0.270, 0.666). As shown in **Figure 10(c)**, 3S3C WOLED with the new CLs covers most of the color space for the DCI standard, which corresponds to 99% color gamut of the DCI standard in CIE1976 (u′v′) color space.

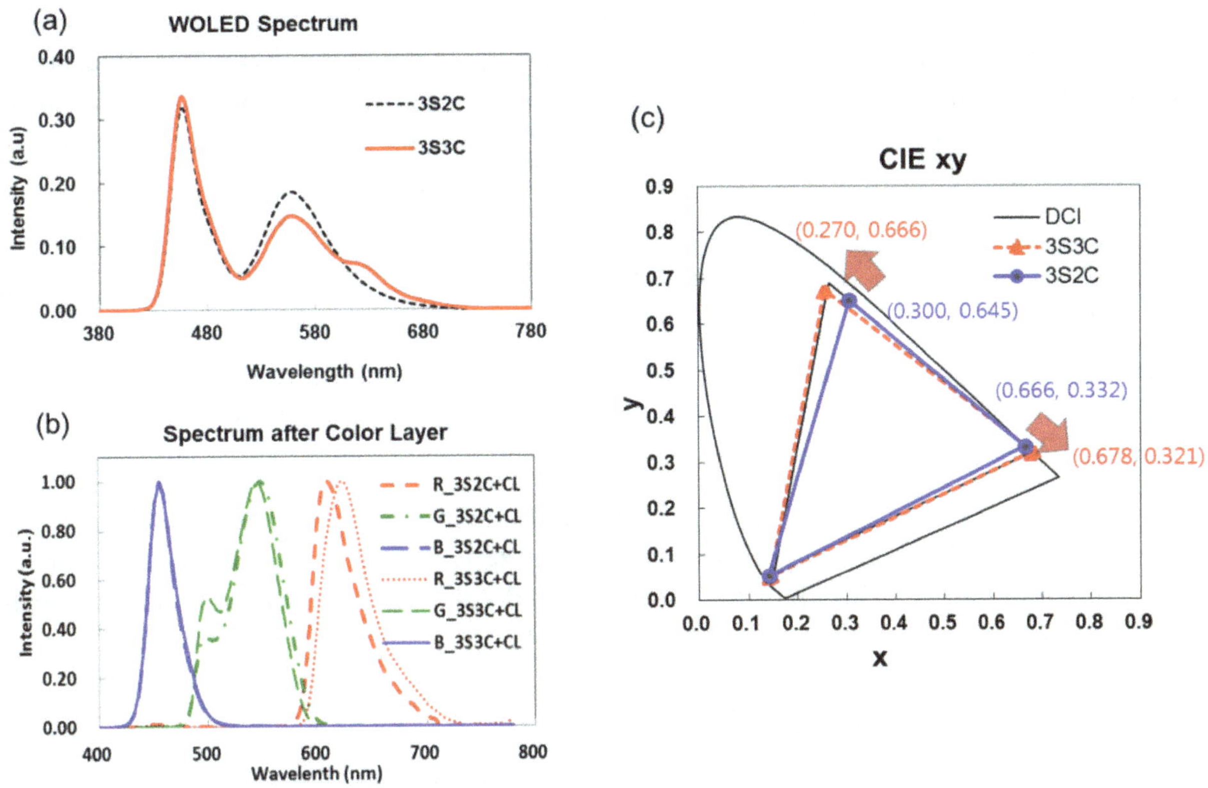

Figure 10. (a) Comparison of emission spectra of 3S2C and 3S3C WOLEDs (b) Emission spectra of 3S2C and 3S3C WOLEDs with color layers (c) Comparison of color gamut of 3S2C and 3S3C WOLEDs.

Table 3 summarizes the brightness and color gamut for 3S2C and 3S3C WOLEDs which are applied to OLED TV made in 2015 and 2016, respectively. As a result of such innovations in WOLED and CL, OLED TV could realize peak brightness of 500 nit and full-window brightness of 150 nit as well as high color gamut, i.e., 129% in sRGB color space and 99% in DCI color space.

	Year 2015	Year 2016
WOLED structure	3-stack 2-color (3S2C)	3-stack 3-color (3S3C)
Brightness (nit)	450/150	500/150
Color gamut (%)	sRGB 114%	sRGB 129%
	DCI 90%	DCI 99%

Table 3. The specification of OLED TVs based on 3S2C and 3S3C WOLEDs.

4. Compensation technologies

4.1. Pixel circuit

OLED displays, having current-driven subpixels, require higher backplane uniformity than LCDs. **Figure 11(a)** shows a simplified pixel circuit diagram of an active-matrix OLED display pixel.

Generally, the image data is supplied as a data voltage via a data line and applied to the gate of the driving transistor (DR) through the switching transistor (SW). The data voltage is stored in the storage capacitor (Cst), which keeps the gate-to-source voltage (Vgs) of DR stable even when the source voltage (Vs) changes according to the current–voltage characteristics of the OLED. The current flowing through DR is determined by Eq. 1:

$$I_{ds} = \frac{1}{2}\mu C_{ox}\frac{W}{L}(V_{gs} - V_{th})^2 \quad (1)$$

The threshold voltage (Vth) determines the x intercept of the $V\text{–}I^{1/2}$ diagram of the transistor, while the mobility (μ), the capacitance per area of the gate insulator (C_{ox}), and the width-to-length ratio (W/L) determine the slope.

These values vary for each pixel because of fluctuations in layer thicknesses, etching biases, etc. and because of TFT degradations such as Vth shifts. The pixel current thus varies for each pixel as shown in **Figure 11(b)**. In order for an OLED display to achieve a high uniformity, the current variation must be compensated in each pixel.

4.2. Internal compensation and pixel circuit

OLED pixels with compensation traditionally employ additional TFTs, capacitors, and lines as in **Figure 12(a)**, which lowers the aperture ratio and increases defects [6], or power line voltage swinging as in **Figure 12(b)**, which is difficult to adopt in large-sized high-resolution panels because of large line loads and a short charging time. In order to achieve mass production of large-sized high-resolution OLED TVs, a simple pixel structure is necessary to reduce defects and improve aperture ratio [11]. Minimizing the number of TFTs in a subpixel can not only reduce defects but also simplify driving signals, which allows a narrow bezel design. We use a single data line and a single gate line for each subpixel, and the four subpixels in a full RGBW pixel share a sensing line and a power line, which reduces line crossings and thus reduces defects.

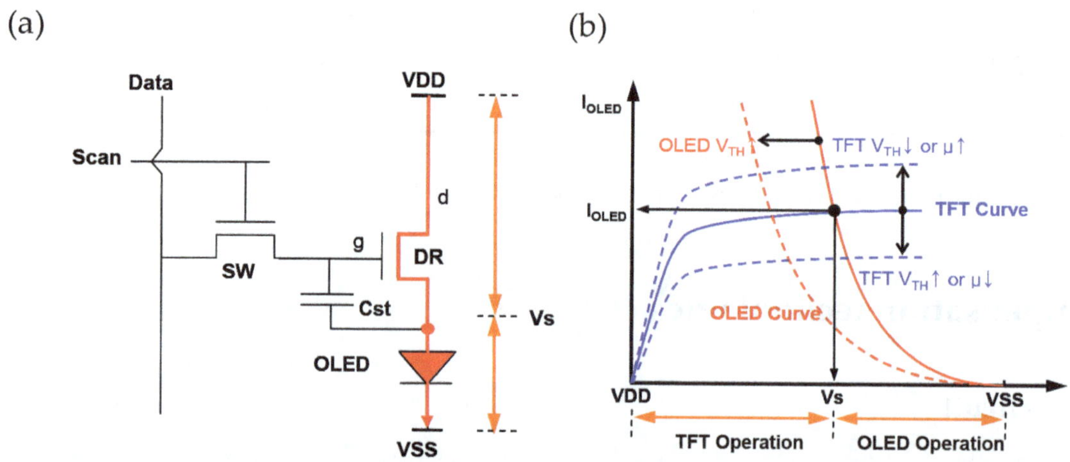

Figure 11. (a) OLED pixel circuit and operation voltages (b) OLED current at various TFT and OLED operation points.

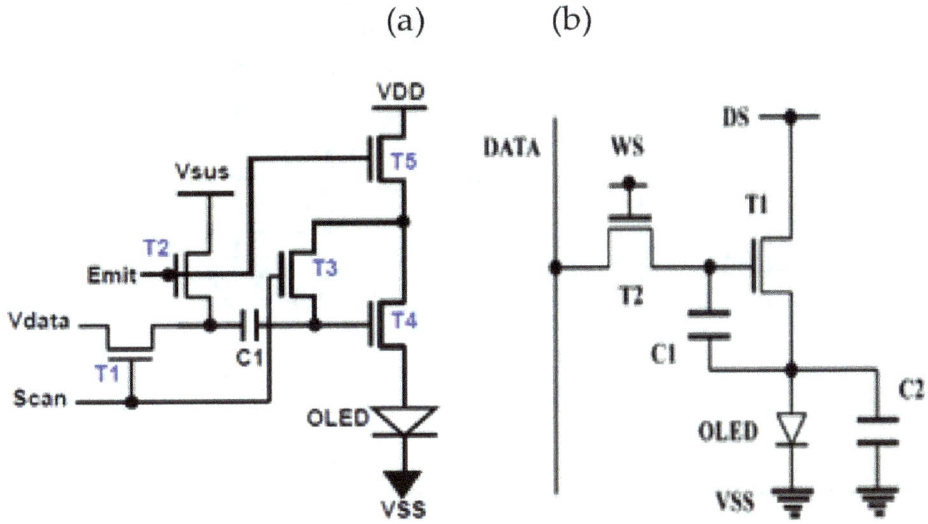

Figure 12. Internal compensation pixel circuits (a) 5T1C voltage programming (b) 2T2C voltage programming with VDD swing.

4.3. External compensation and pixel circuit

Figure 13 shows the concept of our external compensation [29]. From 2003, we developed LGD's unique compensation technology, including simple pixel structure, TFT compensation algorithm, and precise monitoring technology of TFT variation. Our pixel circuit satisfies UHD requirements such as a large screen, high compensation characteristics, and a high productivity. Our external compensation method compensates the threshold voltage and mobility variation of TFT. To get initial good uniformity of the luminance, compensation method is required to compensate variations of threshold voltage and mobility that cause luminance differences. Through the real-time sensing and compensation, our methods compensate threshold voltage shift and mobility shift.

Using this method, we solve reliability issues caused by positive and negative biases, temperature, and current stress. The subpixel itself does not have a compensation circuit, and the external circuit compensates each subpixel correctly. That makes the external circuit more complicated than internal compensation methods, because we need analog-digital converters, a sensing data memory, a compensation algorithm unit, etc. It has its own advantages: we can optimize compensation methods and parameters, not only for Vth but also for mobility, by using refined algorithms.

Figure 14 shows the sensing results of TFT Vth and mobility using external compensation pixel circuit. In our case, we sense the Vth and the mobility of the driving transistor in the source follower method shown in **Figure 14(b)**, where the data voltage for sensing (Vg) is applied to the gate of the driving transistor and the source voltage (Vs) is sensed by an external circuit. The difference between Vg and Vs is the Vth of the driving transistor and is stored to the memory. Using the sensed Vth, a Vth-compensated data voltage shown in Eq. 2 is applied to the gate:

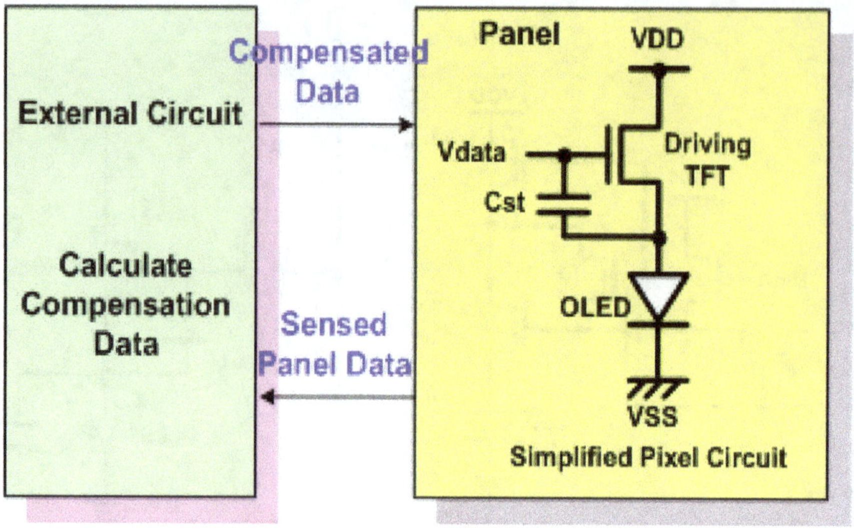

Figure 13. The external compensation method and the simplified pixel circuit.

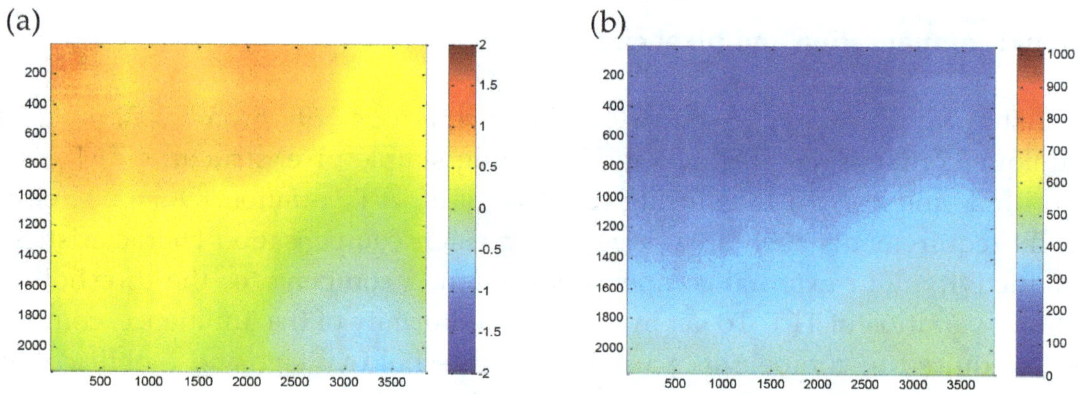

Figure 14. Sensing results of TFT (a) Vth and (b) mobility using extenal compenstion pixel circuit.

$$V'_{data} = V_{data} + V_{th} \tag{2}$$

Then, Vs is sensed, which is in proportion to the product of the TFT characteristics of the pixel except the Vth, which we call *k* here as shown in Eq. 3, where the capacitance of the sensing line (C_{line}) and the sensing time (t) are considered:

$$k = \frac{1}{2} \mu C_{ox} \frac{W}{L} V_{data}{}^{2} C_{line} t \tag{3}$$

Finally, by calculating the average of *k* of all pixels, we compensate each pixel by Eq. 4:

$$V'_{data} = \sqrt{\frac{k_{avg}}{k}} V_{data} + V_{th} \tag{4}$$

The pixel characteristics are sensed and compensated before shipment and in real time.

4.4. OLED degradation compensation

Other quality issues include image sticking. Firstly, being current-driven, OLED pixels generate heat when they emit light, and there may be a luminance change because of high temperature. Secondly, luminance will drop according to total driving time because of OLED degradation, like any other self-luminous device. For the former issue, we have designed a mechanical structure to release heat efficiently, and we use real-time temperature compensation. For the latter issue, we use a known correlation between current efficiency decrease and electric characteristic change of OLED [30]. OLED voltage–current characteristics change according to degradation, and we need a higher voltage to have the same current after OLED usage. We sense voltage for the same predetermined current to estimate OLED degradation at each subpixel and use a lookup table to translate OLED voltage shift to luminance compensation value. **Figure 15** compares time-dependent luminance curves for low stress and high stress, without and with repeated OLED voltage sensing and luminance compensation. Luminance difference can be minimized between high-stress and low-stress subpixels through the OLED degradation compensation. We sense OLED degradation with a predetermined interval because OLED degrades much slower than TFT. We do not compensate OLED luminance completely to maintain initial luminance, because OLED compensation rather accelerates degradation. Instead, we match OLED luminance to target degradation curve. **Figure 16(a)** shows the image sticking

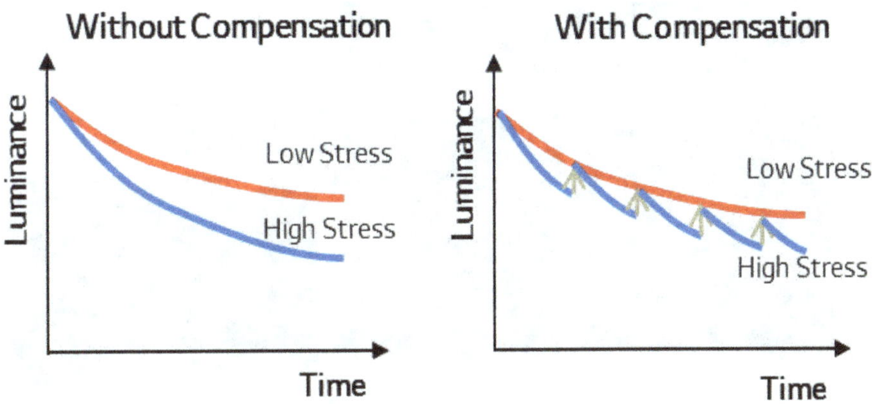

Figure 15. Image sticking compensation method due to OLED degradation.

(a) (b)

Figure 16. Image sticking test results (a) Before compensation, (b) After compensation.

of the OLED panel due to OLED degradation. **Figure 16(b)** demonstrates the drastic disappearance of the image sticking after the OLED degradation compensation.

5. World's first UHD OLED TV products

Figure 17 is a photograph of the 55-, 65-, and 77-inch UHD OLED TVs, which are world's first products. These OLED TVs are employing TFT backplane composed of coplanar a-IGZO TFTs, three-stack three-color tandem WOLED and advanced compensation technologies. Our technology platform allows a panel size scalability, a product reliability, and a high aperture ratio. **Table 4** shows specifications of our UHD OLED TVs with high color gamut, high contrast ratio, and thin thickness.

The reason why OLED TV is known to have the best quality of display is that it realizes high contrast ratio since OLED at each subpixel can be completely and individually turned off by oxide TFTs when the subpixel displays zero signal. Low-leakage current of oxide TFTs contributes to the high contrast ratio of OLED TV. Namely, OLED TV can easily achieve high dynamic range (HDR) without raising up the peak luminance, contrary to LCD TV. Accordingly, OLED

Figure 17. Photograph of LG's 55-, 65- and 77-inch UHD OLED TV products.

Item	Content	Unit
Display type	WRGB OLED	
Panel size	55, 65, 77	inch
Resolution	3840 × 2160 (UHD)	
Brightness	150 (full)/500 (peak)	cd/m^2
Color gamut (DCI* coverage)	99 (DCI)	%
Contrast ratio	> 1000000:1	

Table 4. Specifications of the LG UHD OLED TV (*DCI: Digital Cinema Initiatives).

TV has an advantage to realize wide-span HDR with low power consumption. New compensation technologies can extend life span of OLED TV by relieving image sticking which seems inevitable for the spontaneous emission-type displays. The ultra-slim thickness of OLED TV not only offers premium values in view of design but also entitles OLED TV as green electronics by eliminating several plastic sheets in LCD TV.

6. Conclusion

In this chapter, we have explained the latest progresses in three key technologies, i.e., oxide TFT, WOLED, and advanced compensation circuit and method, applied to brand-new UHD OLED TV. a-IGZO TFTs with coplanar structure have demonstrated excellent electrical and reliability characteristics, which will be a good reference for the future development as well as for mass production of large-sized UHD OLED TVs. New device architectures of three-stack and/or three-color WOLED could realize OLED TV with high luminance and high color gamut. The advanced compensation technologies, including the external compensation method, have enabled a panel size scalability and an enhancement of the lifetime reliability necessary for commercializing large-sized and high-resolution OLED TVs.

Not to mention higher luminance and lower power consumption, OLED TV with higher resolution such as 8 K x 4 K, higher color gamut such as BT2020 and larger size than 80″ should be developed in the future. Although all these are definitely a huge challenge to oxide TFT, WOLED, and advanced compensation technology, it is believed that these will be realized via innovations and breakthroughs coming from momentum of research and development well established in these fields.

Author details

Chang Wook Han*, Hong-Seok Choi, Chanki Ha, Hongjae Shin, Hyun Chul Choi and In Byeong Kang

*Address all correspondence to: hcw@lgdisplay.com

LG Display, Seoul, Korea

References

[1] Kim S et al. Paper No S4.2: Eighth Generation Linear Source for AMOLED Mass Production. In: SID Symposium Digest of Technical Paper (EuroDisplay '15); 2015. p. 18

[2] Han CW et al. 21.2: 55-inch FHD OLED TV employing new tandem WOLEDs. In: SID Symposium Digest of Technical Papers (SID 12). 2012. pp. 279-281

[3] Han CW et al. 53.2: Invited Paper: Advanced Technologies for Large-sized OLED TV. SID Symposium Digest of Technical Papers (SID '14); 2014. pp. 770-773

[4] Nomura K et al. Room-temperature fabrication of transparent flexible thin-film transistors using amorphous oxide semiconductors. Nature. 2004;**432**:488-492. DOI: 10.1038/nature03090

[5] Yabuta H et al. High-mobility thin-film transistor with amorphous InGaZnO4 channel fabricated by room temperature rf-magnetron sputtering. Applied Physics Letters. 2006;**89**:112123. DOI: 10.1063/1.2353811

[6] Oh CH, Shin HJ, Nam WJ et al. Technological Progress and Commercialization of OLED TV. In: SID Symposium Digest of Technical Papers (SID '13); 2013. pp. 239-242

[7] Nag M et al. Characteristics improvement of top-gate self-aligned amorphous indium gallium zinc oxide thin-film transistors using a dual-gate control. Journal of the Society for Information Display. 2017;**25**:349-355. DOI: 10.1002/jsid.558

[8] Bae JU et al. Development of oxide TFT's structures. In: SID Symposium Digest of Technical Papers (SID '13). 2013. pp. 89-92

[9] Kagan CR, Andry P, editors. Thin-Film Transistors. New York: Marcel Dekker; 2003

[10] Sato A et al. Amorphous in-Ga-an-O coplanar homojunction thin-film transistor. Applied Physics Letters. 2009;**94**:133502. DOI: 10.1063/1.3112566

[11] Yoon JS et al. 55-inch OLED TV using optimal driving method for large-size panel based on InGaZnO TFTs. In: SID Symposium Digest of Technical Papers (SID 14). 2014. pp. 849-852

[12] Guo X et al. Investigation on the current nonuniformity in current-mode TFT active-matrix display pixel circuitry. IEEE Transactions on Electron Devices. 2005;**52**:2379-2385. DOI: 10.1109/TED.2005.857936

[13] Chaji GR et al. Electrical compensation of OLED luminance degradation. IEEE Electron Device Letters. 2007;**28**:1108-1110. DOI: 10.1109/LED.2007.909854

[14] In HJ et al. External compensation of nonuniform electrical characteristics of thin-film transistors and degradation of OLED devices in AMOLED displays. IEEE Electron Device Letters. 2009;**30**:377-379. DOI: 10.1109/LED.2009.2014885

[15] Suresh A et al. Bias stress stability of indium gallium zinc oxide channel based transparent thin film transistors. Applied Physics Letters. 2008;**92**:033502. DOI: 10.1063/1.2824758

[16] Oh H et al. Photon-accelerated negative bias instability involving subgap states creation in amorphous in–Ga–Zn–O thin film transistor. Applied Physics Letters. 2010;**97**:183502. DOI: 10.1063/1.3510471

[17] Nomura K et al. Comprehensive studies on the stabilities of a-In-Ga-Zn-O based thin film transistor by constant current stress. Thin Solid Films. 2010;**518**:3012-3016. DOI: 10.1016/j.tsf.2009.09.193

[18] Lee JM et al. Bias-stress-induced stretched-exponential time dependence of threshold voltage shift in InGaZnO thin film transistors. Applied Physics Letters. 2008;**93**:093504. DOI: 10.1063/1.2977865

[19] Nomura K et al. Origins of threshold voltage shifts in room-temperature deposited and annealed a-in-Ga-Zn-O thin-film transistors. Applied Physics Letters. 2009;**95**:013502. DOI: 10.1063/1.3159831

[20] Jeong JK et al. Origin of threshold voltage instability in indium-gallium-zinc oxide thin film transistors. Applied Physics Letters. 2008;**93**:123508. DOI: 10.1063/1.2990657

[21] Chen WT et al. Oxygen-dependent instability and annealing/passivation effects in amorphous in–Ga–Zn–O thin-film transistors. IEEE Electron Device Letters. 2011;**32**:1552-1554. DOI: 10.1109/LED.2011.2165694

[22] Li J et al. Effect of reactive sputtered SiOx passivation layer on the stability of InGaZnO thin film transistors. Vacuum. 2012;**86**:1840-1843. DOI: 10.1016/j.vacuum.2012.04.009

[23] Oh S et al. Suppression of light influx into the channel region of photosensitive thin-film transistors. IEEE Transactions on Electron Devices. 2015;**62**:4057-4062. DOI: 10.1109/TED.2015.2492680

[24] Han CW et al. 11.1: Invited paper: 15-inch RGBW panel using two-stacked white OLED and color filter for large-sized display applications. In: SID Symposium Digest of Technical Papers (SID '10). 2010. pp. 136-139

[25] Lee J et al. Deep blue phosphorescent organic light-emitting diodes with very high brightness and efficiency. Nature Materials. 2016;**15**:92-98. DOI: 10.1038/nmat4446

[26] Lee J et al. Hot excited state management for long-lived blue phosphorescent organic light-emitting diodes. Nature Communications. 2017;**8**:15566. DOI: 10.1038/ncomms15566

[27] Choi HS et al. 45.1: Invited paper: Recent progress of white light-emitting diodes for an application to new models of OLED TV. In: SID Symposium Digest of Technical Papers (SID '16). 2016. pp. 605-608

[28] Choi HS et al. Optimization of 2-stack WOLED structure with consideration on color gamut and power consumption. Journal of Display Technology. 2009;**5**(12):546-551. DOI: 10.1109/JDT.2009.2033822

[29] Jung SH et al. A new AMOLED pixel compensating the combination of n-type TFT and normal OLED device. In: SID Symposium Digest of Technical Papers (SID 09). 2009. pp. 1-4

[30] Shin HJ, Takasugi S, Park KM, et al. Technological progress of panel design and compensation methods for large-size UHD OLED TVs. In: SID Symposium Digest of Technical Papers (SID '14). 2014. pp. 720-723

6

Optimizing of Convolutional Neural Network Accelerator

Wenquan Du, Zixin Wang and Dihu Chen

Abstract

In recent years, convolution neural network (CNN) had been widely used in many image-related machine learning algorithms since its high accuracy for image recognition. As CNN involves an enormous number of computations, it is necessary to accelerate the CNN computation by a hardware accelerator, such as FPGA, GPU and ASIC designs. However, CNN accelerator faces a critical problem: the large time and power consumption caused by the data access of off-chip memory. Here, we describe two methods of CNN accelerator to optimize CNN accelerator, reducing data precision and data-reusing, which can improve the performance of accelerator with the limited on-chip buffer. Three influence factors to data-reusing are proposed and analyzed, including loop execution order, reusing strategy and parallelism strategy. Based on the analysis, we enumerate all legal design possibilities and find out the optimal hardware design with low off-chip memory access and low buffer size. In this way, we can improve the performance and reduce the power consumption of accelerator effectively.

Keywords: convolution neural networks, CNN accelerator, memory access

1. Introduction

Convolution neural networks are developing rapidly in recent years. Due to the outstanding performance in image recognition, CNN are used widely in image classification [1–3]. Moreover, since its great success in image recognition, CNN are studied and applied to many other fields of artificial intelligence, such as speech recognition [4, 5], game play [6], etc.

Increasing the depth of CNN by increasing the amount of layers of CNN is a common and effective method to improve the accuracy of image recognition. For instance, in ILSVRC

2012, the champion work, one kind of CNN model namely AlexNet, achieved the top-5 accuracy of 84.7% in the image classification task, and the CNN model has 5 convolutional layers and 3 fully connected layers [2]. The ResNet, which won the first place in ILSVRC 2015 and achieved 96.43% accuracy exceeding human-level accuracy, consists of 152 layers [3]. Although making CNN model deeper can improve the performance, the computing process of CNN involves an enormous number of computation and data. It brings more pressure to the computing hardware. Traditional CPU became a limitation to CNN. Lacking of parallel computing, using CPU for CNN computing result in poor computing performance and high power consumption. It is necessary to find a better hardware to replace CPU for CNN computing. Therefore, more and more hardware are designed and used for CNN computing, such as FPGA designs [7, 8], GPU designs [9], and ASIC designs [10, 11]. These designs aim to accelerate the computing of CNN, improve the computing performance and reduce the energy consumption. Designing and optimizing a specific CNN hardware accelerator became one of popular topics.

In this chapter, we will propose a typical architecture of CNN accelerator and we will introduce two optimize methods of CNN accelerator, consisting of reducing data precision and data-reusing. Generally, reducing data precision will affect the accuracy of CNN model and data-reusing will bring extra on-chip buffer of CNN accelerator. In this paper, we will focus on these two optimization methods and make a deep analysis of the impact on CNN accelerator. We investigate the impact of reducing data precision to accuracy of CNN from several references [12–16], and found that reducing data precision appropriately almost have no impact on the accuracy of CNN model. It inspires us that reducing data precision is a feasible method to optimize CNN accelerator, which will bring many benefits to CNN accelerator, such as reducing the hardware resource, memory footprint and power consumption, etc. In addition, we will analyze three influence factors of data-reusing, including loops execution order, reusing strategy and parallelism strategy. Based on the analysis, we enumerate all legal design possibilities and find out the optimal hardware design with low off-chip memory access and low buffer size. In this way, we can optimize our CNN accelerator with high performance, small memory footprint and low power consumption.

This chapter is organized as follows:

- Review the history and development of convolution neural networks.
- Describe the architecture and the principle of CNN. A real-life CNN, namely VGG-16 will be presented.
- Present a typical architecture of a CNN accelerator implemented on an embedded FPGA platform and introduce the workflow of a CNN accelerator. Present a typical processing element of CNN accelerator.
- Propose and analyze two optimize methods of CNN accelerator including reducing data precision and data-reusing. Based on the analysis, find out the optimal design space for the CNN accelerator by enumerating all legal design possibilities.

2. Background of convolution neural networks

2.1. CNN development overview

Convolutional neural network (CNN) is one type of Deep Neural Networks. The first CNN model namely LeNet-5 is proposed by LeCun in the paper [1] in 1998 and this model is used in handwriting digit recognition. However, due to the enumerated number of computing in training, CNN has been silence for some time. Until 2012, a breakthrough of CNN occurred. A group from University of Toronto used a deep neural network, namely AlexNet, won the first place in image classification in ILSVRC 2012, and its top-5 error rate achieved 15.3%, compared to the second place which achieved 26.2%, and also dropped the error rate by 10% approximately [2]. They improve the algorithm of CNN model in some aspects, such as deepening the model, using ReLu as the activation function, etc. And they train their CNN model with 2 GPU. Their effort resulted a great leap in Deep Neural Network. Since then, CNN developed rapidly. In 2015, the top-5 error rate of ImageNet champion work namely ResNet achieved 96.43% accuracy and exceeded human-level accuracy [3]. In their work, they continued to deepen the CNN model to 152 layers. In the latest ILSVRC 2017, the champion work top-5 error achieved 2.251% [17]. For the rapid development, CNN is making a great impact on many application areas, such as image and video recognition, speech recognition, game play, etc.

2.2. CNN basic

Convolution neural networks (CNNs) have an inference process for recognition and a back propagation process for training. Since training CNN spends a lot of time, many CNN application usually complete training off-line in advanced, and then use the trained CNN in terminals to execute tasks. Therefore, the inference process for recognition in terminals has more pressing demands on speed and power. In this work, we focus on the inference process of CNN and explore how to speed up the inference process with hardware accelerator.

Figure 1 is the simplified structure of a typical CNN. A typical CNN model usually consists of several parts, including convolutional layer (CONV), nonlinear activation function,

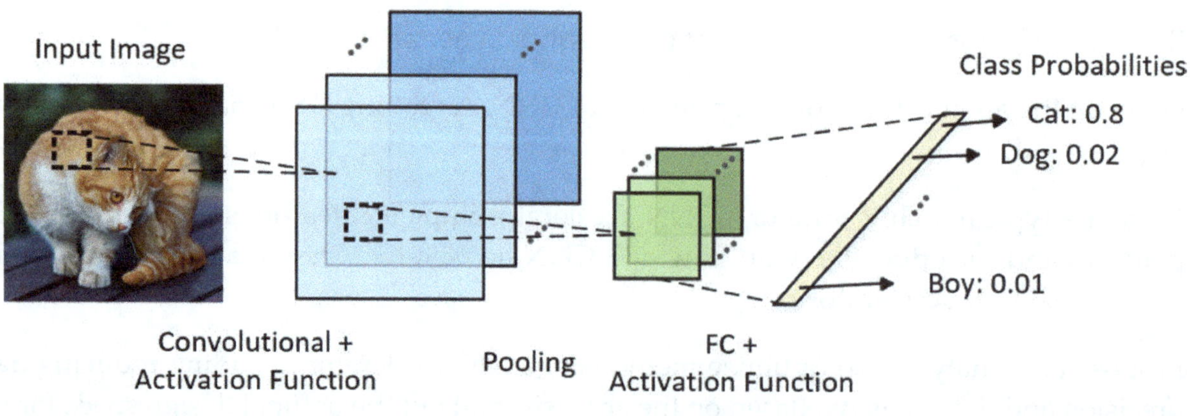

Figure 1. Structure of a typical convolution neural networks.

pooling layer and fully connected layer (FC). Layers are executed layer by layer. CNN reads an input image, and go through a series of CONV layers, nonlinear activation function, pooling layers and generate the output feature maps. These feature maps are turned into a feature vector in the FC layers. Finally, the feature vector is read by a classifier and the input image is classified to the most possible category and the probabilities of each classification are generated.

2.2.1. Convolutional layer

Convolutional layer (CONV) is the most critical layer of CNN. Usually the operations of CONV layers constitute more than 90% of the total CNN operations [18]. For example, in AlexNet, the operations of CONV layers constitute 85–90% of total operations. Therefore, in this work, our discussion and analysis focus on the CONV layers.

CONV layer is like a feature extractor which can extract the features from the input image by the convolutional kernels. When input image read by CONV layer and convolves with weights kernel (generated by training previously), the feature information such as corners, lines are extracted into the output feature maps.

Figure 2 shows a typical CONV layer. The CONV layer takes a set of C feature maps as input and convolves with M sets of N weights kernel to obtain M output feature maps. C is the number of the input channels, M is the number of output channels, H and W are the input feature maps size, R and C are the kernel size, and E and F are the output feature maps size. Usually H is equal to W, R is equal to S and E is equal to F. It shows that the input feature maps convolved by a shifting window with a weight kernel iteratively to get the output feature maps. Actually, the weight kernel shifting with stride S and in most cases the stride S is 1.

Usually the computing process of CONV layer can be expressed with a pseudo code in **Figure 3**. The total computing process of CONV layer can be executed by a loop nest. Every output pixel of output feature maps is the sum of products of Input and Weight. After all loops are executed, all output pixels of output feature maps can be obtained.

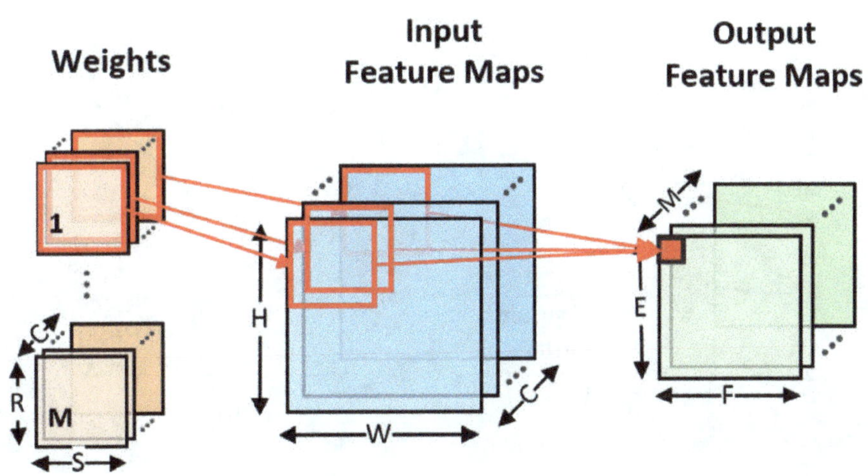

Figure 2. Diagram of convolutional layer.

2.2.2. *Nonlinear activation function*

CNN usually applies nonlinear activation function after each CONV layers or FC layers. The main function of nonlinear activation function is introducing nonlinearity into the CNN. Generally, nonlinear and differentiable are the two conditions that activation functions should to meet. Some conventional nonlinear activation functions in CNN such as sigmoid and tanh are shown in **Figure 4(a)**. However, these activation functions require long training time. In recent years the activation function Rectified Linear Unit (ReLU) [19] becomes more and more popular among CNN models. The expression of ReLU is $f(x) = \max(0, x)$, and it can be shown in **Figure 4(b)**. Compared to the conventional activation functions, the computation process of ReLU is more simple, and it makes training faster. In addition, since a number of output of ReLU are equal to 0, it makes the sparsity of CNN model.

2.2.3. *Pooling layer*

Pooling layer usually attach to the CONV layer. As **Figure 5** shown, usually the type of pooling is maximum or average. The pooling layers read input feature maps, and calculate the maximum or average value of every sub-area of input feature maps so that get low-dimension feature maps. Usually the stride of pooling is equal to the size of pooling.

```
for(x=0;x<F;x++){
 for(y=0;y<E;y++){
  for(k=0;k<C;k++){
    for(i=0;i<R;i++){
      for(j=0;j<S;j++){
        for(m=0;m<M;m++){
            Output[m][x][y] +=
                Input[k][x+i][y+j] ×
                Weight[m][k][i][j]
} } } } } }
```

Figure 3. Pseudo code of the computing process of convolutional layer.

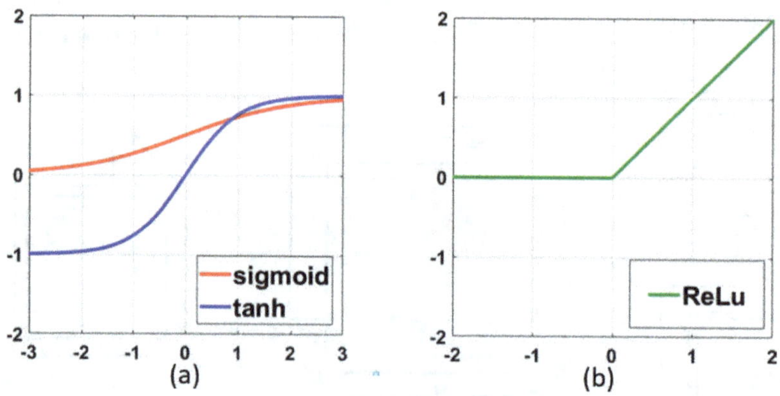

Figure 4. Nonlinear activation function: (a) sigmoid and Tanh; (b) Rectified Linear Unit.

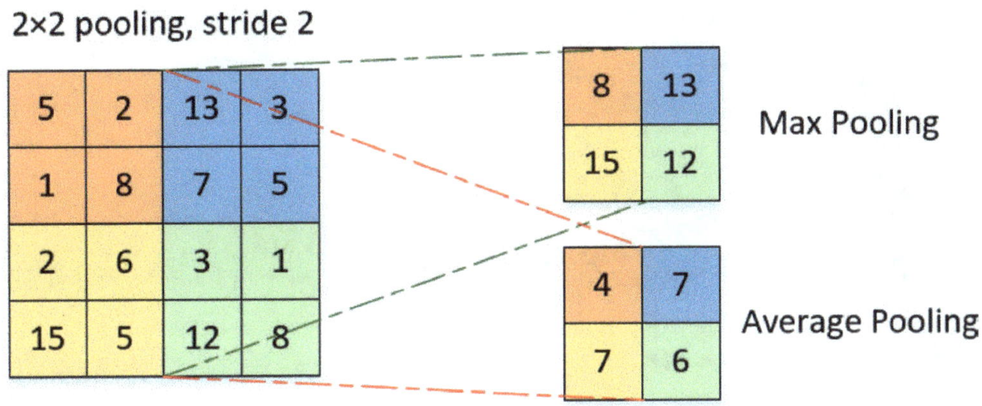

Figure 5. Two type of pooling layer: max pooling and average pooling.

Pooling layer has two major functions. First, pooling can aggregate information from the input feature maps and reduce the size and the dimension of feature maps. It brings a benefit that the size of feature maps and the amount of computation of CNN can be reduced greatly. The second function is that pooling layer can increase the invariance to small shifts.

2.2.4. Fully connected layer

Usually fully connected (FC) Layers are located posterior in the CNN model. The operation of FC layers can be expressed by Eq. (1). Input features (N_i) are fully connected to an output feature vector (N_o) by weights ($N_i \times N_o$). The N_i and N_o are the lengths of the input and output feature vector and the amount of weights is $N_i \times N_o$. $Bias_o$ is the corresponding bias term of the oth output feature vector.

FC layers usually turn the input feature maps into a feature vector. And finally the output feature vector read by a classifier which is a Softmax function and then the probability of each label are generated.

$$Out_o = \sum_{i=1}^{Ni} In_i \times W_{io} + Bias_o \tag{1}$$

2.3. A real-life CNN

Next, we will introduce a real-life CNN namely VGG-16 [20]. In ILSVRC 2014, VGG-16 won the second place in image classification task and its top-5 error achieved 7.4%. **Figure 6** [21] shows the architecture of VGG-16 model. VGG-16 is consisting of 13 CONV layers in 5 groups, 5 max pooling layers, and 3 FC layers. The network receives three 224 × 224 input images and the input images go through a series of layers. Finally, the network generates a output feature vector with the depth of 1000, which represent the likelihoods of 1000 categories. Compared to other CNN models, one of the characteristics of VGG-16 is that all kernel size of each CONV layer is 3 × 3.

Actually, each CONV layer of VGG-16 involves a large amount of operations and data. For a more visualized understanding, we make a detailed statistic for VGG-16. **Figure 7(a)** shows the amount of Input data, Weights and Output data for each CONV layer of VGG-16.

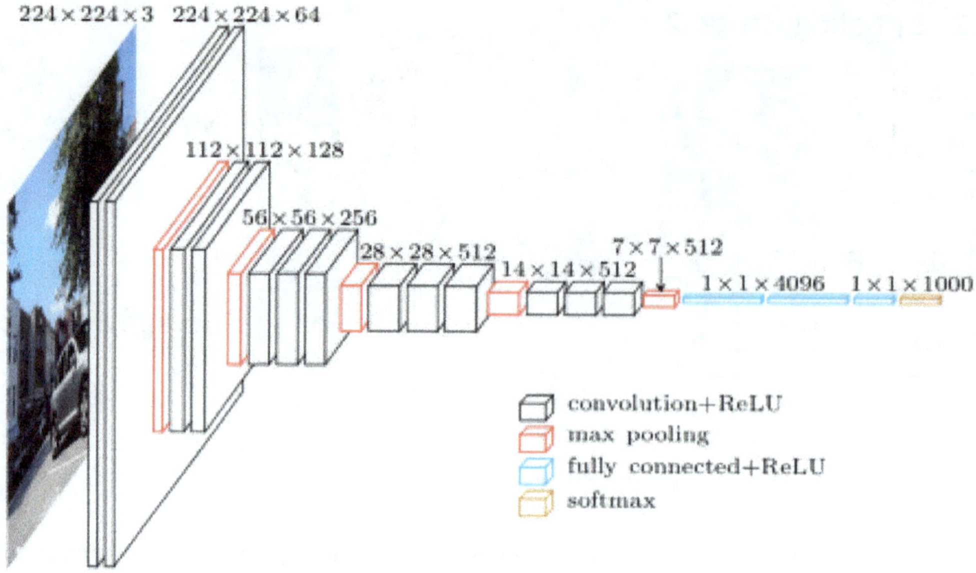

Figure 6. Architecture of VGG-16 model [21].

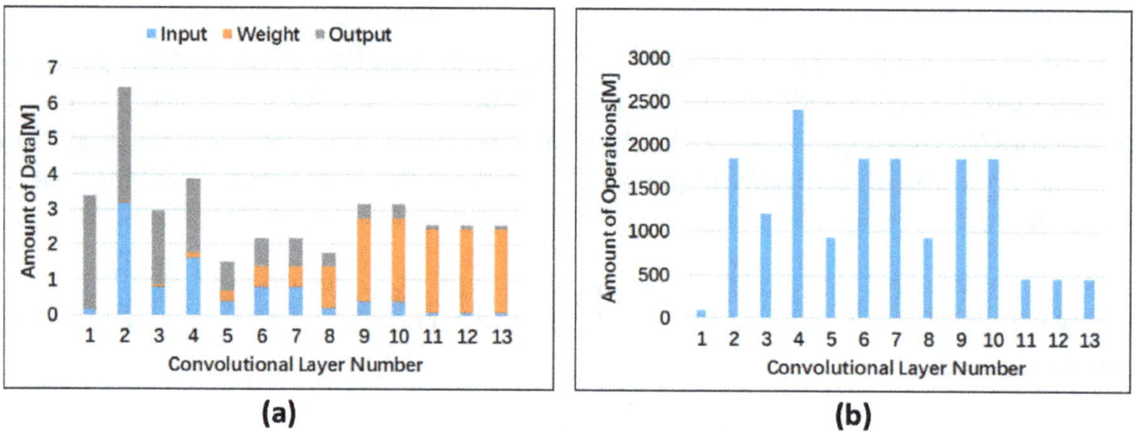

Figure 7. Statistic for each CONV layer of VGG-16: (a) amount of Input, Weight and Output for different CONV layers; (b) amount of operations for different CONV layers.

The 13 bars represent the CONV layer from layer 1 to 13. The height of each bar represents the total amount of data received and generated for each CONV layer, consisting of Input data, Weights and Output data. We can find that in different CONV layers, the amount of Input, Weight or Output vary greatly. For example, the CONV2 requires about 0.035 million weights, and the CONV9 requires 2.25 million weights, which is approximately 64 times to CONV2. **Figure 7(b)** shows the total amount of operation (multiply-add operation) for each CONV layer of VGG-16. The height of each bar represents the total amount of operation for each CONV layer. For example, there are about 87 million times multiply-add operations in the CONV1 and about 1764 million times in the CONV2. We can see that although the operation of CONV is similar, the amount of data involved and operation vary greatly between different CONV layers.

3. CNN accelerator

In this section, we will introduce the CNN accelerator and the challenges CNN accelerator face. As previous section shown, the operations of convolutional layers constitute a large proportion of whole CNN. Therefore, following in our work, the discussion and optimization of CNN accelerator focus on the CONV layers of CNN.

Due to deep convolution neural networks involve an enormous amount of data and operations as shown in Section 2, it is necessary to accelerate the CNN computation by hardware. Actually, many previous works developed the hardware platforms to accelerate the CNN and the implementations obtained good performance and high energy efficiency. Generally CNN can be accelerated by hardware platforms such as FPGA platforms [7, 8], GPU platforms [9], and ASIC platforms [10, 11]. These works have a same feature: they obtained high computational performance for CNN due to the parallelism of the hardware accelerator. In this work, we mainly discuss and analyze the CNN accelerator based on an embedded FPGA platform.

3.1. Overall architecture

Figure 8 is a typical architecture overview of CNN accelerator based on an embedded FPGA platform, and it is a CPU + FPGA architecture. The whole system is consisting of two parts including the Processing System (PS) and the Programmable Logic (PL).

Processing System (PS) mainly consists of CPU and off-chip memory. Due to the enormous amount of input data and weights, it is impossible to store data and weights in on-chip memory. Therefore, usually data and weights are stored in the off-chip memory like DDR3 at the beginning. CPU can configure the control module of accelerator, so that adjust the accelerator to accommodate different scale of CONV layers. In addition, CPU can realize

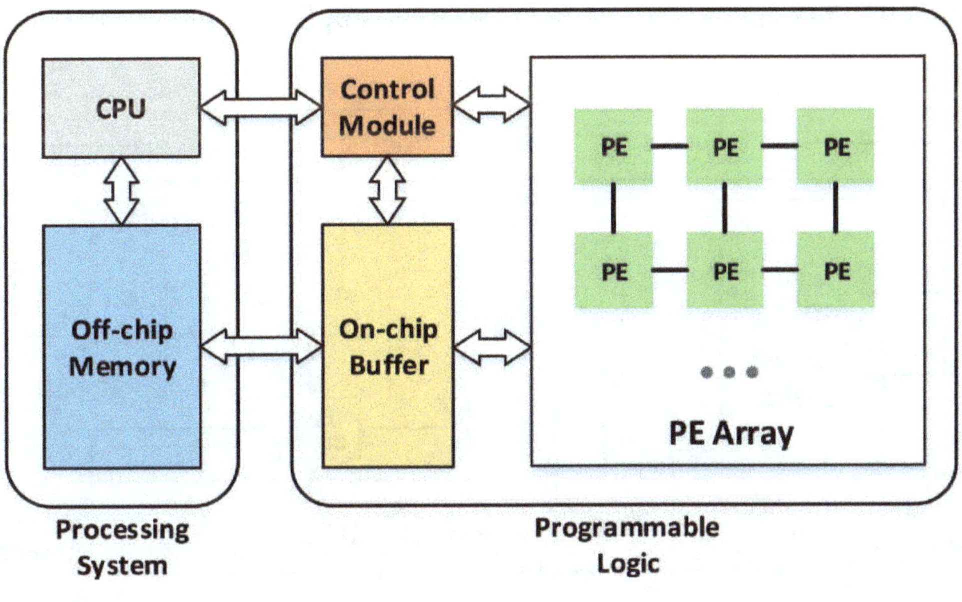

Figure 8. Overview of a typical architecture of CNN accelerator based on an embedded FPGA platform.

some simple operation such as the Softmax function of CNN model. We know that the operation of CONV layers usually constitute more than 90% of the total CNN operations. Accelerating the operations except the CONV layers have little performance improvement. Therefore, we can use CPU to handle the operations except the operation CONV layers such as Softmax function.

Programmable Logic (PL) actually is a FPGA chip and we can program the PL to meet our requirement. PL consists of several parts including processing element (PE) array, control module and on-chip buffer. PE array is consisting of a certain number of PE. PEs are the computation unit for convolution and usually the number of PEs decide the computational performance of CNN accelerator. Data can be interchanged between PEs so that data can be reused without accessing buffer. On-chip buffer is used to cache data and weights for PEs and store the results. Since data and weights of CONV layers are reused repeatedly, buffering and reusing data can reduce the off-chip memory access and it will be introduced in Section 4. Control module receives configuration information from PS, and control the computational process and dataflow of PE array.

The whole working process of CNN accelerator of an implementation can be divided into three steps. First, before system working, all data and weights are stored in the off-chip memory. Next, CPU starts to configure the control module of accelerator and then control module control the on-chip buffer to fetch data from off-chip memory. So that PE array can read data and weights from on-chip buffer and start computation. Finally, PE array finishes all computation and returns the results to off-chip memory so that CPU can read the results.

3.2. PE architecture

Processing element (PE) is the computation unit for CNN accelerator. **Figure 9** shows a typical architecture of PE. It is consisting of multipliers, adder trees, ReLU module, pooling module and registers for Input, Weights and Psum. The multipliers and adder trees are used to complete the convolutional operation. The ReLU module applies a ReLU function to the psum

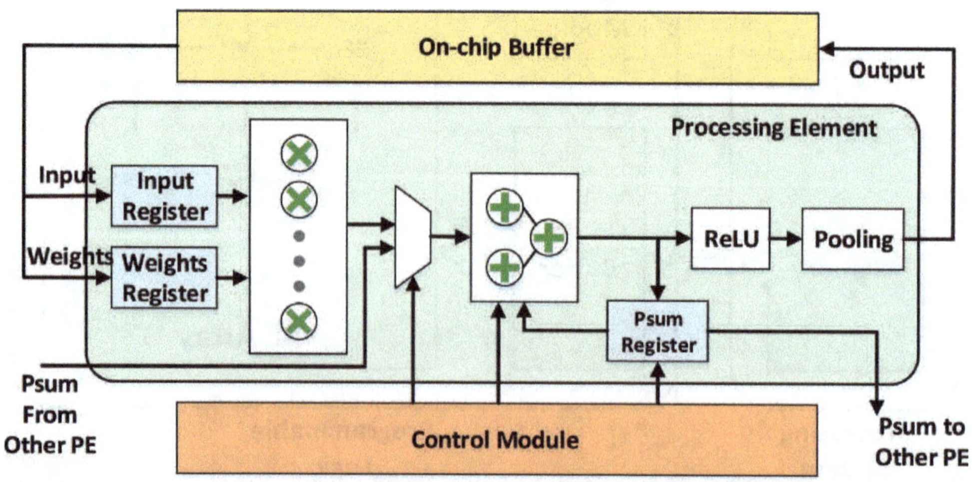

Figure 9. Typical architecture of processing element.

from adder trees. Pooling module realizes the max or average pooling operation and generates the result to on-chip buffer. Register is used to store the Input, Weights and Psum so that PE can reuse the data. PE can interchange data with other PEs. It improves the convenience of data accessing for PE, and it reduces the amount of on-chip buffer access.

Before PE working, control module receives the configuration information and configures the PE. After finishing configuration, PE starts to operation. Input and Weights register read and store data. Multipliers and adder trees read data from register or other PEs, and psum are generated and store in psum register. After finishing convolutional operation, psum are sent to ReLU module and Pooling module and Output are generated and sent to on-chip buffer. PE operation is done.

3.3. Challenges

Although the CONV layer algorithm is simple, due to the enormous amount of data and computation, hardware accelerators face some grave challenges. One of the challenges is the limitation of off-chip memory bandwidth. Generally, CNN accelerator computes with high parallelism by increasing the processing elements (PEs), which can improve the computational performance of accelerator. However, it is accompanied by the pressure of the bandwidth caused by the large amount of data access. Another challenge is that the large amount of off-chip memory access consumes a lot of energy.

Figure 10 is a diagram from reference [22]. It shows the normalized energy cost of each level memory hierarchy relative to the computation of one multiply-accumulate (MAC) operation and the data are extracted from a commercial 65-nm process. We can see that the energy cost of DRAM access is much higher than the energy cost of on-chip buffer access and MAC operation. Therefore, large amount of DRAM access will cause the high power consumption. Besides, large memory footprint caused by enormous amount of data is also an inevitable challenge.

To solve these problems, previous works proposed some optimization methods. One of the optimization methods is reducing data precision. Several researches show that reducing the data precision appropriately of CNN model almost has no impact on image recognition

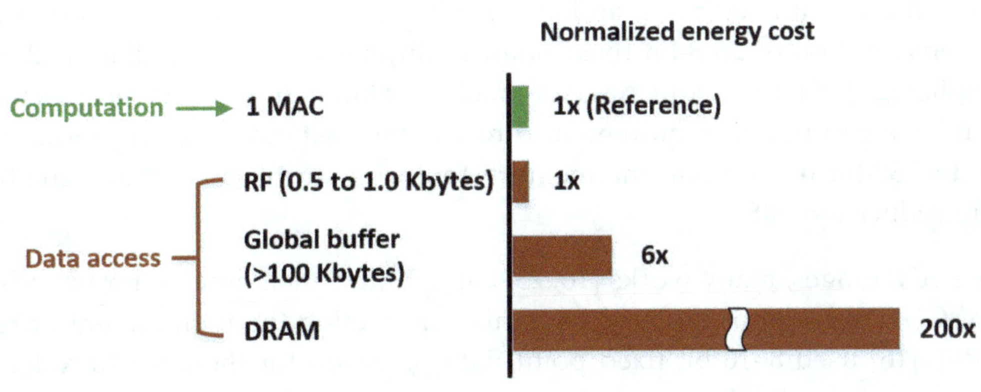

Figure 10. Normalized energy cost of each level memory hierarchy [22].

accuracy. It inspires us that reducing data precision is an efficient and significant method to reduce memory footprint and computational hardware resource. In addition, data-reusing [7, 8, 23] is also an optimization direction for CNN accelerator. Due to most of data in CNN are used repeatedly, data-reusing is also an efficient method to reduce memory access, consequently reducing the memory bandwidth and power consumption. In the next section, we will introduce these two optimization methods in detail.

4. Accelerator optimization

4.1. Data precision

Generally, to guarantee high recognition accuracy, 32-bit floating point data and weights are used to train CNN model. However, such high data precision brings more pressure to hardware because high data precision usually requires more computational resources and larger memory footprint. More and more studies indicated that reducing data precision appropriately almost has no impact on the accuracy of CNN model [12–14]. In Ref. [15], authors used MNIST dataset to test the impact of data precision on accuracy. The test result showed that using 16-bit fixed point data instead of 32-bit floating point in the inference process and using 32-bit floating point in the training process, the accuracy almost had no reduction. If using 16-bit fixed point data in the inference process while using 32-bit fixed point in the training process, the error rate reduced from 99.18 to 99.09%. In the reference [16], authors explored the precision requirements of CNN using AlexNet and VGG models. 8-bit fixed point for the convolution weights are used to test in the process of inference. Compared to full precision weights, the result showed accuracy reduced less than 1%.

These studies show that in many cases there is not necessary for CNN model to use high data precision such as 32-bit floating point. It inspires us reducing data precision is a feasible method to optimize CNN accelerator. Reducing data precision brings a lot of benefits to CNN accelerator. Firstly, for the reason that computation of floating point requires more computational resource than the computation of fixed point, using fixed point data in CNN accelerator will save a large number of computational resource. In addition, using lower precision data will make CNN accelerator more energy-efficiency. For instance, the power consumption of an 8-bit fixed point adder is 30× less than a 32-bit floating point adder and the power consumption of an 8-bit fixed point multiplier is 18.5× less than a 32-bit floating point multiplier [24]. Finally, reducing data precision of data or weights reduces the memory footprint and the bandwidth requirement directly. For instance, if using 16-bit fixed point data instead of 32-bit fixed point, the memory footprint shrink by half and the bandwidth requirement reduce by half.

Due to these advantages, many works [16, 25, 26] optimize their accelerator by reducing data precision of CNN accelerator under the premise of meeting the requirement of recognition accuracy. Ref. [16] used 8–16 bit fixed point data precision for the both AlexNet and VGG models. The accuracy reduction caused by the fixed point operations in the FPGA implementation is less than 2% for top-1 accuracy and less than 1% for top-5 accuracy.

4.2. Data-reusing

Data-reusing is an important optimization method of CNN accelerator to reduce the memory accesses. The main idea of data-reusing is use on-chip buffer to buffer the data which will be used repeatedly. When next time we use these data again, they can be accessed in the on-chip buffer, so that we do not need to access off-chip memory.

Figure 11 shows the computational process of convolution. In this process, a 3 × 3 weight kernel shift and convolve with 9 pixels of input feature map in sequence. Between two times convolving as the red frame and green frame shown in **Figure 11**, the value of weight kernel does not change, and the red area of input feature map will be used repeatedly. So that the weight kernel and the part of input feature map can be reused.

Table 1 shows the amount of data, of total data accesses and the average access times per data of Input, Weights and Output for CONV1 of VGG-16 model. The amount of data represents the amount of actual Input, Weights or Output. The amount of total data access represents the amount of data access of Input, Weights or Output. The average amount of using times per data is calculated by the amount of total data access dividing the amount of data and it represents the using times of each Input, Weights or Output. From **Table 1**, we can find that in the computing process of convolutional layers, although a large amount of Input, Weights and Output are involved, actually these data are always used repeatedly. Some data even can

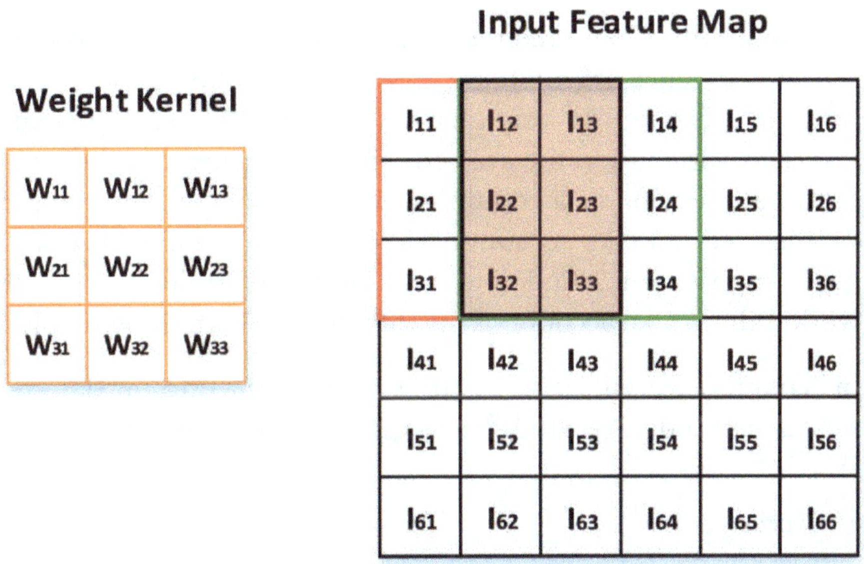

Figure 11. Data-reusing in the process of convolution.

	# of data(K)	# of total data access(K)	Average # of using times per data
Input	150	86,704	576
Weights	1.728	86,704	50,176
Output	3211	86,704	27

Table 1. Statistics of CONV1 of VGG-16 model.

be used over thousands of times. Therefore, data-reusing is an available optimized method to reduce the data accesses of off-chip memory.

Data-reusing greatly reduce off-chip memory accesses, and the benefits are obvious. First, it can relieve the pressure of the off-chip memory bandwidth. Second, it can shorten the operation time of accelerator since the time of accessing off-chip memory is more longer than accessing on-chip buffer. Finally, as **Figure 10** shown, since the energy cost of off-chip memory access is much higher than the energy cost of on-chip buffer access, data-reusing will make accelerator more energy-efficiency.

Although data-reusing is an available optimized method to reduce the data accesses of off-chip memory, it needs extra on-chip memory as a price to buffer data. Usually the benefits and the cost brought by data-reusing vary greatly from different data and CONV layers. From **Table 1** we can easily infer that reusing Weights reduce more data accesses of off-chip memory than reusing Input or Output, for the reason that the average amount of using times per Weights is far larger than the Input and Output. In addition, the size of buffer for reusing Weights in CONV1 is larger than CONV13, for the reason that the amount of Weights in CONV1 is far larger than CONV13 as **Figure 7(a)** shown. From the above discussion, although data-reusing is an available method, due to the different benefits and cost caused by reusing different data and reusing in different CONV layers, data-reusing is a complicated optimization method involving many factors. Therefore, it is necessary to explore the strategy of data-reusing.

Actually, some previous works explored the optimization of memory access by reusing data. Work [7] explored the influence of loop tiling and transformation to data-reusing, which has an impact on computing throughput and memory bandwidth. Work [8] explored the influence of multiple design variables to the accelerator design. These works took account of the influence factors to data-reusing in a certain extent, such as the loop tiling and loop interchange; however, they did not take account of the influence of parallel computation to data-reusing. Work [27] proposed a methodology on how to determine the parallelism strategy. However, this work did not take account of the influence of loop tiling and loop interchange.

Following in this chapter, we will discuss the factors impacting on data-reusing in deep, including loops execution order, reusing strategy and parallelism strategy.

4.2.1. Loop execution order

The computational procedure of convolutional layer can be expressed by a convolution loop nest as **Figure 3**. The similar iteration loop dimensions can be merged to a new loop dimension, and the loops execution order can be transformed, as **Figure 12(a)** and **(b)** shown. **Figure 12(b)** is transformed by **Figure 12(a)**. The loops execution order of two loop nests is different but the result is same.

For the reason that the weights will be accessed repeatedly in the loop ***a***, we can buffer the weights which will be accessed repeatedly in loop ***a*** and reuse them. Assuming reusing weights in loop ***a***, we need corresponding on-chip buffer to store the weights. All weight which will be accessed in loop ***a*** should be buffered. For example, in the loops execution order

```
L1: a   for(x=0;x<F;x++){
          for(y=0;y<E;y++){
L2: b       for(k=0;k<C;k++){
L3: c         for(i=0;i<R;i++){
                for(j=0;j<S;j++){
L4: d             for(m=0;m<M;m++){
                    Output[m][x][y] +=
                      Input[k][x+i][y+j] ×
                      Weight[m][k][i][j]
        } } } } } }
```

(a)

```
L1: b   for(k=0;k<C;k++){
L2: c     for(i=0;i<R;i++){
            for(j=0;j<S;j++){
L3: d         for(m=0;m<M;m++){
L4: a           for(x=0;x<F;x++){
                  for(y=0;y<E;y++){
                    Output[m][x][y] +=
                      Input[k][x+i][y+j] ×
                      Weight[m][k][i][j]
        } } } } } }
```

(b)

Figure 12. Two different loops execution order: (a) order a; (b) order b.

Figure 13. Reusing weight in two situations with different loops execution order: (a) situation a; (b) situation b.

of **Figure 12(a)**, when executing loop ***a***, the weights in the inner loops such as loop ***b***, ***c*** and ***d*** will be accessed. Thus, if we want to reuse the weights in loop ***a***, we need to buffer the weights which accessed not only in loop ***a***, but also in inner loop ***b***, ***c*** and ***d***. Therefore, the on-chip buffer size depends on the amount of the weights which will be accessed when executing loop ***a***. And the amount of the weights accessed depends on the relative position of loop ***a*** in the whole loops nest.

Compare the loop execution order of two cases of **Figure 12(a)** and **(b)**. In **Figure 12(a)**, if we want to reuse the weight which is represented by the red pixel in **Figure 13(a)** in loop ***a***, between two times for accessing the red pixel weight, the other weights as the blue frame shown in **Figure 13(a)** will be accessed. Thus, we need to buffer these weights which are shown as the blue frames in **Figure 13(a)** and the amount of these weight is $C \times R \times S \times M$. Assuming the bit width of weights is 2 bytes, the buffer size for reusing weights is $2 \times C \times R \times S \times M$ bytes.

In **Figure 12(b)**, due to the loop ***a*** is the innermost loop of the loop nest, if we want to reuse the weight which is shown as the red pixel in **Figure 13(b)**, between two times for accessing the red pixel weight, there is no other weights need to access. So that we just need to buffer the weight which is shown as the blue frame in **Figure 13(b)** and the amount of weight is only 1. Thus the buffer size for reusing weights is 2 bytes. By comparing these two cases, we can find that the loops execution order has a critical influence on the on-chip buffer size for reusing data.

Similarly, the Input and Output can be reused either and the size of the buffer for reusing Input and Output can be calculated like above. When consider the influence on the buffer size for reusing data, Input, Weight and Output all should be taken into account, and find out the optimal loops execution order.

4.2.2. Reusing strategy

In the above, we introduce the influence of the loop execution order on the buffer size. Actually reuse strategy is also an important influence factor to the data reuse. Reuse strategy is a concrete scheme indicating which data will be reused in which loop dimension. For instance, in **Figure 12(a)**, we can reuse or not reuse the Weight in loop ***a***. Similarly, we can choose in loop ***a*** or loop ***d*** that we reuse Input. Thus, there are a lot of reusing strategies combination we can choose. Given a loops execution order, different reuse strategy will make buffer size and memory access different. When we consider the data-reusing, reusing strategy should be taken into account and find out the corresponding optimal reusing strategy.

4.2.3. Parallelism strategy

Parallel computing is an effective method to improve computational performance of CNN model. Actually parallelism is also a method reusing data indirectly. In some cases, parallel computing involves the same data, thus when parallel computing we just need to access the data once, compared to serial computing, which we need to access data repeatedly. However, parallelism also brings some problems such as higher hardware overhead, higher bandwidth requirement, extra buffer size, etc. Generally, a legal parallelism strategy of CONV layer for **Figure 3** must meet the following constraints:

$$\begin{cases} 0 < P_a \times P_b \times P_c \times P_d \leq (\#\text{ of PEs}) \\ 0 < P_a < E \times F \\ 0 < P_b < C \\ 0 < P_c < R \times S \\ 0 < P_d < M \end{cases} \tag{2}$$

In Eq. (2) P_x represent the parallelism degree in loop dimension $\boldsymbol{x}$. Eq. (2) indicates the number of PEs is the limitation of total parallelism. And the parallelism degree in each loop dimension cannot exceed the loop iterations.

Under the constraints as Eq. (2), there are many combinations of the parallelism strategy. For the reason that the computational performance of CNN accelerator is generally proportional to the total parallelism degree, if the total parallelism degree is similar, the computational performances of taking different parallelism strategies are similar. However, the memory access of taking different parallelism strategies differs greatly. Assuming two cases, the first case is that take a parallel computing in the loop dimension ***b*** and the parallelism degree is Pm. The diagram of the computing process is shown as **Figure 14(a)**. The red frame represents the data involved in a parallel computing. The computing diagram shows that when parallel computing in the loop dimension ***b***, there are Pm Input and Pm Weight, totally 2Pm data are involved in parallel computing per time. Another case is that take a parallel computing in the loop dimension ***d*** and the parallelism degree is also Pm, as **Figure 14(b)**

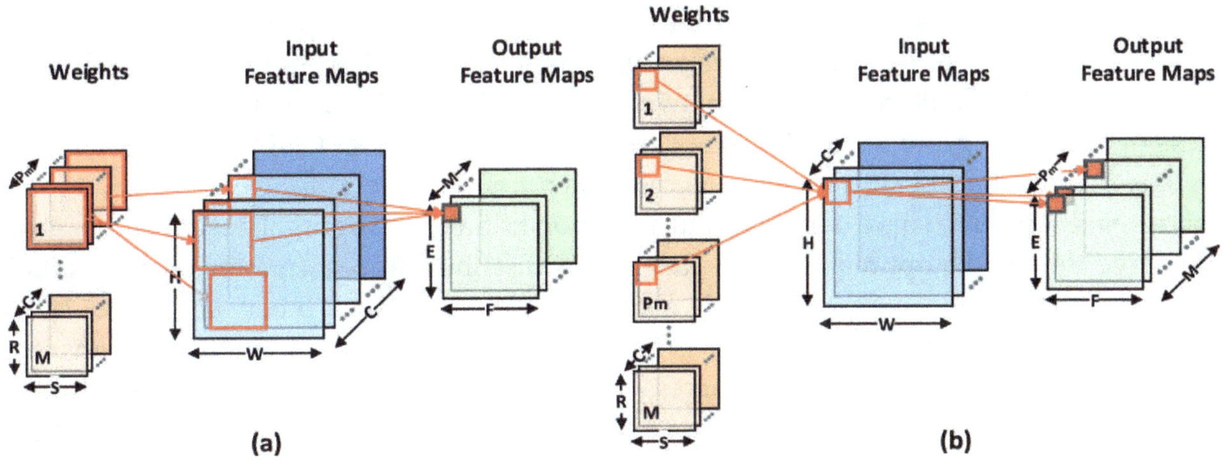

Figure 14. Computing process of two parallelism strategies: (a) parallelism in loop dimension ***b***; (b) parallelism in loop dimension ***d***.

shown. Similarly, we can easily get that there are 1 Input and Pm Weight, totally Pm + 1 data are involved in parallel computing per time. Obviously, the amount of data accessed by the second parallelism strategy is less approximately by half than the first parallel strategy. By comparison between these two cases, we can find that different parallelism strategies have different influence on the amount of data accessed. Therefore, when we design the CNN accelerator and take the parallel computing for the accelerator, a good parallelism strategy can reduce the amount of data accessed, and reduce the amount of memory accesses accordingly.

Usually parallel computing will bring extra on-chip buffer. When calculating the on-chip buffer size, we can regard parallel computing as the combination of a number of (it depends on the parallelism degree) independent computing. In that circumstance, the total buffer size can be calculated by the buffer size of each independent computing. Thus we can calculate the buffer size of each independent computing and get the total buffer size for parallel computing.

4.3. Design space evaluation

In the above, we introduce the influence of data precision and data-reusing to CNN accelerator. In this section we combine all the design factors of accelerator and make a comprehensive analysis. Here we use the buffer size and off-chip memory access as the evaluation parameters of CNN accelerator. Enumerate all design possibilities which are the combinations of different loops execution order, reusing strategy and parallelism strategy. And calculate the buffer size and the off-chip memory access of all design possibilities with the analysis in the previous section and generate a series of groups consist of design space, buffer size and off-chip memory access, and depict all these groups on a graph.

Based on the CONV1 of the VGG-16 model and using 16-bit fixed point data precision, under the constraint that the number of PEs less than 500 which means the total parallelism degree less than 500, we enumerate all design possibilities and obtain a series groups of all design space. To simplified analysis, we remove the overlapping and obviously poor design groups and remain a series of the optimal design groups. **Figure 15** depicts these optimal design groups and it shows the relationship between the on-chip buffer size and off-chip memory access of all design groups. The "x" axis denotes the buffer size of CNN accelerator and the "y" axis denotes the off-chip memory access. Each point in the graph represents a kind of design space. The blue curve depicts the distribution trend of the optimal design space. As the blue curve shown, the optimal design space points distribute approximately on a hyperbola. It indicates that when we design a CNN accelerator, the relationship between buffer size of accelerator and off-chip memory access is an inverse relationship as a whole. It means if we want to reduce the off-chip memory access, corresponding on-chip buffer are required for reuse data. Reduce the on-chip buffer size, usually accompanied by the increment of off-chip memory access.

Based on **Figure 15**, we can choose the optimal design space according to the requirement of hardware. For example, if our FPGA platform only provides 1 KB on-chip buffer, we can choose the design space of point A in **Figure 15** to design our accelerator. If we want to reduce the memory bandwidth and the power consumption of accelerator and we do not concern about the buffer size, we can choose the design space of point C. Moreover, if we want to make a balance between on-chip buffer size and off-chip memory access, we can choose the design space of point B. In a word, we can obtain the appropriate and optimal design space of CNN accelerator referring to **Figure 15**.

4.4. Power consumption evaluation

In the previous section, we mentioned that the power consumption between different memory accesses differs greatly. According to **Figure 10**, under the commercial 65-nm process, the power consumption of DRAM access is 33 times more than the global buffer access. Therefore,

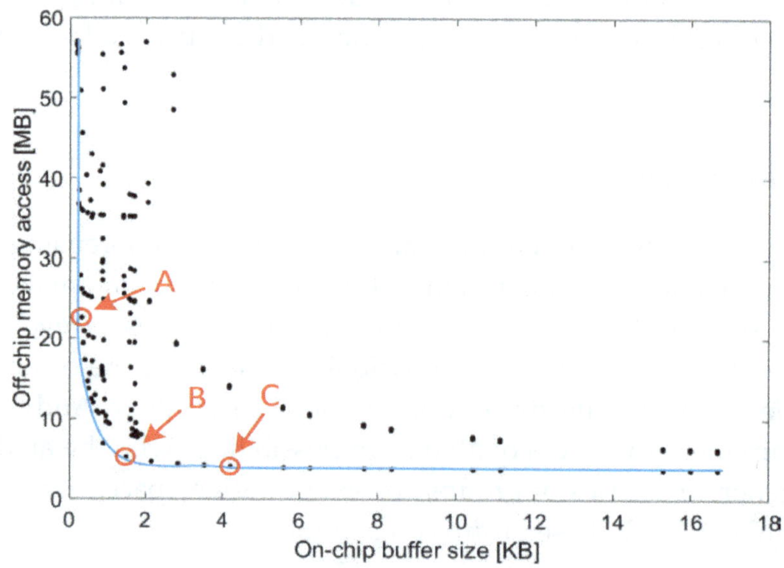

Figure 15. The relationship between the on-chip buffer size and off-chip memory access of all design groups for CONV1 of VGG-16.

we can consider that the power consumption of DRAM access is the main part of power consumption of memory access. Storing and reusing data with on-chip buffer can reduce the DRAM access, so that it can reduce the power consumption of the computing process of CNN accelerator.

As **Figure 15** shown, the off-chip memory access of point B is 4.5 times more than point A, while the on-chip buffer size of point B is only 2 KB approximately more the point A. It means we can increase a small on-chip buffer (2 KB) to obtain the 4.5 times reduction of the power consumption of memory access of CONV1.

Although it is theoretical estimate of power consumption, it shows that we can make a balance between on-chip memory size and power consumption. Our design space evaluation method can help us to choose the optimal design space of CNN accelerator which can reduce the power consumption by increasing small on-chip buffer size. Facing the tendency that miniaturization and low power consumption of IOT, our evaluation method is an effective design strategy and match the concept of green electronics.

5. Summary

Convolution neural network (CNN) has been developing rapidly and used widely for many computer science domains in the past few years, such as image recognition, speech recognition, game play, etc. In the image recognition filed, the recognition accuracy of ResNet exceeded human accuracy in 2015. The outstanding performance makes CNN more and more popular in the artificial intelligence applications. Many researches increase the depth of CNN model to improve the accuracy; in the meanwhile, it brings great pressure to the hardware. Therefore, many specific CNN accelerators are designed and used for CNN computing, including FPGA designs, GPU designs, and ASIC designs, which aim to improve the computing performance and reduce the energy consumption of CNN computing.

In this chapter, we reviewed the history of CNN and introduce the basic and principle of CNN. Following we presented a real-life CNN model, namely VGG-16. We illustrated several CNN accelerators and then we introduced and analyzed two optimization methods of CNN accelerators, including reducing data precision and data-reusing. Based on the analysis, we enumerated all legal design possibilities of CNN accelerator and the optimal design space for CNN accelerator can be obtained. By depicting the design space groups on a graph, we obtained the appropriate design space of CNN accelerator according to our design requirements.

Acknowledgements

This work was supported by the Project Science and Technology of Guangdong Province of China (2015B090912001, 2016B010123005, and 2017B090909005).

Author details

Wenquan Du, Zixin Wang and Dihu Chen*

*Address all correspondence to: stscdh@mail.sysu.edu.cn

Sun Yat-Sen University, Guangzhou, China

References

[1] Lécun Y, Bottou L, Bengio Y, Haffner P. Gradient-based learning applied to document recognition. Proceedings of the IEEE. 1998;**86**(11):2278-2324

[2] Krizhevsky A, Sutskever I, Hinton GE, Editors. ImageNet classification with deep convolutional neural networks. International Conference on Neural Information Processing Systems; 2012

[3] He K, Zhang X, Ren S, Sun J, editors. Deep Residual Learning for Image Recognition. Computer Vision and Pattern Recognition; 2016

[4] Sainath TN, Mohamed AR, Kingsbury B, Ramabhadran B, editors. Deep convolutional neural networks for LVCSR. In: IEEE International Conference on Acoustics, Speech and Signal Processing; 2013

[5] Hinton G, Deng L, Yu D, Dahl GE, Mohamed AR, Jaitly N, et al. Deep neural networks for acoustic modeling in speech recognition: The shared views of four research groups. IEEE Signal Processing Magazine. 2012;**29**(6):82-97

[6] Silver D, Huang A, Maddison CJ, Guez A, Sifre L, Driessche GVD, et al. Mastering the game of go with deep neural networks and tree search. Nature. 2016;**529**(7587):484

[7] Zhang C, Li P, Sun G, Guan Y, Xiao B, Cong J, editors. Optimizing FPGA-based accelerator design for deep convolutional neural networks. In: Acm/Sigda International Symposium on Field-Programmable Gate Arrays; 2015

[8] Ma Y, Cao Y, Vrudhula S, Seo JS, editors. Optimizing Loop Operation and dataflow in FPGA acceleration of deep convolutional neural networks. In: Acm/Sigda International Symposium on Field-Programmable Gate Arrays; 2017

[9] Chetlur S, Woolley C, Vandermersch P, Cohen J, Tran J, Catanzaro B, et al. cuDNN: Efficient primitives for deep learning. Computer Science. 2014;arXiv:1410.0759

[10] Chen YH, Krishna T, Emer JS, Sze V. Eyeriss: An energy-efficient reconfigurable accelerator for deep convolutional neural networks. IEEE Journal of Solid-State Circuits. 2016;**PP**(99):1-12

[11] Luo T, Liu S, Li L, Wang Y, Zhang S, Chen T, et al. DaDianNao: A neural network supercomputer. IEEE Transactions on Computers. 2017;**66**(1):73-88

[12] Holi JL, Hwang JN. Finite precision error analysis of neural network hardware implementations. Computers IEEE Transactions on. 1993;**42**(3):281-290

[13] Draghici S. On the capabilities of neural networks using limited precision weights. Neural Networks. 2002;**15**(3):395

[14] Chen T, Du Z, Sun N, Wang J, Wu C, Chen Y, et al. DianNao: A small-footprint high-throughput accelerator for ubiquitous machine-learning. Acm Sigplan Notices. 2014; **49**(4):269-284

[15] Luo T, Luo T, Liu S, He L, He L, Wang J, et al., editors. DaDianNao: A machine-learning supercomputer. In: Ieee/Acm International Symposium on Microarchitecture. 2014

[16] Suda N, Chandra V, Dasika G, Mohanty A, Ma Y, Vrudhula S, et al. Throughput-Optimized OpenCL-based FPGA Accelerator for Large-Scale Convolutional Neural Networks. In: Proceedings of the 2016 ACM/SIGDA International Symposium on Field-Programmable Gate Arrays. Monterey, California, USA. 21 – 23 February, 2016. pp. 16-25

[17] Hu J, Shen L, Sun G. Squeeze-and-excitation networks. Computer Science. 2017;arXiv: 1709.01507

[18] Desoli G, Chawla N, Boesch T, Singh SP, Guidetti E, Ambroggi FD, et al., editors. 14.1 A 2.9TOPS/W deep convolutional neural network SoC in FD-SOI 28nm for intelligent embedded systems. In: Solid-State Circuits Conference; 2017

[19] Nair V, Hinton GE, editors. Rectified linear units improve restricted boltzmann machines. In: International Conference on International Conference on Machine Learning; 2010

[20] Simonyan K, Zisserman A. Very deep convolutional networks for large-scale image recognition. Computer Science. 2014;arXiv:1409.1556

[21] Deep CNN and Weak Supervision Learning for Visual Recognition [Internet]. 2016. Available from: https://blog.heuritech.com/2016/02/29/a-brief-report-of-the-heuritech-deep-learning-meetup-5/

[22] Chen YH, Emer J, Sze V. Using dataflow to optimize energy efficiency of deep neural network accelerators. IEEE Micro. 2017;**37**(3):12-21

[23] Alwani M, Chen H, Ferdman M, Milder P, editors. Fused-layer CNN accelerators. In: Ieee/acm International Symposium on Microarchitecture; 2016

[24] Sze V, Chen YH, Yang TJ, Emer JS. Efficient processing of deep neural networks: A tutorial and survey. Proceedings of the IEEE. 2017;**105**(12):2295-2329

[25] Gupta S, Agrawal A, Gopalakrishnan K, Narayanan P. Deep learning with limited numerical precision. Computer Science. 2015;arXiv:1502.02551

[26] Qiu J, Wang J, Yao S, Guo K, Li B, Zhou E, et al., editors. Going deeper with embedded FPGA platform for convolutional neural network. In: Acm/Sigda International Symposium on Field-Programmable Gate Arrays; 2016

[27] Li H, Fan X, Jiao L, Cao W, Zhou X, Wang L, editors. A high performance FPGA-based accelerator for large-scale convolutional neural networks. In: International Conference on Field Programmable Logic and Applications; 2016

7

Integrated p-NOI Structures on Nanoporous Material Designed for Biodetection

Cristian Ravariu, Elena Manea, Alina Popescu and
Cătălin Pârvulescu

Abstract

Pesticides are utilized to protect the crops, destroying or controlling any pest. Unfortunately, pesticides pollute the entire environment: plants, organisms, soil, and water. This chapter describes a paraoxon pesticide biosensor that includes nanostructures and porous materials integrated on silicon (Si), as convergent objectives of the green microelectronics strategy. The transducer element is in an interdigitated capacitive electrode that recently highlighted a special nanostructure—the planar nothing on insulator (p-NOI)—included in the capacitive detection system. The biodetection is based on the hydrolysis of the acetylcholinesterase (AChE) enzyme as biosensor receptor. So, the final application is an enzymatic biosensor that utilizes the nanoporous Si layer for the enzyme adsorption, with p-NOI capacitive transducer, for the environmental monitoring.

Keywords: biomaterials, electronics, nanotechnologies, biosensor tool, green environment

1. Introduction

Green industries imply few research directions: (i) an industry based on recycling the unwanted resources that take care from the beginning of the company construction to conceive a sustainable fabrication process flow; (ii) a traditional industry that becomes sensitive to environment protection, minimizing the waste and pollution; or (iii) a top industry that provides useful tools for the environmental conservation and monitoring.

The nano-biotechnology-based industry which falls into the last category can produce pesticide biosensors, wastewater detectors [1], micro-nano-filters for air-water-soil cleaning [2], and pathogen biosensors using nanomaterials such as metal nanoparticles, quantum dots, magnetic nanoparticles, carbon nanotubes, and graphene due to their surface properties, excellent electron transfer, and a large ratio of surface area to volume, making them particularly attractive for use in labels or transducing platforms for optical or electrochemical sensors and biosensors.

The examples of such biosensors are the organophosphorus pesticides using liposome-based nano-biosensors [3]. Gold nanoparticles for pesticide detection using cyclic voltammetry [4], organophosphorous pesticide (OP) biosensor based on quenching of the fluorescence from CdTe QDs [5]. Acetylcholinesterase action is monitored using a localized surface plasmon resonance (LSPR) fiber optic biosensor [6]. AuNP-AChE conjugates for paraoxon electrochemical biosensor [7]. AuNP-AChE onto chemically reduced graphene nanosheets (cr-Gs) [8], graphene oxide/Nafion (RGON) nanohybrids electrochemical biosensor platform to detect organophosphorus hydrolase as an enzyme for the hydrolysis of Ops [9], pathogen detection in soils using nanobiosensors [10, 11].

Unfortunately, the pesticides used in this field not only spoil the soil but also infest the entire food chain. Less toxic new generations of pesticides may reduce the risks transmitted to people and environment, especially by water contamination. The pesticides reduce the nitrogen fixation in plants, consequently decrease the biodiversity, destroy habitats, and threaten jeopardized species [12]. Integrated biosensors usually contain onto the same chip of the semiconductor solid-state support, the transducer as an electronic device, and the biological detector as an enzyme [13] or an antibody [14].

This chapter describes a pesticide biosensors fabricated using nanoporous Si materials to entrap the receptor element, along with the transducer element consisting of an interdigitated capacitive electrodes to detect pesticides, like paraoxon. The novel detection scheme is using interdigitated capacitive electrodes which highlighted a special nanostructure called as the planar nothing on insulator (p-NOI) [15, 16]. The biodetection is based on the hydrolysis under the acetylcholinesterase (AChE) enzyme action, as biosensor-specific receptor [17]. The final product is an integrated biosensor that is constructed by microtechnological processes aided by biotechnological enzyme processing steps, having a nanoporous Si layer coupled to a p-NOI capacitive transducer, which is sensitive to the pesticide concentration.

The p-NOI structure that is integrated inside the biosensor transducer has another facet in this work: the first p-NOI structure still exists between top metal on insulator placed on silicon, and the second p-NOI structure is present between two adjacent lateral metal fingers. The first one must accomplish an isolation through the bottom nanoporous material. The second one has the distance between fingers high enough versus a nanometric p-NOI that allows a tunnel current flow [15]. Hence, the tunneling conduction is missing in this case. But the liquid droplet that connects two adjacent fingers by an ionic conductor offers a novel conduction route.

2. The work principle and simulation results for a p-NOI structure

Recently, the nothing on insulator (NOI) device, as the succession n-Si/Vacuum nanocavity/n-Si (nVn) on insulator, was proposed [18] and timely updated [19]. The horizontal variant implementation for the NOI transistor is unknown at the actual technology level, etching a straight cavity in Si from 10 to 20 nm depth of only 2 to 3 nm width, without pipes which seem to be impossible [20].

If the nVn succession is used as the device body, we speak about a NOI (nothing on insulator) transistor [20]. If oxide (O) is used instead of the vacuum (V) cavity and, additionally, a metal is used instead of one semiconductor zone, we speak about a mOn succession as metal/oxide/n-Si. This structure still conserves the NOI work principle [21]. Also, if oxide is replaced by any insulator (I) placed between two metals placed on a Si wafer surface, we speak about MIM structure [22]. The mOn and MIM structures use the same thin insulator tunneling principle but benefit on materials placed on the front plan of the Si wafer. Both of them are associated with the planar variant of a NOI device, simply noted by p-NOI device [15].

Therefore, a vertical implementation of the p-NOI variant is more suitable for the integration of the biosensor transducer. The insulator can be oxide or sandwich of insulators of 10 nm up to 50 nm thickness to prevent the substrate tunneling [23]. The oxide is grown by the Si planar technology. Therefore, the presented p-NOI structure is a vertical simplified NOI variant, with the advantage to be inherent integrated on the Si wafer during the biosensor metallic electrode configuration on insulator. On the other hand, the Fowler-Nordheim tunneling through the bottom insulator is poor. Hence, more than 50 nm oxide thickness ensures an excellent dielectric insulation that is suitable for the biosensor transducer purposes. The explanation comes from two Fowler-Nordheim tunneling ways in this transducer: (i) the useful one that acts the p-NOI device at the surface of the device and (ii) the parasitic tunneling toward substrate that must be avoided. The transducer successfully interacts with bio-liquid on the top of the wafer, generating the capacitance variation, while efficiently prevents the leakage current toward substrate, for thick-enough oxide layer. However, a principle that must be checked is to simulate an exponential I–V dependence for a vertical p-NOI, to put in agreement the Fowler-Nordheim tunneling principle with the p-NOI conduction mechanisms [24].

Figure 1 presents the proposed vertical p-NOI structure with substrate as back-gate. The usual anodes play the gate role, and the cathodes are the source or drain. Therefore, the notations are kept as in a transistor configuration. Essentially, in **Figure 1** there are three vertical

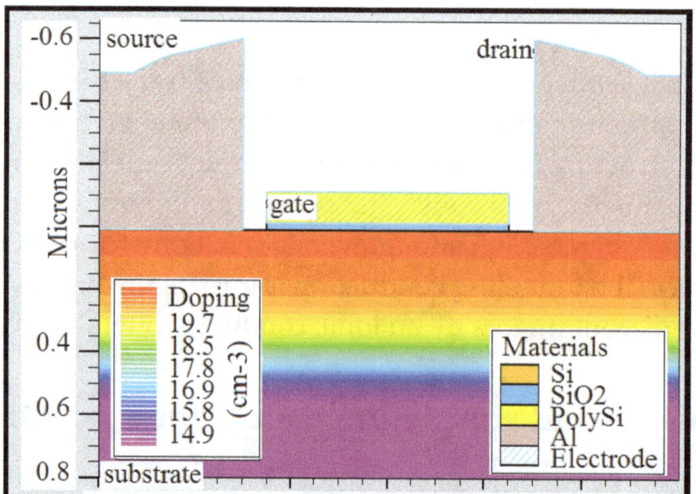

Figure 1. The basic p-NOI device in the planar configuration.

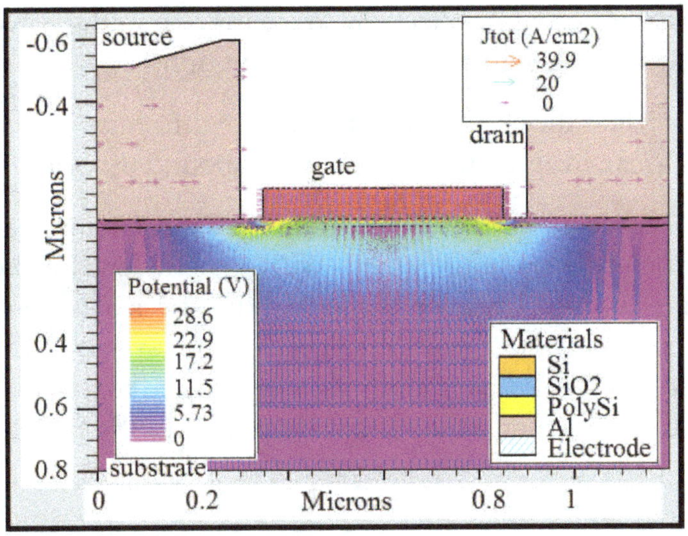

Figure 2. The p-NOI device biased at +29 V.

metal-oxide-semiconductor-metal structures as simultaneous three vertical p-NOI structures, similar to the adjacent fingers included in the next studied biosensor.

In **Figure 1** the p-NOI structure size and doping concentration are presented. On the polysilicon terminal, a voltage of 29 V is applied, and the other metallic contacts are grounded. **Figure 2** shows the potential distribution in the central p-NOI device and the current vectors through the structure, after ATLAS Running. The maximum current density is 39.9A/cm^2. In **Figure 3a** and **3b**, the current-voltage characteristics, I_G-V_G-type, are shown through the p-NOI structure when the gate voltage has increased from 0 to 29 V. It is demonstrated in this case that a tunnel current arises, after the exponential shape of the curves at linear and logarithmic scale, being a good start-up result.

For a p-NOI structure with adjacent metals, the applied voltage on the polysilicon electrode can be more favorable when the semiconductor is superficially doped with 10^{20} cm^{-3}.

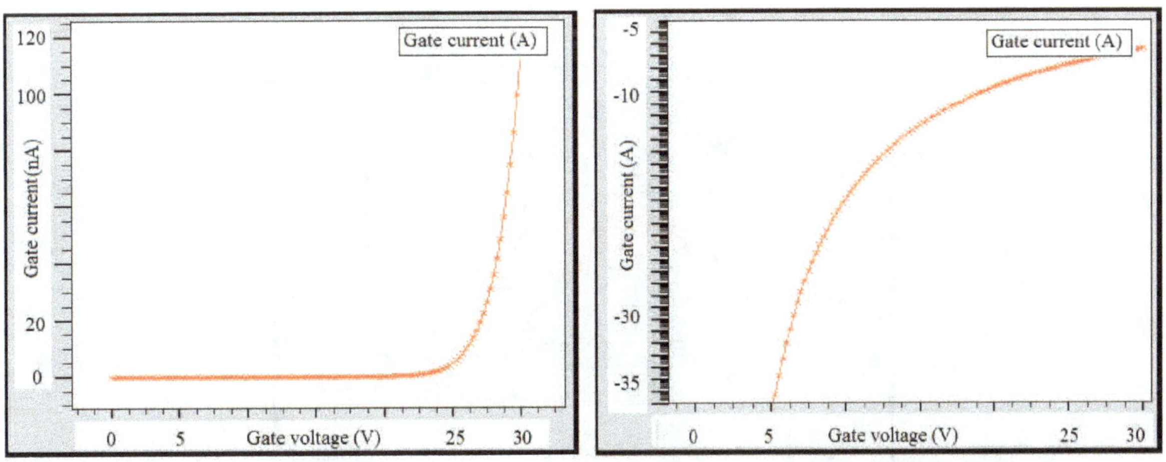

(a) (b)

Figure 3. The simulated characteristics of the vertical p-NOI at scale: (a) linear and (b) logarithmic.

Figure 4 shows the distribution of the electron concentration in the p-NOI structure biased at 29 V. There does not seem to be significant carrier depletion.

It can be seen in **Figure 5** that the electric field inside the ultrathin oxide has an increased value up to approximative 8 V/cm, in agreement to the Si-SiO_2 boundary conditions with $\varepsilon_{Si}/\varepsilon_{oxide} \sim 2.7$.

The explanation is still associated to a strong tunneling for the main electrode and weak tunneling for the adjacent electrode, in agreement with the Fowler-Nordheim tunneling theory applied for the NOI device [24]. However, the higher distances between two adjacent metallic fingers inside the next pesticide biosensors foster rather the capacitive effect of p-NOI than the conductive effect. Therefore a capacitive analysis is performed for the extreme case of an ultrathin oxide thickness of 5 nm, when a AC signal is applied to the p-NOI structure (**Figure 6**).

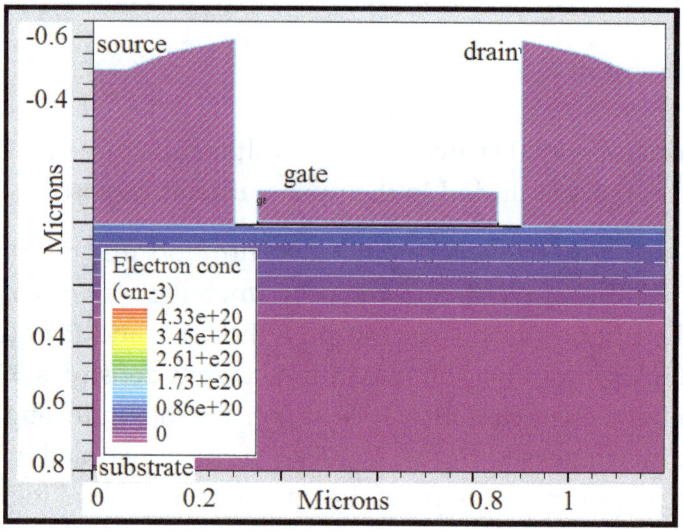

Figure 4. Distribution of the electron concentration in the p-NOI structure at 29 V.

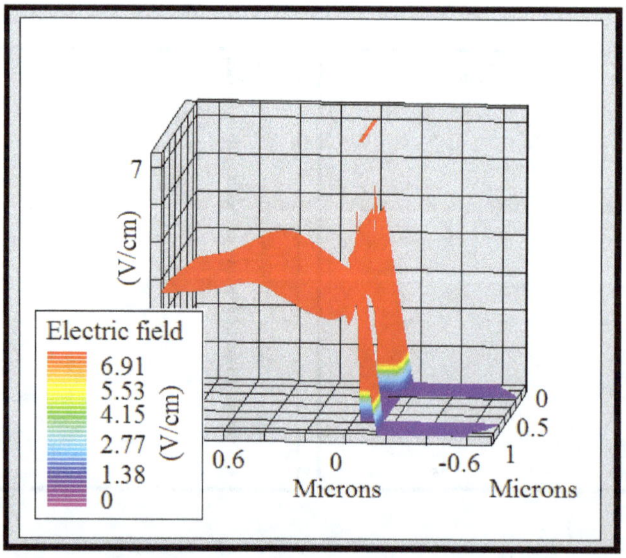

Figure 5. The electric field distribution in the p-NOI structure at 29 V.

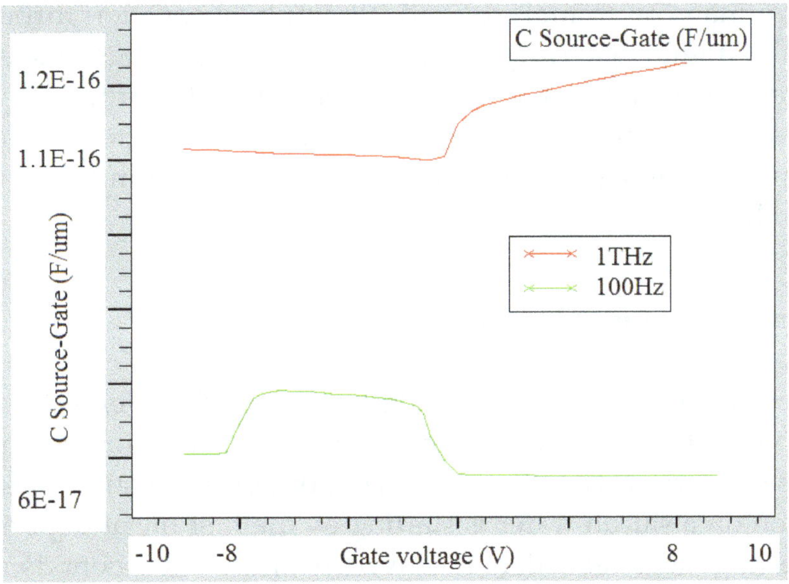

Figure 6. The CV analysis for a AC sweep for V_G between −10 V and +10 V for two adjacent metal fingers like source and gate of a p-NOI structure with 5 nm film thickness and similar size as in **Figure 1**, for low frequency (100 Hz) and high frequency (1 THz).

Obviously, the capacitive range is in agreement with the sizes of the p-NOI structure form (**Figure 1**) and varies between 6×10^{-17} and 1.2×10^{-16} F/μm (**Figure 6**).

3. The work principle of the pesticide biosensor with nanoporous Si layer

3.1. Work principle

The proposed pesticide biosensor works with acetylcholinesterase noted by AChE, with the code—EC 3.1.1.7. This receptor element is used to degrade agonists of the acetylcholine (AcH) neurohormone. In the living matter, AcH is present at neuromuscular junction and in the cholinergic nervous system, modulating the electrical pulse transmission at synaptic spaces, as other neurotransmitters [25]. The AChE has a very high catalytic affinity for acetylcholine and for its agonists as parasympathomimetic pesticides. This property opens the door of pesticide-selective detection by AChE-based enzyme biosensors [26].

The pesticides are intensively used in agriculture usually as organochlorines, carbamates, and organophosphate. Paraoxon belongs to the organophosphate class, being an oxon and the active metabolite of the parathion pesticide. Their working principle on the pests is based on the inhibition of AChE, allowing acetylcholine to transfer nerve impulses indefinitely and causing paralysis. Paraoxon is a novel generation of pesticide, which reacts as an inhibitor of AChE. Pesticides from this group act directly by stimulating the nicotinic receptors or indirectly by the inhibition of cholinesterase, as an acetylcholinesterase inhibitor, abbreviated as AChEI. Paraoxon is one of the most potent acetylcholinesterase inhibitor available in insecticide [27]. In water solvent, it stands for a high risk of poisoning for humans or animals, due to its simply absorption through teguments in contact with the contaminated water from environment. As pesticide, the parathion is dissolved in water and usually is applied by treatment. It is frequently sprayed to rice and fruits. The usual concentrations are 0.05 and 0.1% [28]. After the rain, the pesticide is accumulated in water and soil. This parathion degradation in time produces multiple water-soluble products.

3.2. Nanoporous materials for enzyme entrapping

This section depicts the paraoxon biosensor starting from a Si wafer technology. Some intermediate nanoporous materials are used in the biosensor construction, for the enzyme entrapping. Among these materials, TiO_2 [29], Al_2O_3 [13], or porous Si still exists [30]. The porous material integration on a silicon wafer is starting by the first metal deposition, followed by subsequent processing steps, in order to convert them into compounds and finally into a porous matrix. The main steps of porous Si layer formation are:

- The Si start wafer is n-type, <100> orientation, and 2–9 Ωcm resistivity.
- The first process is a thermal oxidation that allows the pattern configuration.
- The next process is the boron ion implantation through the patterned mask in a high dose on the front wafer to convert the upper Si layer into p-type.
- Then, the Si p-type layer is converted in porous Si by anodization in the electrolyte $HF{:}CH_3COOH{:}H_2O$ with 180:60:60 ratios, at current density more than 2 mA/cm^2.
- The last process consists in annealing at 550°C in H_2 and 850°C in N_2 to increase the film stability.

This porous Si technology provides usual porous Si layers with a porosity of 56, suitable for the enzyme entrapping process [30].

These intermediate porous materials augment the capillary, allowing the biomaterials entrapping in a liquid phase, during the pre-deposition technological stage. At the same time, the porous layer must be grown onto the Si wafer in order to be strongly anchored to substrate and in order to avoid accidental detachments. Nanoporous Si can be easily converted from a Si thin upper layer. Having a closer lattice constant with Si, the porous Si stands for an efficient intermediate material for the next technological steps. The nanoporous Si material preparation by anodization is a perfect compatible method with the microelectronics technology. The pore sizes can be simply adapted in respect with the anodization reaction parameters, changing the electrolyte composition. Due to an increased area, offered by the nanoporous Si material against the monocrystalline Si, an enhanced miniaturization with capacitive electrodes can be performed. Therefore, the porous Si was selected as intermediate layer for the AChE enzyme entrapping. This solution is also in agreement with the nowadays tendency.

4. Results of some key technological steps

4.1. Nanoporous Si characterization

The morphology of the synthesized porous Si film is characterized by SEM microscopy (**Figure 7**). When the anodization process is tested for 60 mA/cm^2, some multiple pores are crowded inside a single larger pore rather with a crater shape. When the anodization process is tested for 300 mA/cm^2, some huge pores reached 7 μm size. However, the current density of 50 mA/cm^2 offered the optimum porosity for the enzyme entrapping [22].

Figure 8 presents the designed mask for the nanoporous Si region configuration.

Then, **Figure 9** presents the next mask used for the metal deposition on nanoporous Si material.

4.2. Electrode design and processing

The biosensor transducer is represented by an electrical capacitance. From the design stage, the capacitor has a constant armature surface and a fix distance between electrodes, so that any variation in capacitance reflects the electrical permittivity change of the material. This permittivity variation is proportional with the quantity of ions accumulated after the enzymatic-assisted reaction of the pesticide hydrolysis. Therefore, in order to increase the sensor sensitivity, as high as possible, the active area is demanded high. In this sense, the electrodes are designed with an interdigitated geometry [31].

Now, **Figure 10** shows multiple metallic traces for a global view of the biosensor transducer.

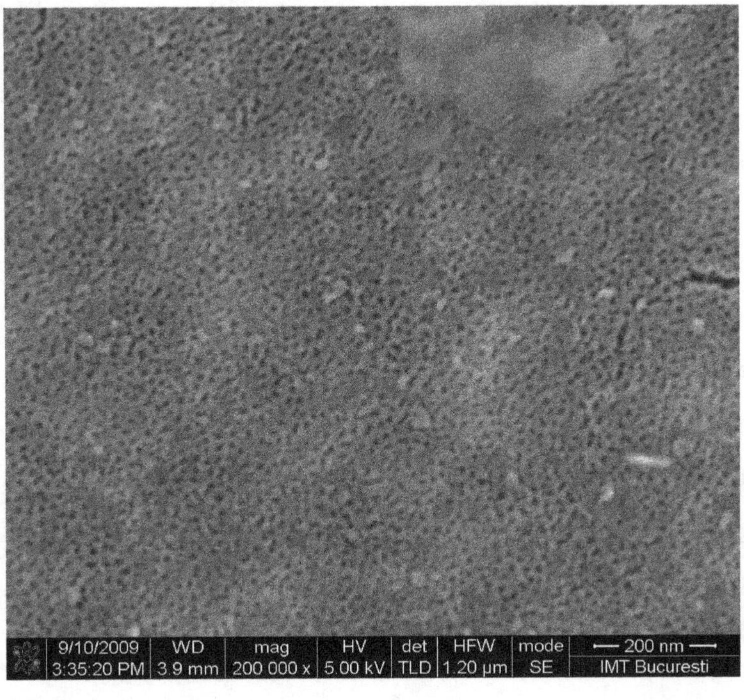

Figure 7. The SEM image reveals the nanoporous material.

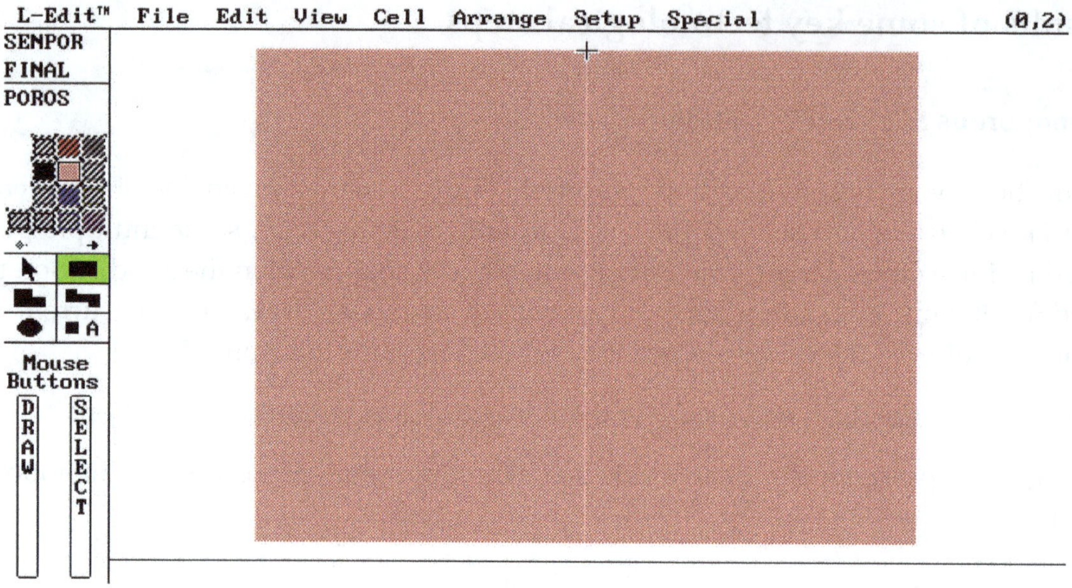

Figure 8. The porous Si mask.

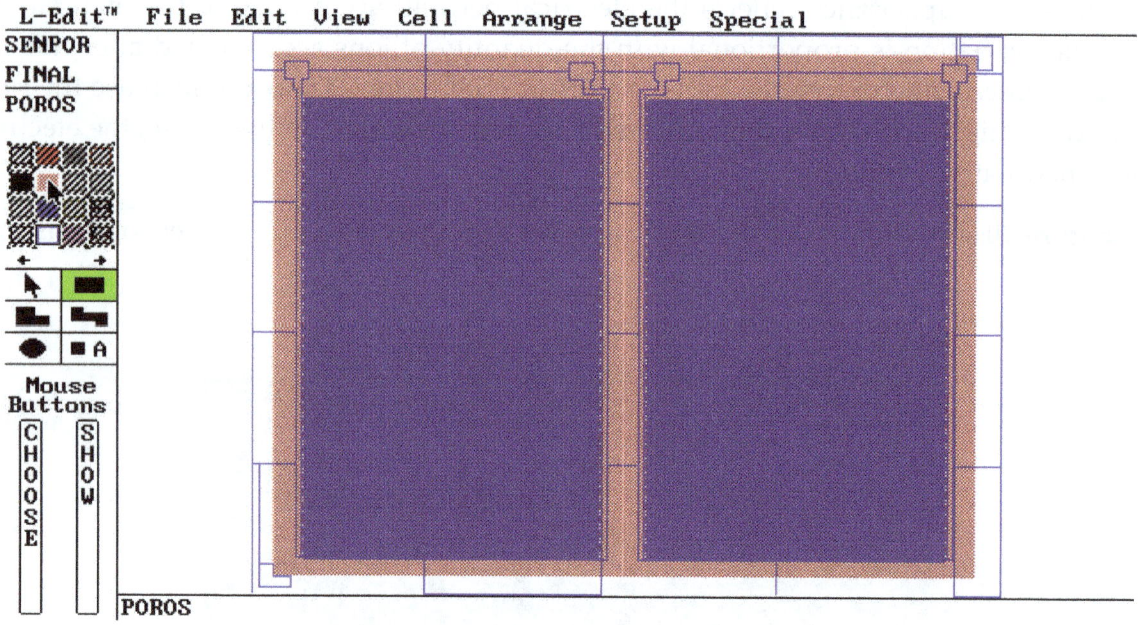

Figure 9. Mask of the metal overlap onto nanoporous Si material.

4.3. The enzyme processing step

In the first step, the powder of the acetylcholinesterase enzyme is blended with serum albumin, all being of Sigma provenience. A phosphate buffer solution keeps pH = 7.1 as constant pH. The mixture has helped to be entrapped on the Si wafer surface by the adsorption method. In this scope, the glutaraldehyde, in solution of 2.5% concentration, is added as cross-link agent.

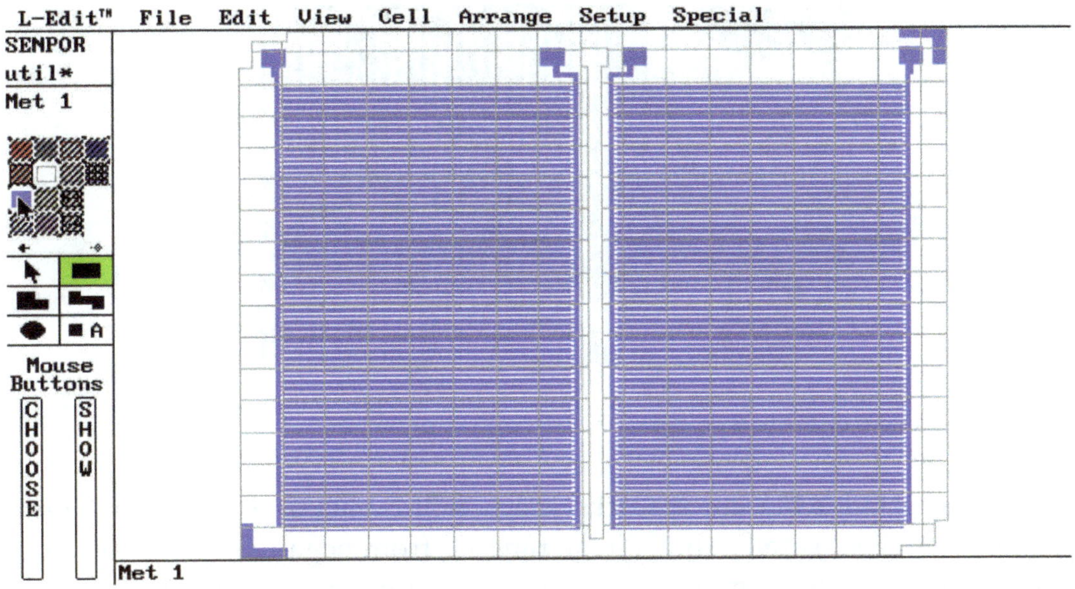

Figure 10. Final overlap of the metallic electrodes.

Figure 11. The top view of a piece of the enzymatic membrane.

The enzymatic area overlaps on the nanoporous Si area that facilitates the immobilization. The structure is introduced in drying stove for 24 hours. A top view of a piece of the processed AChE membrane is presented in **Figure 11**.

5. The final product: a tool for the environment monitoring

The final interdigitated structure comprises 98 horizontal metallic traces, which are starting from the central pillar for each electrode, detail in **Figure 12**.

After the enzyme entrapping, the integrated biosensor was tested by capacitance-voltage (C-V) analysis. The capacitance-voltage experimental curve is measured for the final product, after the enzyme entrapping, but in the absence of pesticide in measured solution.

The capacitance ranges from 45 nF up to a minimum of 7 nF, in agreement with p-NOI electrode structures with 8μm width and 10 μm gap space. The shape of the C-V curve

Figure 12. Detail of electrodes on the final product.

proves that our capacitive biosensor works as a p-NOI capacitor, with enzyme on nanoporous Si material on Si substrate. The voltage ranging from negative values toward positive values brings the capacitor from the accumulation regime through the depletion regime—the middle decreasing part of the curves—toward the inversion regime. The C-V curve is almost reproducible that indicates an adequate enzyme entrapping onto the silicon surface occurred.

6. Conclusions

The chapter presented a biosensor generated by micro-nanotechnology and biotechnology to serve as a monitoring tool to ensure a green environment condition. The fabrication steps of an integrated pesticide biosensor with AChE enzyme on nanoporous Si structure were presented. The structure of interdigitated electrodes was investigated as physical phenomena simulations inside the novel proposed planar nothing on insulator (p-NOI) structure. The contributions and purpose of this work were:

- Fabrication of nanoporous Si layer onto the Si surface and characterization by CV
- The enzyme membrane immobilization technique by adsorption onto the porous Si layer

The growing technological process of the nanoporous Si layer onto the Si surface was processed by the conversion of the n-type wafer into a p-type at the surface, followed by anodization. The enzyme membrane immobilization technique was by adsorption onto the porous Si layer and fixed with glutaraldehyde by the cross-link method. Finally, the preparation of the capacitive electrodes as an interdigitated structure comprised 98 horizontal metallic p-NOI traces. The primary C-V curves checked the sensor functionality.

Acknowledgements

This work was partially supported by a grant of the Romanian National Authority for Scientific Research and Innovation, CNCS/CCCDI UEFISCDI, project number PN-III-P2-2.1-PED-2016-0427, within PNCDI III, project number 205PED/2017, for IMT part.

This work was also partially supported by a grant of Ministry of Research and Innovation, CNCS-UEFISCDI, project number PN-III-P4-ID-PCE-2016-0480, within PNCDI III, project number 4/2017, for UPB part.

Conflict of interest

There are no conflicts of interest known.

Author details

Cristian Ravariu[1]*, Elena Manea[2], Alina Popescu[2] and Cătălin Pârvulescu[2]

*Address all correspondence to: cristian.ravariu@upb.ro

1 Department of Electronic Device, Circuits and Architectures, Faculty of Electronics Telecommunications and Information Technology, "Politehnica" University of Bucharest, Bucharest, Romania

2 National Institute for Research and Development of Microtechnology – IMT Bucharest, Voluntari, Ilfov, Romania

References

[1] Lucarelli F, Palchetti I, Marrazza G, Mascini M. Electrochemical DNA biosensor as a screening tool for the detection of toxicants in water and wastewater samples. Talanta. 2002;**56**:949-957. DOI: 10.1016/S0039-9140(01)00655-5

[2] Warkiani ME, Lou CP, Liu HB, Gong HQ. A high-flux isopore micro-fabricated membrane for effective concentration and recovering of waterborne pathogens. Biomedical Microdevices. 2012;**14**(4):669-677. DOI: 10.1007/s10544-012-9647-6

[3] Vamvakaki V, Chaniotakis NA. Pesticide detection with a liposome-based nano-biosensor. Biosensors and Bioelectronics. 2007;**22**(12):2848-2853. DOI: 10.1016/j.bios.2006.11.024

[4] Kim G-Y, Kang M-S, Shim J, Moo S-H. Substrate-bound tyrosinase electrode using gold nanoparticles anchored to pyrroloquinoline quinone for a pesticide biosensor. Sensors and Actuators B: Chemical. 2008;**133**(1):1-4. DOI: doi.org/10.1016/j.snb.2008.01.055

[5] Zheng Z, Zhou Y, Li X, Liu S, Tang Z. Highly-sensitive organophosphorous pesticide biosensors based on nanostructured films of acetylcholinesterase and CdTe quantum dots. Biosensors and Bioelectronics. 2011;**26**(6):3081-3085. DOI: doi.org/10.1016/j.bios.2010.12.021

[6] Lin T, Huang K, Liu C. Determination of organophosphorous pesticides by a novel biosensor based on localized surface plasmon resonance. Biosensors and Bioelectronics. 2006;**22**(4):513-518. DOI: doi.org/10.1016/j.bios.2006.05.007

[7] Lu D, Shao G, Du D, Wang J, Wang L, Wang W, Lin Y. Enzyme entrapped nanoporous scaffolds formed through flow-induced gelation in a microfluidic filter device for sensitive biosensing of organophosphorus compounds. Lab on Chip. 2011;**11**(12):381-384. DOI: doi.org/10.1039/C0LC00337A

[8] Wang Y, Zhang S, Du D, Shao Y, Li Z, Wang J, Engelhard M H, Li J, Lin Y J: Self assembly of acetylcholinesterase on a gold nanoparticles–graphene nanosheet hybrid for organophosphate pesticidedetection using polyelectrolyte as a linker. Journal of Material Chemistry. 2011;**21**(2):5319-5325. DOI: doi.org/10.1039/C0JM03441J

[9] Choi BG, Park H, Park TJ, Yang MH, Kim JS, Jang S, Heo NS, Lee SY, Kong J, Hong WH. Solution chemistry of self-assembled graphene nanohybrids for high-performance flexible biosensors. ACS Nano. 2010;**4**(5):2910-2918. DOI: 10.1021/nn100145x

[10] Madiyar FR, Syed LU, Arumugam P, Li J. Chapter 6. Electrical capture and detection of microbes using dielectrophoresis at nanoelectrode arrays advances in applied nanotechnology for agriculture. In: Advances in Applied Nanotechnology for Agriculture. ACS Symposium Series. Vol. 1143. ACS Publisher (American Chemical Society); 2013. pp. 109-124. DOI: 10.1021/bk-2013-1143.ch006

[11] Wang C, Madiyar F, Yu C, Li J. Detection of extremely low concentration waterborne pathogen using a multiplexing selfreferencing SERS microfluidic biosensor. Journal of Biological Engineering. 2017;**11**(9):1-11. DOI: 10.1186/s13036-017-0051-x

[12] Malaj E, Von der Ohe PC, Grote M, Kühne R, Mondy CP, Usseglio-Polatera P, Brack W, Schäfer RB. Organic chemicals jeopardize the health of freshwater ecosystems on the continental scale. Proceedings of the National Academy of Sciences of United States of America. 2014;**111**(26):9549-9554. DOI: 10.1073/pnas.1321082111

[13] Ravariu C, Popescu A, Podaru C, Manea E, Babarada F. The nanoporous Al_2O_3 material used for the enzyme entrapping in a glucose biosensor. In: Proceedings of the XII Mediterranean Conference on Medical and Biological Engineering and Computing (MEDICON'2010); 27-30 May 2010; Chalkidiki, Greece: Springer Publisher; Vol. 29. 2010. pp. 459-462

[14] Crosson C, Rossi C. Quartz crystal microbalance immunosensor for the quantification of immunoglobulin G in bovine milk. Biosensors and Bioelectronics. 2013;**42**:453-459. DOI: 10.1016/j.bios.2012.11.010

[15] Ravariu C, Mihaiescu DE, Babarada F, Manea E, Idu M, Vladoianu L. Vertical variants of PIN and p-NOI tunnel electronic devices and potential applications. In: Proceedings of the 5th International IEEE Symposium on Electrical and Electronics Engineering (ISEEE'2017); 20-22 October 2017; Galati. Romania: IEEE; 2017. pp. 36.1-36.6

[16] Ravariu C, Babarada F. Resizing and reshaping of the nothing on insulator NOI transistor. Advanced Nano-Bio-Materials and Devices. 2017;**1**(1):18-23

[17] Kavitha CV, Gaonkar SL, Narendra Sharath Chandra JN, Sadashiva CT, Rangappa KS. Synthesis and screening for acetylcholinesterase inhibitor activity of some novel 2-butyl-1,3-diaza-spiro[4,4]non-1-en-4-ones: Derivatives of irbesartan key intermediate. Bioorganic & Medicinal Chemistry. 2007;**15**(23):7391-7398. DOI: 10.1016/j.bmc.2007.07.014

[18] Ravariu C. A NOI – Nanotransistor. In: Proceedings of the IEEE International Conference of Semiconductors (CAS'2005); 11-13 October 2005; Sinaia. Romania: IEEE; 2005. pp. 65-68

[19] Ravariu C. The implementation methodology of the real effects in a NOI nanostructure aided by simulation and modelling. Simulation Modeling Practice and Theory. 2010;**18**(9):1274-1285. DOI: 10.1016/j.simpat.2010.05.002

[20] Ravariu C. Compact NOI nano-device simulation. IEEE Transactions on Very Large Scale Integration (VLSI) Systems. 2014;**22**(8):1841-1844. DOI: 10.1109/TVLSI.2013.2278474

[21] Ravariu C. Semiconductor materials optimization for a TFET device with nothing region on insulator. IEEE Transactions on Semiconductor Manufacturing. 2013;**26**(3):406-413. DOI: 10.1109/TSM.2013.2258411

[22] Suzuki M, Sagawa M, Kusunoki T, Nishimura E, Ikeda M, Tsuji K: Enhancing electron-emission efficiency of MIM tunneling cathodes by reducing insulator trap density. IEEE Transactions on Electron Devices 2015;**59**(8):2256-2262. DOI: 10.1109/TED.2012.2197625

[23] Karthick S, Ajayan J, Vivek K, Kavin-Arasan K, Manikandan A. A review on performance comparison of SOI MOSFET with STS-SOI MOSFET. In: Proceedings of the 2nd International Conference on Electronics and Communication Systems (ICECS'2015); 26-27 February 2015; Coimbatore, India: IEEE; 2015. pp. 1401-1406

[24] Ravariu C. Deeper insights of the conduction mechanisms in a vacuum SOI nanotransistor. IEEE Transactions on Electron Devices. 2016;**63**(8):3278-3283. DOI: 10.1109/TED.2016.2580180

[25] Ravariu C, Botan R. The electrical transport mechanisms investigation in adrenergic synapses using a parallel BioOI biodevice. In: Proceedings of the 19th IEEE International Conference of Biosignal (EURASIP'2008); 29-30 June 2008; Brno. Czech: IEEE-EMBS; 2008. pp. 63.115-63.118

[26] Viswanathan S, Radecka H, Radecki J: Electrochemical biosensor for pesticides based on acetylcholinesterase immobilized on polyaniline deposited on vertically assembled carbon nanotubes wrapped with ssDNA. Biosensors and Bioelectronics. 2009;**24**:2772-2777. DOI: 10.1016/j.bios.2009.01.044

[27] Jeanty G, Marty JL. Detection of paraoxon by continuous flow system based enzyme sensor. Biosensors and Bioelectronics. 1998;**13**(2):213-218. DOI: 10.1016/S0956-5663(97)00035-3

[28] Husain K, Ansari RA, Ferder L. Pharmacological agents in the prophylaxis/treatment of organophosphorous pesticide intoxication. Indian Journal of Experimental Biology. 2010;**48**(7):642-650

[29] Ravariu C, Manea E, Babarada F. Masks and metallic electrodes compounds for silicon biosensor integration. Journal of Alloys and Compounds, Elsevier. 2017;**697**:72-79. DOI: 10.1016/j.jallcom.2016.12.099

[30] Dhanekar S, Jain S. Porous silicon biosensor: Current status. Biosensors and Bioelectronics. 2013;**41**:55-64. DOI 10.1016/j.bios.2012.09.045

8

Low-Temperature Solution-Processable Functional Oxide Materials for Printed Electronics

Phan Trong Tue

Abstract

Over a decade since the first printed oxide transistor has been reported, printed oxide electronics is now becoming emerging technologies for realization of flexible, large-scale, low-cost electronic devices and systems. This chapter summarizes recent progress in the development of low-temperature solution-processable functional oxide materials and devices, and it also addresses critical challenges for the fundamental understanding and practical implementation of complex oxides in devices. The first part of this chapter gives an overview of the development of functional oxide inks such as semiconductors, conductors, and dielectrics. The second part discusses high-resolution printing technologies and some applications of printed electronics to exemplify their potential.

Keywords: oxide materials, thin-film transistors, oxide semiconductors, oxide dielectrics, transparent conducting oxides, solution process, low-temperature process, printed electronics

1. General introduction

Oxide materials have become high-tech functional materials beyond their traditional role as dielectrics. They show a rich variety of complex emergent behaviors, such as memristive effect [1], catalytic activity [2], and complex multiferroic effects [3]. The discovery of new metal oxides with interesting and useful properties continues to drive much research in chemistry, physics, and materials science. The physicochemical properties of oxides can be tuned through variation of factors such as composition, temperature, pressure, strain, external fields, defects, film orientation, and nanoparticle size [4–6]. Structure-property detailed analysis and understanding of the physicochemical properties of the oxides are prerequisites to improve their properties and to spur development of new oxide materials.

Among the various process technologies, solution process has many advantages, not only simple and low-cost process but also homogeneity and excellent composition control and high throughput [7]. Oxide solutions are generally synthesized using functional metal precursors in solvents and deposited on substrates by various coating methods. The coated oxide gels were pre-annealed to remove the solvents and post-annealed to develop active layers. Design of metal oxide precursor solutions (metal composition, metal precursor, chelating agents, etc.), treatment of intermediate oxide gels, and annealing techniques are of paramount importance for controlling and improving structural and opto-electronic properties of final oxides.

In this chapter, we review the progress in solution-processed functional oxide thin films produced at a low temperature and highlight the critical challenges for the fundamental understanding and practical implementation of complex oxides in devices.

2. Low-temperature solution-processed oxide dielectric thin films

2.1. Introduction

In this section, we focus on the low-temperature solution deposition of high-quality oxide dielectric thin films for the fabrication of active electronic devices such as thin-film transistors (TFTs). Highly stable oxide dielectric materials are of paramount importance as implemented through sophisticated additive processing with the other components of a TFT. However, compared to vacuum-based deposited oxide films, solution-processed counterparts are generally found to be inferior due to their deposition, metal precursors, morphological characteristics, and performance limitations. Although realizing capabilities to print and integrate solution-processed device-quality oxide dielectrics poses a very significant challenge, success is very likely to open many opportunities in fabricating active electronics as well as in providing new approaches to the production of various unique optical and optoelectronic devices. In addition, solution processing of oxide dielectric thin films at a temperature well below 300°C, which likely opens the door to flexible electronics, has remained challenging. By various approaches including tailoring precursor solution or functional solution chemistry, deposition, annealing/heating profile, activation of amorphous film, such as photo-assisted annealing (UV), solvothermal synthesis, and so on, low-temperature solution-processed oxide dielectric thin films have become possible now.

2.2. Materials selection for oxide dielectrics

Conventionally, silicon dioxide (SiO_2) has been used as dielectric layer in metal-oxide-semiconductor (MOSFET) due to its excellent insulating property and perfect Si/SiO_2 interfacial properties. Advanced technology has enabled fabrication of sub-22 nm channel-length MOSFET, which requires the thickness of SiO_2 to be less than 1 nm [8–10]. However, it is not possible to maintain small leakage currents, which increase dramatically through tunneling for very thin films. In this regard, it is useful to consider the important film features and performance metrics that will produce an optimal gate dielectric. The following are general requirements for a good dielectric film: large relative dielectric constant (>10) and small leakage current density (<10 nA/cm^2 at 1 MV/cm), low-dielectric loss (tanσ < 0.01), and large breakdown field (>4 MV/cm) to preserve the device function as shown in **Figure 1(a)**. In addition, high-mechanical strength, low-thermal expansion, low-water adsorption quality, and high-chemical inertness properties are highly desired. It is worth noting that the transistor mobility strongly depends on the quality of semiconductor/dielectric interface. A dielectric with a rough surface would result in irregular semiconductor/dielectric interface, impeding the flow of charges through the semiconductor. Thus, atomically flat dielectric film surface is essentially required. To achieve the necessary leakage current and breakdown field, films must be as dense as possible and exhibit no pores or cracks. Both from the perspective of surface smoothness and the need for high-breakdown field and low-leakage current, films with amorphous structure are generally preferred for the fabrication of gate dielectric layers.

Because of the challenges in producing such insulators through solution methods, most solution-processed oxide TFTs [11–17] have been fabricated by using binary oxide gate insulators formed through vacuum-based depositions. Although binary oxides will continue to be used for TFT gate dielectric applications, they do not represent an optimal approach to realizing high-performance devices. Generally, binary oxides tend to crystalize [18–20] at relatively low-process temperatures, resulting in enhanced impurity interdiffusion and high-leakage currents due to formation of grain boundaries. Two most important prerequisites of oxide

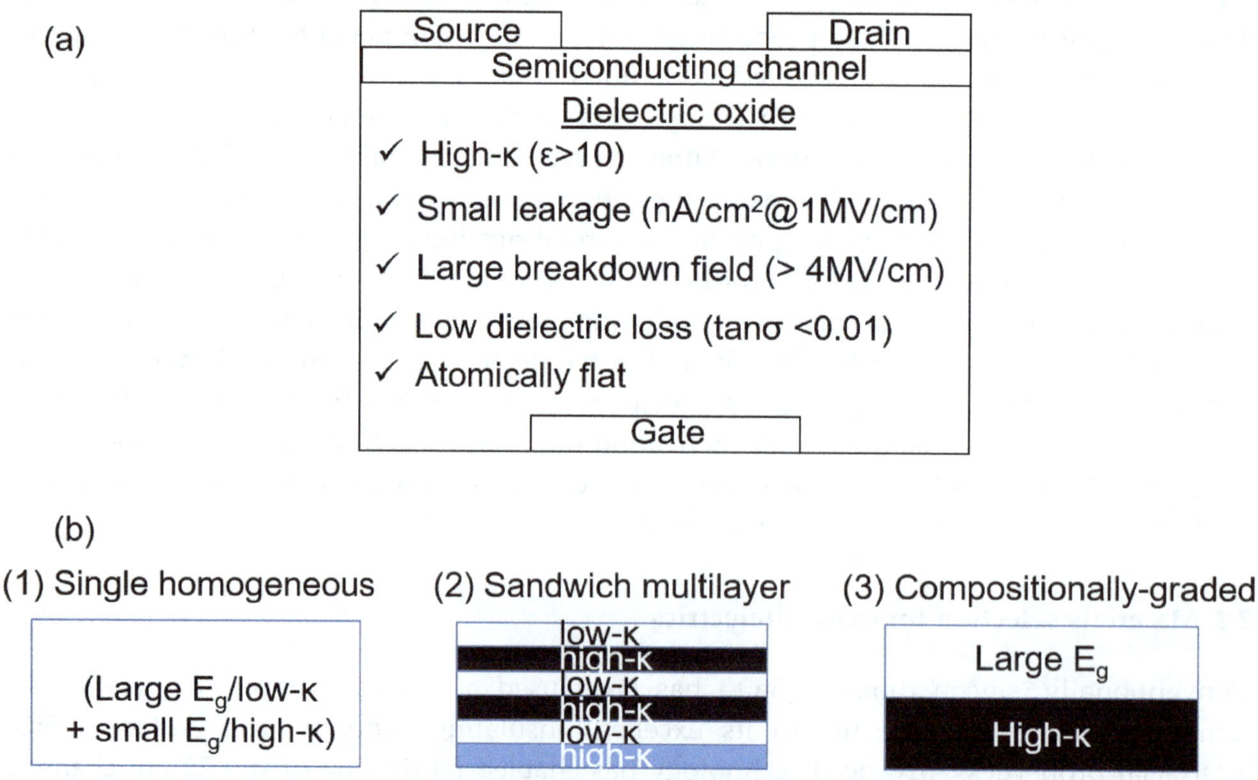

Figure 1. (a) General requirements of a dielectric layer and (b) major designs of device-quality dielectric layer.

dielectrics are high-dielectric constant and large breakdown field [21]. However, selection of a binary oxide as an insulator generally involves a compromise between these two characteristics. That is because binary oxides with high-dielectric constants normally have small band gaps (HfO_2, Ta_2O_5, ZrO_2, La_2O_3, and TiO_2) and vice versa (SiO_2 and Al_2O_3).

One approach to the production of high-performance dielectrics relies on the use of mixed multiple-component oxides. These oxides provide convenient means for controlling the dielectric constant and breakdown field through incorporation of components that specifically contribute to either dielectric constant or breakdown. Furthermore, amorphous phase can be stabilized by mixing multi-components, resulting in the films with extremely flat surfaces. Common binary oxides used for tuning these properties are listed in **Table 1** [20, 22–26].

To meet the performance requirements of gate dielectrics in TFTs, multi-component oxides can be produced by strategies as illustrated in **Figure 1(b)**. A single homogeneous dielectric can be produced by combining selected wide band-gap materials with those exhibiting smaller gaps and higher dielectric constants. For example, the mixtures of Hf-Si-O [27–29] and HfO_2-Al_2O_3 [30, 31] have been extensively studied as gate dielectrics in Si CMOS devices. Alternatively, wide- and small-gap materials can be interleaved to form multilayered structures, as demonstrated by stacked layers of TiO_2 and Al_2O_3 produced through atomic layer deposition. The presence of sharp dielectric interfaces in such structured materials provides a means to improve dielectric-breakdown fields. Finally, a compositionally graded material dominated by a high-dielectric-constant material at the metal-insulator interface and a high-band-gap material at the dielectric-semiconductor interface provides an additional alternative.

2.3. Producing high-quality films from solution

The fundamental challenges in depositing oxide thin films from solution are associated with the processes of conversion of soluble precursors into dense solids. Thus, understanding the structure of metal-organic precursors in solution and their effects on processability and on the final structure and properties of the oxide is the key to the production of high-quality oxides. Although improvements of solution-processed oxide dielectrics reported so far are impressive, many of them exhibit porous structures with coarse morphologies indicating that proper

Material	κ	E_g (eV)	CBO (eV)	VBO (eV)
SiO_2	3.9	9.0	3.2	4.7
Al_2O_3	9.0	8.8	2.8	4.9
La_2O_3	30	6.0	2.3	2.6
ZrO_2	25	5.8	1.5	3.2
HfO_2	25	5.8	1.4	3.3
Ta_2O_5	22	4.4	0.35	2.95
TiO_2	80	3.5	0	2.4

Table 1. Summary of some relevant characteristics of major binary oxides.

chemistries have not been utilized. By placing greater emphasis on conversion pathways from precursor to oxide, low-energy reactions should be devised that allow condensation to proceed uniformly. Especially for electronic applications, thin oxide films must retain density, homogeneity, and uniformity during condensation.

It was found that *hybrid clusters, having inorganic cores coordinated by organic ligands*, are the typical form of metal-organic precursor structures [32]. The study has shown that solvothermal synthesis of the precursor results in significantly improved insulating properties (e.g., two orders lower leakage current) of high-temperature annealed oxide films. We put emphasis on the structural analysis of the cluster precursors and annealed solids and relate the results to the significant improvement of properties by solvothermal treatment of solutions. A change in the cluster core toward structural unification can be brought about by the solvothermal treatment, leading to higher uniformity and higher stability of clusters. The final structure of the material maintained the features of the core structure in solution, even after annealing at high temperatures. These results demonstrate the key role played by designing cluster structure in solution. In addition, improved synthesis of the cluster precursor under solvothermal conditions leads to low-temperature deposition of oxide insulating films at below 200°C. Based on these strategies, we have designed and succeeded in producing various high-quality oxide dielectric films including lanthanoids-zirconium-oxide systems (Ln-Zr-O, Ln = La, Ce, Nd, Sm, Gd, Ho, Tm), Hf-Zr-O, Y-Zr-O, Hf-Ta-O, La-Ti-O, and Hf-La-O. As a typical example, detailed studies on La-Zr-O system will be shown in the following part with a focus on the structural analysis of the cluster precursor under solvothermal conditions.

2.3.1. Hybrid cluster precursors of the La-Zr-O insulator for transistors

Both lanthanum oxide and zirconium oxide are typical high-κ materials, having dielectric constant values in the range of 20–30. However, lanthanum oxide is hygroscopic, and both oxides are polycrystalline. Addition of Zr to La-O to create insulating LZO system with a dielectric constant in the range of 20–25 exhibits diverse chemistries in solution and resists crystallization in the solid phase. The LZO dielectric has shown some excellent properties in all-oxide TFTs [33–35], but leakage needs to be further suppressed and a processing temperature that is compatible with plastic substrates is highly desirable. Analyses of structures of solutions, gels, and solids by various characterizations have revealed a close structural relationship between the clusters in the solutions and the final solids even after annealing at high temperatures [32].

The synthesis of LZO precursor solution is summarized in **Figure 2**. First, lanthanum (III) acetate (La(OAc)) and zirconium (IV) butoxide solution (Zr(BtO)) were each dissolved in appropriate amounts of propionic acid (PrA) to produce La and Zr solutions. After that, the two solutions were mixed to obtain LZO mixtures with La/Zr molar ratios of 3/7 (LZ37) or 5/5 (LZ55). For solvothermal treatment, the LZO mixtures were sealed in an autoclave (AC) container and heated at 160–180°C for 2–5 h with magnetic stirring. The precursor solutions were spin coated on Pt/Ti/SiO_2/Si substrates, followed by annealing at 200–500°C in oxygen.

The thermal behaviors of precursor solutions were analyzed by thermal-gravity differential thermal analysis (TG-DTA) (**Figure 3**). Comparing LZO solutions with and without solvothermal treatment, we have observed three key features: (1) the decomposition

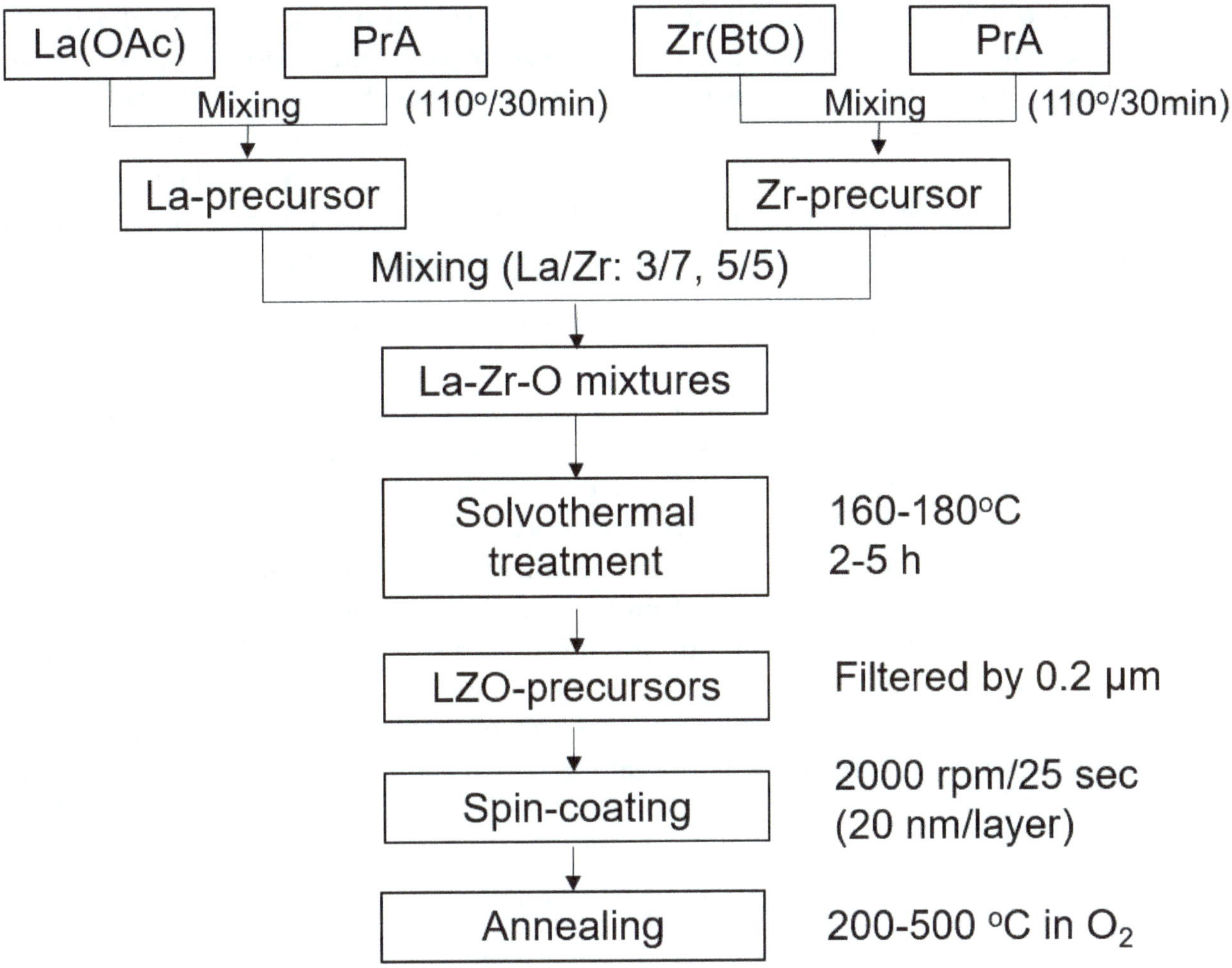

Figure 2. Schematic diagram of synthesis of the LZO precursor solution.

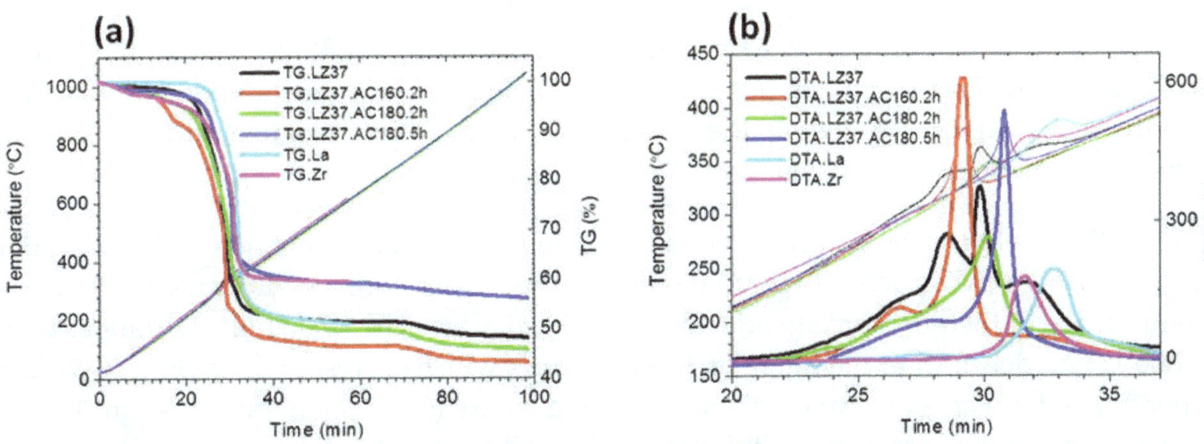

Figure 3. TG-DTA analyses of different LZO precursors solution.

temperature of the organic components was affected by the solvothermal treatment, i.e., evolution from multiple, different cluster structures toward a single cluster structure, leading to a more uniform and stable cluster structure. (2) Both the La-only and Zr-only precursors showed decomposition temperatures higher than those of LZO precursors, indicating that the La and Zr components were not simply mechanically mixed, even before solvothermal treatment. (3) The residual masses after TG analysis were quite different for solvothermally treated precursor solutions, representing the variation in the inorganic and organic contents in the precursors. Therefore, we speculate that the cluster development underwent two stages. The first stage is likely to be associated with the decomposition of the initial Zr clusters during their transition into La-Zr clusters. The second stage involved further growth/condensation of the clusters.

Interestingly, the solvothermal treatment impacted not only on the solution structure but also on the final material structure. It was turned out that solvothermal treatment led to a uniform structure in the precursor, which was even inherited by the final annealed oxide. Without pre-formation of a uniform cluster structure, the final material remained non-uniform even if the metal components in the precursors were well mixed on the cluster scale but not inside the clusters. This suggests that the inorganic core of the cluster in the precursor remained the structural unit after annealing, during which the organic ligands around it decomposed and the cores compacted, without significantly reacting with the surrounding components. Therefore, the solvothermally treated La-Zr solutions resulted in more uniform oxide solids with improved insulating and dielectric properties. The reorganization of the clusters under solvothermal conditions also enhanced UV light absorption and enabled film deposition under UV light at low temperature, which is presented in more detail in the following part.

2.3.2. *Lowering processing temperature for the La-Zr-O dielectric*

As mentioned earlier, a cluster core containing two or more metal elements with a structure similar to that of desired final oxide can be used as a building block for film deposition. In this method, the preferential decomposition of one metal compound over the others that causes compositional segregation is not expected to occur because the different metal elements have already been combined into one core and thermal dynamically stabilized. Hence, the decomposition and densification are similar to those occurring in binary metal oxide system. The structure of the clusters, as well as their uniformity and stability, influences the insulating properties of low-temperature deposited film.

The solvothermally treated solutions were found to have a highly enhanced UV absorption ability because of the structural reorganization of the cluster cores, which may facilitate the decomposition of their organic ligands under UV irradiation. The LZ37 films UV-annealed at 200–300°C had densities in the range of 4.3–4.5 g cm^{-3}, which corresponds to only 70–74% of the density of LZO crystals. An excellent insulating ability (~10^{-8} A cm^{-2} at 2 MV cm^{-1}) of the film was achieved (**Figure 4(b)**). Without solvothermal treatment of the solution, the current density is several orders higher, indicating significant effect of solvothermal treatment on the improvement of the film properties. In addition to the improvement of the cluster core, the solvothermal treatment of solutions also enhances the organic ingredient to be stabilized by the UV annealing, which facilitates formation of C—O bonding. Improvements of both the cluster core and the organic ingredient greatly contribute to the enhanced dielectric properties of LZO films.

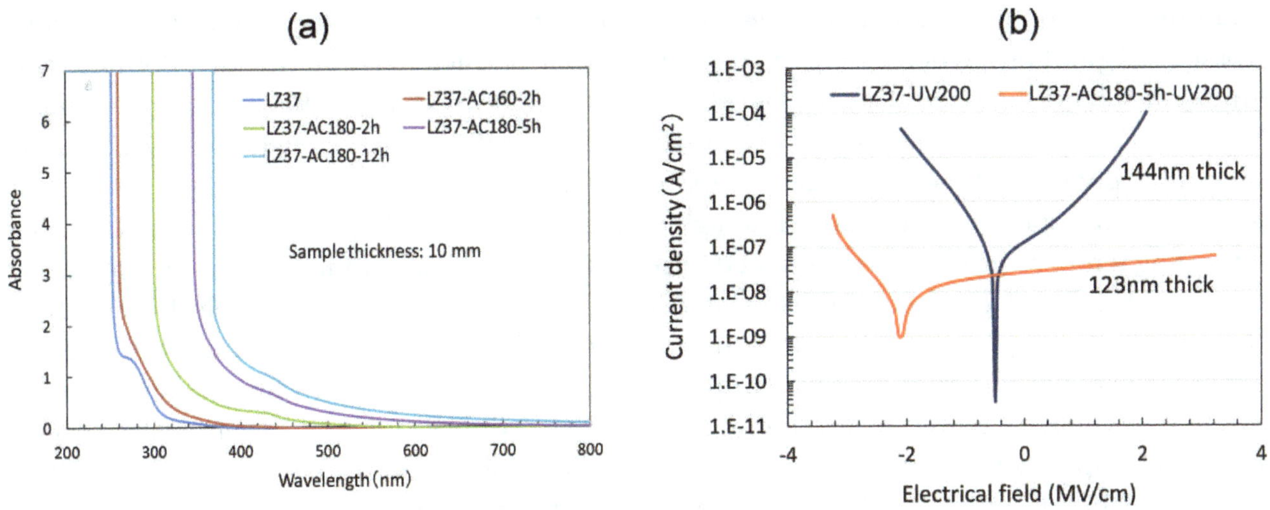

Figure 4. (a) UV-vis absorption spectra of LZO precursor solutions and (b) eakage currents of LZO films with and without solvothermal treatment.

Finally, using the LZ37 film UV/O_3-annealed at 200°C, we fabricated a TFT with a bottom-gate top-contact structure. The TFT exhibited a low-gate leakage current of less than 10 pA at an operating voltage of 15 V, a large "on/off" ratio of near 10^6, a field-effect mobility (μ) of 0.37 $cm^2\ V^{-1}\ s^{-1}$, and a subthreshold swing factor (*SS*) of 0.61 V $decade^{-1}$. The off current of the drain (3–30 pA) and the gate leakage (~10 pA) were extremely low and comparable to those of TFTs with the thermally grown SiO_2 insulator [36], indicating excellent insulating property of the low-temperature-processed LaZrO. The *SS* value is similar to those of high-temperature-processed In-Zn-O/LaZrO [35] and In-Zn-O/SiO_2 (channel/gate insulator) TFTs, suggesting the similar channel/gate insulator interface properties.

3. Low-temperature solution-processed oxide semiconductor thin films and transistors

3.1. Introduction

Inorganic semiconductors including silicon and chalcogenides have been solution-processed into high-performance semiconductors that have better mobility (>10 $cm^2\ V^{-1}\ s^{-1}$) and stability in comparison with organic semiconductors [37, 38]. These solution-processed inorganic semiconductors, however, generally require high annealing temperature, necessary for generating crystalline phases, impurity-free, and dense structures for device-quality films.

On the contrary, in oxide semiconductors, the amorphous phases are capable of exhibiting electron mobilities comparable to those of their crystalline phase counterparts [39–41], which allow the exclusion of a high-temperature annealing process for obtaining crystalline phases. The conduction band minimum (CBM), which constitutes the electron conduction pathway, is composed of vacant metal cation s-states, and the spatial expanse of these s-states is greater than the inter-cation distances. The s-state spatial overlap is primarily determined by the principal quantum number (n), and therefore, heavy post-transition metal cations with (n–1)d^{10} ns^{o} electronic configurations, where n ≥ 5, are ideal oxide semiconductor candidates [42]. Both In^{3+} and Sn^{4+}, which have the same [Kr]($4d^{10}$)$5s^{o}$ electronic configuration, meet this requirement, and a highly dispersed CBM is also found in ZnO due to the small interaction distances [43]. To date, most amorphous oxide semiconductors (AOSs) are deposited on rigid substrates, such as glass and metal foil, and are processable at a high temperature (>400°C). Recent efforts to lower the annealing temperature have shown that device-quality AOSs, exhibiting high performance and device stability that are not easily achievable in organic semiconductor-based electronics, have been successfully grown on plastic substrates. In the following part, we summarize the recent advances in the development of solution-processed AOS at a low temperature. In particular, we discuss the chemical pathways (colloidal-based process, sol-gel routes, auto-combustion chemistry, and impurity-free precursor-based approach) and physical approaches by newly developed annealing techniques (photo-assisted, microwave, and high pressure), which effectively enable the fabrication of low-temperature, solution-processable, high-performance AOS-TFTs.

3.2. Approaches to low-temperature solution-processed AOS-TFTs

3.2.1. Chemical pathways

3.2.1.1. Nanomaterials-based process

The nanoparticles (NPs)-based chemical approach is, in principle, the promising pathway for low-temperature annealed, high-performance semiconducting layers. Nevertheless, whereas the metal NPs, smaller than a few tens of nanometers, can be melted even below 200°C due to the dramatic lowering of the melting point [44, 45], the oxide NPs are not capable of undergoing the structural transformation into a granular film morphology at low temperatures, since the melting point of oxide NPs is not decreased depending on the particle size. This unique physical property of oxide NPs leads to the large surface area, porous, and poorly interconnected particulate film at low annealing temperatures, which limits device performance (low mobility and poor stability). For example, In_2O_3 NPs, less than 10 nm in size and spherical in shape, have been generated through chemical synthesis and implemented into device structures by ink-jet printing even at room temperature. However, the μ was confined to 0.8 cm^2 V^{-1} s^{-1} [46–48], which is associated with the inefficient carrier transport at junctions between neighboring NPs.

3.2.1.2. Sol-gel chemistry

In metal salt-based sol-gel chemistry, the precursor solution is synthesized by dissolving metal salt precursors in solvents with stabilizing agents and water. The resulting metal complexes undergo a hydrolysis reaction through the loss of a proton by one or more of the water molecules that surround the metal cations in the first solvation shell. As a consequence, the aquo ligand molecule, water, bonded to the metal cation is transformed into either a hydroxo ligand, OH^-, or an oxo ligand, O^{2-}. Subsequently, a condensation reaction occurs due to the oxolation generating an oxo-bridge, M-O-M, allowing the formation of a metal oxide skeleton framework with a dense microstructure.

Elimination of impurities such as hydroxides, anions, and carbon along with the formation of oxygen vacancies is critical for determining properties of AOS because they not only hinder the sol-gel oxide framework formation reaction but also interrupt the efficient transport of charge carriers [49]. To enable low-temperature processable products, the thermal decomposition of metal precursors should be completed at a low temperature, as evidenced by the use of metal nitrate precursors, the anion of which is almost completely decomposed even at 250°C, instead of other metal precursors (chloride, acetate) [50]. In addition, the chemical structure should be tailored toward a framework containing less hydroxide, which can be realized by doping an element that has a high electronegativity [51].

A "sol-gel on chip" hydrolysis approach from soluble metal alkoxide precursors was reported [52], which affords unprecedented high μ of ~10 $cm^2\ V^{-1}\ s^{-1}$, reproducible and stable turn-on voltage (~0 V) at maximum process temperature as low as 230°C. The approach uses the *in-situ* hydrolysis and condensation of transition metal alkoxides when they are exposed to an aqueous environment by nucleophilic substitution, thus affording the M-O-M framework at low temperature. The process is applicable to a broad range of AOS that are of immediate interest in TFT applications.

3.2.1.3. *Impurity-free precursor-based approach*

The presence of impurities has a negative impact on the performance of oxide semiconductors. In this regard, the metal-hydroxide nanocluster is a viable alternative to metal-salt precursors due to the absence of impurity-containing chemical species. The metal hydroxide is converted into a metal oxide framework by a thermally activated reaction. Aqueous Zn hydroxide solution was shown to drastically lower the annealing temperature to 150°C, producing ZnO-TFT with a μ of 0.4 $cm^2\ V^{-1}\ s^{-1}$ [53]. Another approach for the impurity-free precursor is the use of aqueous carbon-free metal-oxide precursors such as zinc oxide hydrate dissolved in ammonium hydroxide to generate Zn ammonium complex precursor [54]. This route allows for the growth of ultra-thin (4–5 nm), high-quality polycrystalline ZnO films on arbitrary substrates. Transistors fabricated using this simple process at 180°C showed an electron mobility of up to 11.0 $cm^2\ V^{-1}\ s^{-1}$.

Formation of the ammine-hydroxo complex and its low-temperature conversion to corresponding oxide, however, are generally not convenient, except for Zn. Thus, the design of a new chemical complex with the easy accessibility to other elements and the low-temperature processability for chemical transformations are essential in the impurity-free precursor-based approach.

3.2.1.4. *Auto-combustion synthesis*

A redox-based combustion synthetic approach was applied to oxide thin films using acetylacetone or urea as a fuel and metal nitrates as oxidizers [55]. The self-energy-generating combustion chemistry provides a localized energy supply, eliminating the need for high, externally applied processing temperatures. Furthermore, the atomically local oxidizer supply can efficiently remove organic impurities without coke formation in combustion reactions with balanced redox chemistry. The semiconducting In_2O_3 annealed at temperature as low as 200°C yielded TFTs with μ approaching 1.0 and 13.0 $cm^2\ V^{-1}\ s^{-1}$ on SiO_2 and Al_2O_3 gate dielectrics, respectively.

3.2.2. *Physical pathways*

3.2.2.1. *Microwave annealing (MW)*

The use of microwave-assisted annealing enabled quick fabrication of ZnO-TFT at 140°C with a μ of 1.7 $cm^2 V^{-1} s^{-1}$, which is improved to three- to sixfolds in comparison with TFTs processed by conventional hot-plate annealing [56]. This enhanced TFT performance is attributed to the improved development of ZnO grains suitable for reducing the length of transport paths across grain boundaries and the shorter Zn-Zn distance enabling better overlap between the Zn s-orbitals, thus creating chemical structures appropriate for high-performance AOS.

3.2.2.2. *Deep-ultraviolet assisted annealing (DUV)*

Much efforts have also been focused on the treatment of sol-gel films by photochemical activation using deep-ultraviolet (DUV) irradiation [36, 57–61]. When the as-deposited film is exposed to DUV irradiation, high-energy DUV photons induce photochemical cleavage of alkoxide groups and activate metal and oxygen atoms to facilitate M-O-M network formation (**Figure 5**, step 1, condensation). Further irradiation induces a gradual removal of oxygen and carbon (and, thereby, near-complete condensation) and a transition to a film densification (**Figure 5**, step 2, densification). This DUV technique is applicable to numerous AOS. The μ of the room temperature DUV-IGZO is as high as 14.0 $cm^2 V^{-1} s^{-1}$.

3.2.2.3. *High-pressure annealing (HPA)*

HPA influences thermodynamics of solution-processed AOS films, facilitating denser film formation by reducing its thickness under a high pressure (2 MPa). In addition, high-oxygen pressure changes the Gibbs free energy of the system and strengthens the bonding between metal ions and oxygen. As a result, TFT processed at 220°C exhibited a μ of 1.7 $cm^2 V^{-1} s^{-1}$ and high-bias stability [62].

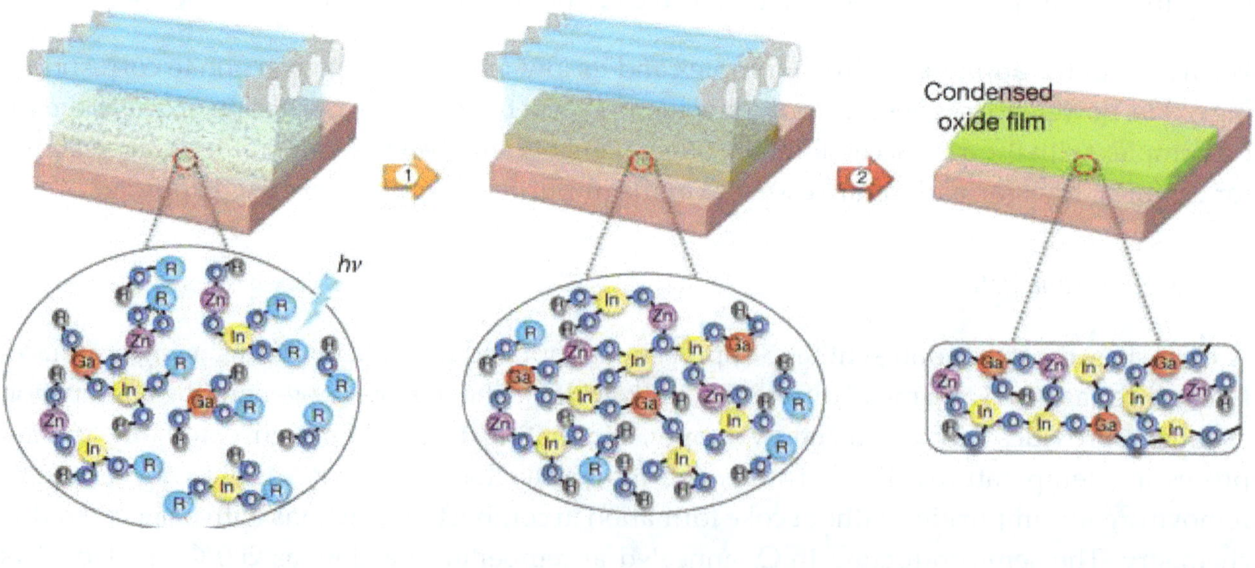

Figure 5. Schematic diagram for the condensation mechanism of metal-oxide precursors by DUV irradiation.

3.2.2.4. *Laser-assisted annealing*

A femtosecond (fs) laser annealing was applied for an effective and rapid fabrication of AOS thin films [63]. The method led to improvements in μ (9.0 cm^2 V^{-1} s^{-1}) and in "on/off" ratio due to the efficient removal of impurities and enhanced metal-oxide composition.

A summary of solution-processed AOS-TFTs fabricated at a low temperature (<300°C) is given in **Table 2**.

3.3. Conclusions and outlook

During the past decade, there has been significant active research regarding the development of high-performance semiconductors that can be generated by a low-cost, large-scale solution process. The recent progress with solution-processed, high-performance AOS confirms their potential feasibility in electronic applications. For practical realization of high-performance electronic devices even on plastic substrates, the high mobility at annealing temperature below 150°C

Precursor type	Semi.	Ann. temp. (°C)	Insulator	Mobility (cm^2 V^{-1} s^{-1})	On/off	Note
Nanoparticle	In_2O_3	RT	—	0.8	2×10^3	Colloidal [48]
Nanorod	ZnO	230	SiO_2	0.6	<10^6	Colloidal [64]
Ammine-hydroxo	ZnO	150	SiO_2	0.4	10^6	Impurity-free [53]
Ammonia complex	Li-ZnO	300	SiO_2	11.45	10^7–10^8	Alkali doping [65]
Metal salt + fuel	In_2O_3	200	Al_2O_3	13.0	10^3	Combustion [55]
			SiO_2	1.0	10^5	
Alkoxide	IZO	210	Al_2O_3	6.0	10^8	Sol-gel on chip [52]
Metal salt	In_2O_3	300	SiO_2	22.1	10^5–10^6	O_2/O_3 anneal [66]
		200		0.85		
Hydroxo	ZnO	140	SiO_2	1.7	10^7	MW [56]
Metal nitrate	In_2O_3	140	Al_2O_3	2.0	10^8	Precursor soaking [67]
Metal salts	IGZO	<80	Al_2O_3	8.76	10^8–10^9	DUV [36]
	In_2O_3			11.29		
Metal nitrate	In_2O_3	180		3.2	>10^6	Far UV [68]
		200		7.5		
Aqueous carbon-free	ZnO	80–180	Al_2O_3/ZrO_2	11.0	10^4–10^5	DUV [54]
Metal salt	IZO	220	SiO_2	1.78	10^6	HPA [62]
Metal salt	IZO	—	—	9.0	—	Laser anneal [63]

Table 2. Summary of low-temperature (<300°C) solution-processed oxide semiconductors.

is an inevitable requisite, which is considered achievable in solution processable AOS when the chemical/physical challenges addressed. The development of new chemical precursors for impurity-free AOS, the incorporation of dopants that can improve the electrical characteristics, and bias stability, and advanced annealing techniques for the generation of high-quality AOS would significantly improve the electrical performance of AOS processed by a solution process at a low temperature, thus making various practical electronic applications possible.

4. Conducing oxide thin films prepared from low-temperature solution processing

4.1. Transparent conducting oxides (TCOs) in general

The use of highly conductive and highly transparent thin films in the visible range of the spectrum is of great importance for a variety of optoelectronic device applications such as displays, solar cells, opto-electrical interfaces, and circuitries. Transparent conducting oxides (TCOs), in contrast to glass fiber, silicon, and compound semiconductors, are highly flexible intermediate states, whose conductivity can be tuned from insulating through semiconducting to conducting as well as their transparency adjustable. Furthermore, main carriers can be switched between n-type and p-type, opening a wide range of new technological applications.

So far, TCOs including binary, ternary, and quaternary oxide systems are mainly based on indium (III) oxide (In_2O_3), tin (IV) oxide (SnO_2), zinc (II) oxide (ZnO), and their mixtures with some dopants to tune structural and opto-electrical properties (**Figure 6**). The plot of In_2O_3 includes results for Sn-doped c (ITO) and other dopants, and the plot for various deposition methods is shown in **Figure 7** [69]. The slopes of the plots for improvement versus time in lowering doped In_2O_3 and doped SnO_2 resistivities appear to plateau at approximately 10^{-4} and 2×10^{-4} Ω cm, respectively. Thus, it is likely that further technological improvement in lowering resistivity of these materials is limited. However, the doped ZnO plot still presents a descending slope of improvement versus time. Therefore, it can be considered that further improvement of conductivity of ZnO-based systems would be possible.

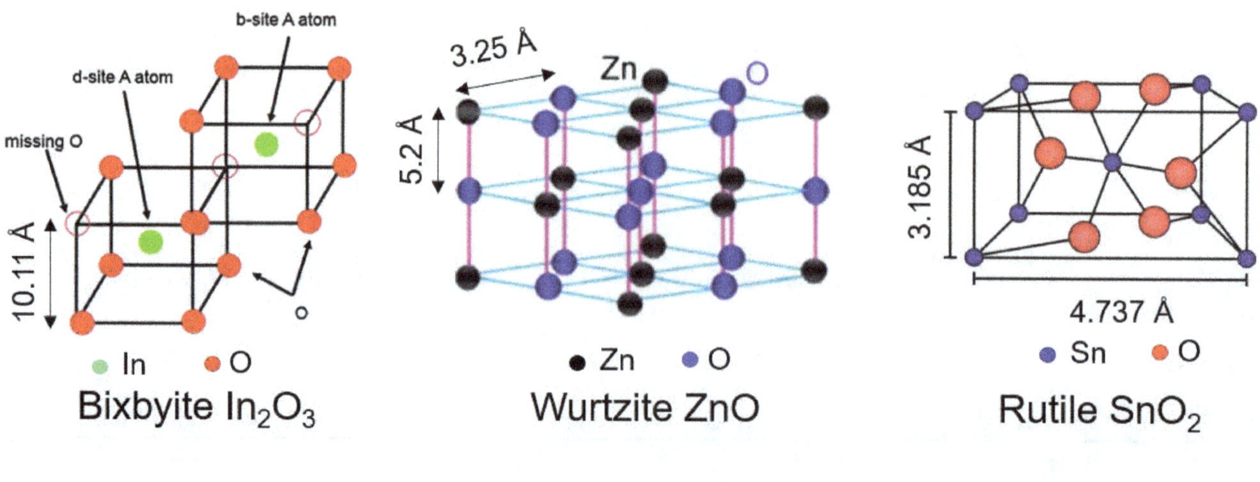

Figure 6. Unit structure of typical binary oxides: bixbyite In_2O_3, wurtzite ZnO, and rutile SnO_2.

4.2. N-type conductive oxides

4.2.1. Indium-tin-oxide (ITO)

ITO is the most widely used TCOs because of its two key characteristics: conductivity and optical transparency. Conventionally, ITO films are produced by vacuum-based deposition techniques. The best resistivity of ITO film deposited by sputtering was reported as low as 1.2×10^{-4} Ω cm [70]. As an alternative to vacuum-based deposition, recently much efforts have been focused on production of solution-based deposition for high-performance ITO films using various approaches such as nanomaterials-based [71–75], aqueous metal salts solution [76], combustion synthesis [77, 78], and advanced annealing techniques (microwave annealing [79] and ultraviolet laser annealing [80]. A summary of properties of ITO films prepared by various solution-based deposition processes is given in **Table 3**.

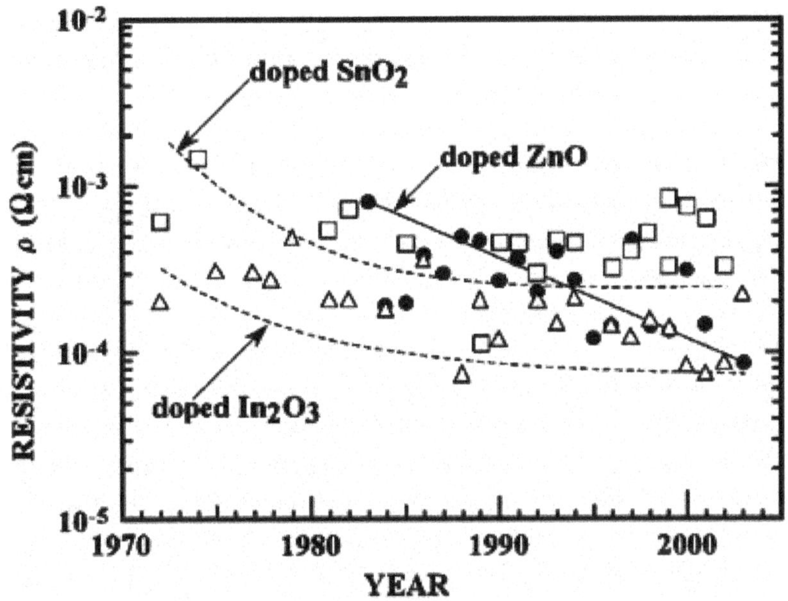

Figure 7. The plot of resistivity improvement versus time for various TCOs.

ρ (10^{-4} Ω cm)	R_s (Ω)	*t* (nm)	*T* (%)	Φ_H (10^{-2} Ω^{-1})	Deposition	Ref
45.7	47.5	144	85.3	0.420	Electrospray	[79]
42.3	188	225	80	—	Dip coating	[76]
2.5	42	60	~85	0.469	Dip coating	[81]
2.1	7.1	295	~78	1.171	Dip coating	[82]
310	—	75	>90	—	Spin coating	[80]
0.52	356	146	93	0.136	Spin coating	[74]
4.2	81	70	—	—	Spin coating	[77]
7.2	30	241	90.2	1.191	Spin coating	[72]

Table 3. Properties of ITO films prepared by various solution-deposition processes.

A redox-based combustion synthetic approach was applied to ITO thin film using acetylacetone as a fuel and metal nitrate as an oxidizer, which enabled production of high-quality ITO film at a temperature as low as 250°C [77]. The redox reaction between the fuel and oxidizer in the precursor solution generates internal heat, facilitating conversion from metal precursor to metal oxide at low temperature. The SCS-ITO thin film exhibited high-crystalline quality, atomically smooth surface (RMS ~ 4.1 Å), and low-electrical resistivity (4.2×10^{-4} Ω cm). The TFT using SCS-ITO film as the S/D electrodes showed excellent electrical properties with negligible hysteresis. The obtained "on/off" ratio, *SS*-factor, V_{th}, and μ were 5×10^7, 0.43 V/decade, 0.7 V, and 2.1 $cm^2 V^{-1} s^{-1}$, respectively (**Figure 8**). The performance and stability of the SCS-ITO TFT are comparable to those of the sputtered-ITO TFT, emphasizing that the SCS-ITO film is a promising candidate for totally solution-processed oxide TFTs.

4.2.2. Alternatives to ITO

There are increasing emphases on not only developing a TCO with higher conductivity but also with a material other than ITO. Two major concerns with ITO are the cost and the potentially limited resource.

Current candidate materials for a higher conductivity TCO are summarized in **Table 4**, which include the conventional binary oxides of CdO, SnO_2, In_2O_3, and ZnO, with alternative dopants, and combinations of these binaries, plus Ga2O3, in ternary, quaternary, and even quinary combination of these five oxides.

As a promising binary TCO material, ZnO is considered the prime candidate for a higher conductivity and to be low cost. However, ZnO is more sensitive to oxygen than ITO, and process control is more difficult. Work continues on improving ZnO with alternative dopants, especially Al (AZO) and Ga (GZO). As for sol-gel AZO film, the lowest reported resistivity is 1.5×10^{-4} Ω cm [83], which is close to that of film prepared by sputtering [84].

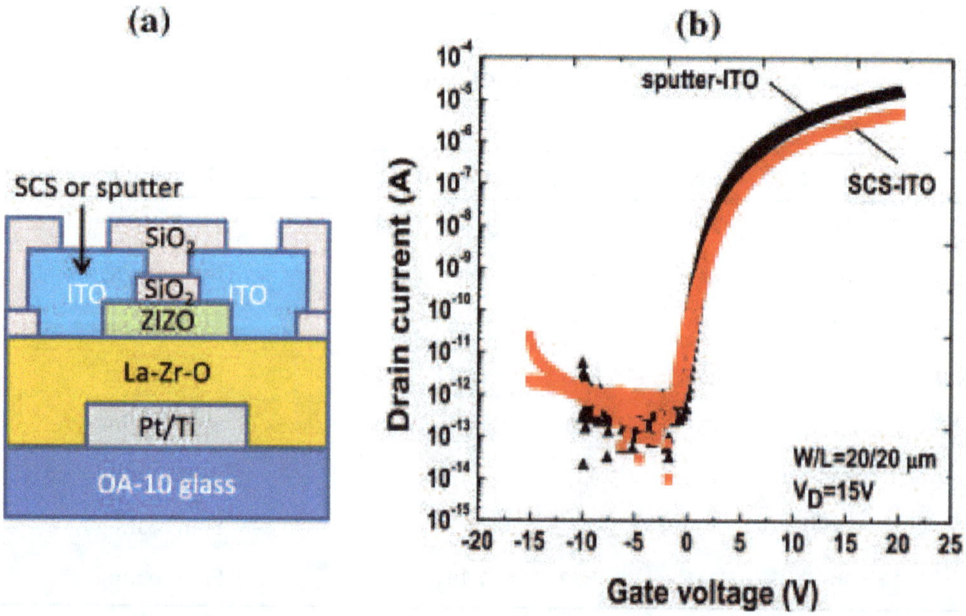

Figure 8. (a) Cross-sectional structure and (b) transfer curves of ZIZO-TFT using SCS-ITO as S/D electrodes.

Material	Dopant or compound
In_2O_3	Sn, Ge, Mo, F, Ti, Zr, Hf, Nb, Ta, W, Te, Mg
SnO_2	Sb, F, As, Nb, Ta
ZnO	Al, Ga, B, In, Y, Sc, F, V, Si, Ge, Ti, Zr, Hf
CdO	In, Sn
ZnO-SnO_2	Zn_2SnO_4, $ZnSnO_3$
ZnO-In_2O_3	$Zn_2In_2O_5$, $Zn_3In_2O_6$
Ga-In-O	Sn, Ge
$CdSb_2O_6$	Y
Cd-In-Sn-O	$CdIn_2O_4$-Cd_2SnO_4

Table 4. Conductive oxides based on multinary metal oxide systems other than ITO.

4.3. P-type conductive oxides

Highly conductive p-type oxides serve as critical components for various technological developments such as efficient charge injection layers for light-emitting devices [85], solar cells with better band-matching current collectors [86, 87], invisible circuits, and applications in near-infrared optoelectronics where n-type TCOs provide poor optical transmission. In contrast to widespread use of n-type TCOs such as ITO, p-type TCOs have not been commercialized yet due to their significantly low-carrier mobility and electrical conductivity. In fact, research toward p-type amorphous TCOs is highly challenging even by using vacuum deposition techniques, which result in very limited materials: Zn-Rh-O (resistivity 0.5 Ω cm) and Zn-Co-O (0.05 Ω cm). The difficulties originated from electronic structure of p-type TCOs, which exhibits the strong localization of the upper edge of the valence band to oxide ions.

The Hosono group has initiated breakthrough research toward p-type oxides. They found a series of transparent p-type Cu oxides with delafossite structure such as $CuGaO_2$ and $CuAlO_2$ [41, 88–90], and amorphous Zn-Rh-O, Zn-Co-O systems [41, 88, 91]. However, most of these p-type TCOs were produced by vacuum-based deposition techniques. Solution-processed p-type conductive oxides have remained very challenging.

Recently, misfit-layered $Ca_3Co_4O_9$ thin film synthesized through solution processing is shown to be high-performing p-type TCOs [92]. The synthesis method consists of sol-gel chemistry, spin coating, and heat treatment at 650°C. A resistivity and visible range transparency of 57 mΩ cm and 67%, respectively, were obtained. However, the required high-annealing temperature over 600°C may limit its practical uses.

In order to produce conductive amorphous p-type oxides from solution at low temperature, we focused on Ru oxides [93]. Crystalline RuO_2 is a well-known electrode material that exhibits metallic conduction. Initially, we found that the amorphous phase of RuO_2 is unstable, and only crystalline phase can be obtained through solution processing even at a low temperature below 300°C. We then used lanthanide elements (Ln, except Ce) to stabilize the amorphous phase. As a result, we found a series of solution-processable amorphous p-type conductive Ln-M-O (a-Ln-M-O, where M = Ru, Ir, and Ln is lanthanide elements except Ce) having low resistivity (10^{-3}–10^{-2} Ω cm) (**Figure 9**). The resistivity increases with increasing Ln atomic number (Z) and changes evidently faster for Z > 64 (Gd), which may be more related to the influence of 4f electrons (4f subshell is more than half-filled at Z > 64) than to the decreasing size of Ln ions. These oxides are thermally stable to a high degree, being amorphous up to 800°C, and processable below 400°C. These oxides have three pronounced features: first, their valence electrons have open-shell configurations (t_{2g}^4 in Ru 4d and t_{2g}^5 in Ir 5d), which is distinctively different from the closed-shell or pseudo-closed-shell configurations in other p-type oxides (e.g., Cu $3d^{10}s^o$ in Cu oxides, Sn $5s^2$ in SnO, Rh t_{2g}^6 in Zn-Rh-O, and Co t_{2g}^6 in Zn-Co-O). Second, the conduction of amorphous La-Ru-O (semiconducting) is completely different from that of crystalline phases (e.g., $La_3Ru_3O_{11}$ and $La_4Ru_6O_{19}$, which are both metallic). Third, the resistivity of solution-processed a-La-Ru-O is lower than that of the sputtered sample. Hence, detailed understanding of the electronic structure of these materials and the mechanism underlying processing-composition-conduction correlation is of fundamental importance for future research.

The unique properties of amorphous p-type oxides have high potential for use in printed electronics, e.g., as gate electrode in n-type oxide TFTs [33, 94], and may be modified to use as p-channel in TFTs or as an active layer in solar cells. The narrow band-gap is advantageous for light absorbing in solar cells. The electronic configurations of these amorphous oxides are apparently not analogous to those of known p-type oxides. This deserves theoretical investigation and suggests that more p-type amorphous oxides are ahead of us.

It is known that the conduction of ruthenium pyrochlore, $A_2Ru_2O_7$, is sensitive to the type of A element. These compounds are semiconducting for A = Y and the lanthanides Pr-Lu, and weakly metallic for A = Bi and Pb. The Ln of the aforementioned oxides was thus replaced with Bi and Pb, leading to low-resistivity amorphous oxides produced at low

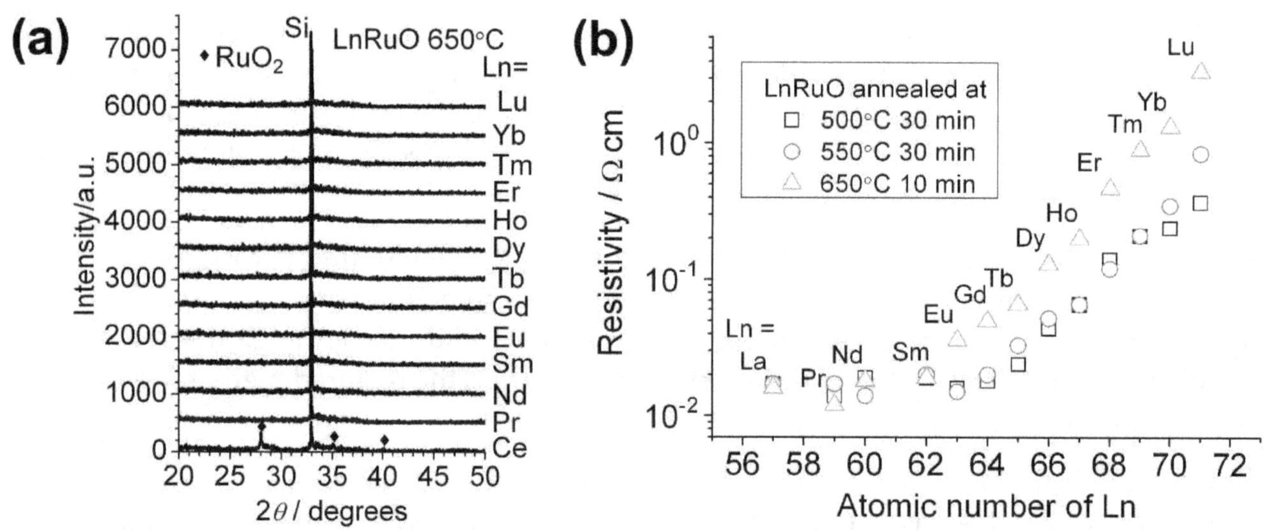

Figure 9. (a) XRD patterns and (b) resistivity of LnRuO films.

Material	Method	T (°C)	ρ/σ	μ (cm² V⁻¹ s⁻¹)	E_g (eV)	Ref.
$CuAlO_2$	So-gel	900	250 Ω cm	—	—	2002 [92]
$CuCrO_2$, $CuAl_{0.5}Cr_{0.5}O_2$	Sol-gel	600	11–16 Ω cm	—	—	2009 [96]
$CuFeO_2$	So-gel	700	0.36 S cm^{-1}	0.7	3.05	2012 [97]
$CuCrO_2$:Mg	CSD	600	0.31 Ω cm	0.7		2012 [98]
Ln-M-O (M = Ru, Ir, and Ln is a lanthanide)	CSD	<400	10^{-3}–10^{-2} Ω cm	—	—	2012 [93]
A-B-O (A = Bi, Pb; B = Ru, Ir)	CSD	240	1.3–3.8 mΩ cm	—	0.2	2012 [95]
$Ca_3Co_4O_9$	CSD	650	57 mΩ cm	—	—	2014 [92]
$CuCrO_2$	Combustion	180	0.14 S cm^{-1}	0.23	3.0	2018 [99]
$CuCrO_2$	Spray pyrolysis	345	12 S cm^{-1}	6.4×10^{-3}	2.4	2015 [100]
Y:$CuAlO_2$	Electro-hydrodynamic	1100	580 Ω cm	–	–	2015 [101]
Li:ZnO		500	15.06 Ω cm	0.918	–	

Table 5. Electro-optical characteristics of some solution-processed p-type conductive oxides.

temperatures [95]. The lowest necessary temperatures for a-BiRuO, a-PbRuO, and a-BiIrO were 240, 290, and 350°C, respectively, resulting in RT DC resistivities of 3.8, 1.7, and 3.8 mΩ cm, respectively. The resistivity of a-BiRuO film has nearly reached the value of crystalline bulk $Bi_3Ru_3O_{11}$ (1.4 mΩ cm), which suggests that the films are of high quality. In p-type oxides, the low-resistivity values of these amorphous oxides are matched only by epitaxial LaCuOSe:Mg annealed at a high temperature of 1000°C (1.1 mΩ cm). A summary of electro-optical characteristics of some solution-processed p-type conductive oxides is given in **Table 5**.

5. Applications of solution-processed functional oxides for TFTs and display

5.1. Low-temperature all solution-derived amorphous oxide TFTs

TFTs, in which all the layers were fabricated using simple chemical solution-processed, vacuum-free routes, followed by thermal annealing at 400°C, were demonstrated [34]. A ruthenium oxide (RuO_2) was used for both gate and source/drain electrodes. Amorphous LZO and zirconium-indium-zinc oxide (ZIZO) films were used as the gate insulator and channel layer, respectively, which enabled the fabrication of a TFT with the desired performance at a sufficiently low temperature. Transfer characteristics of the low-temperature all solution-processed TFT are shown in **Figure 10**. The obtained "on/off" ratio, *SS*-factor, and μ were ~6×10^5, 250 mV/decade, and 5.80 cm^2 V^{-1} s^{-1}, respectively.

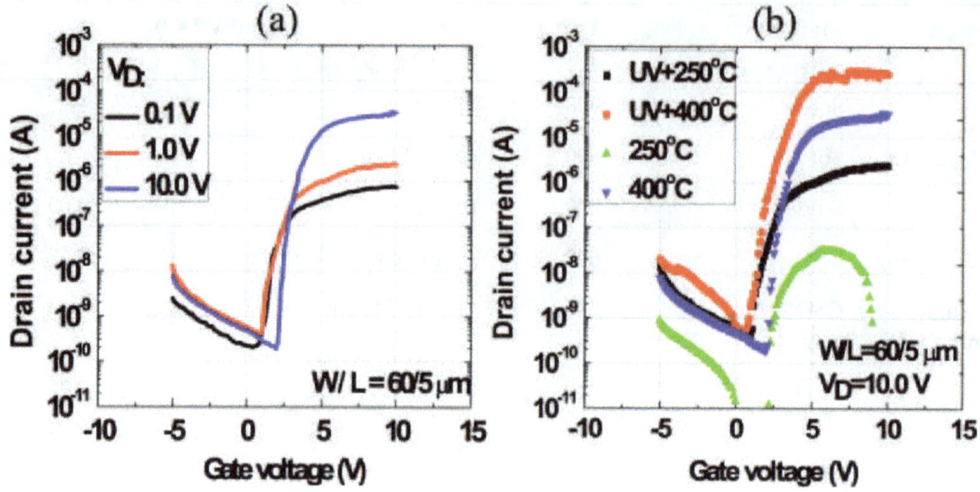

Figure 10. Transfer curves of low-temperature all solution-processed oxide TFT.

5.2. Electrophoretic display driven by all solution process active-matrix oxide TFTs

Precursor solutions and a solution process were developed for the fabrication of active-matrix amorphous oxide-TFTs [102]. The gate lines (GE), gate insulator (GI), and channel layer (CH) comprised RuO_2, LZO, and ZIZO films, respectively. The polysilazane-based silicon dioxide (PSZ) film was used for the channel stopper (CS) layer. The source and drain (S/D) electrode had a double-layer structure comprising ITO and RuO_2 films. The silsesquioxane-based silicon dioxide (SQ) and RuO_2 films were used for the passivation layer (PV) and pixel electrodes (PE), respectively (**Figure 11(a)**). Details of the precursor solutions and annealing conditions for the film formations are summarized in **Table 6**. The TFTs exhibited a μ of 2.68 $cm^2\ V^{-1}\ s^{-1}$, a *SS*-factor of 1.09 V/decade, a V_{th} of 3.06 V, and an "on/off" ratio of 10^5. Electrophoretic displays (EPDs) with a resolution of 101.6 ppi were successfully fabricated using the all solution-processed active-matrix TFTs (**Figure 11(b** and **c)**). Bi-stable black/white images were retained after cutting of the power supply and video signal in these TFT-EPDs (**Figure 11(d)**). It is considered that low-cost electronic paper can be realized using this technology in the near future.

5.3. Direct printing of oxide TFT by nano-rheology printing

Among various solution-based approaches, a direct printing is a promising low-cost technique for fabricating oxide TFTs. The printing technique offers several advantages in manufacturing electronics such as a direct writing of materials, reduction of materials waste, and reproducibility with high resolution, which are not affordable from other solution-based approaches [103–105]. While many printed organic TFTs have been reported, relatively fewer studies, associated with directly printed oxide TFTs, have been pursued [17, 33, 106]. Furthermore, the printing has mainly been applied for metal-oxide semiconductor as the channel layer, while other layers such as GE, S/D, CS, and PV were fabricated by other techniques.

Recently, a new printing method, so-called nano-rheology printing (nRP), for metal oxide patterns within sub-micrometer range, was introduced [33]. The nRP method, which is a type of direct thermal imprinting, uses viscoelastic transformation like glass transition of oxide

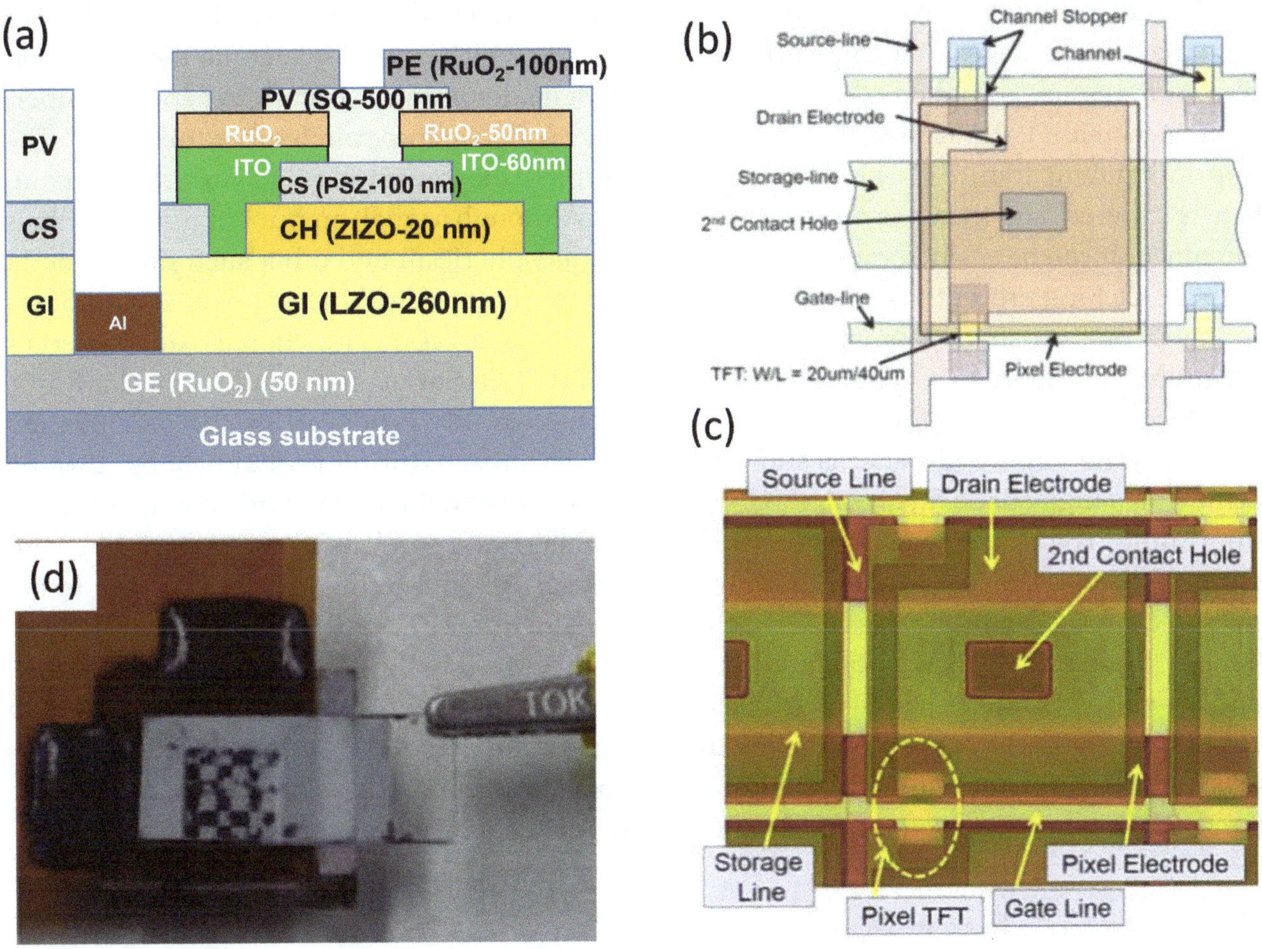

Figure 11. Cross-sectional structure of TFT (a), diagram of the pixel layout (b), and photograph of the pixels (c) in the TFT-driven electrophoretic display. The pixel pitch was 250 × 250 μm^2, which corresponds to a resolution of 101.6 ppi. The aperture ratio was 84.6%.

Layer	Materials	Precursor material	Solvent	Annealing condition
GE	RuO_2	$Ru(NO)(OAc)_3$	MEA, PrA	500°C/10 min
GI	LZO	$La(O_2C_2H_3)_3$, $Zr(OC_4H_9)_4$	PrA	550°C/O_2/10 min
CH	ZIZO	$Zr(OC_4H_9)_4$, $In(OCCH_3CHOCCH_3)_3$, $ZnCl_2$	PrA, 2ME	550°C/O_2/20 min
CS	SiO_2	Polysilazane		500°C/H_2O/30 min
S/D	RuO_2/ITO	$Ru(NO)(OAc)_3$, $In(NO_3)_3{\cdot}3H_2O$, acacH, NH_4Ac, $SnCl_2$	MEA, PrA, 2ME	500°C/air/10 min
PV	SiO_2	Silsesquioxane		500°C/O_2/10 min
PE	RuO_2	$Ru(NO)(OAc)_3$	MEA, PrA	250°C/air/5 min

MEA, 2-aminoethanol; PrA, propionic acid; 2ME, 2-methoxyethanol; acacH, acetylacetonate; NH_4Ac, ammonium acetate.

Table 6. The precursor solutions and annealing conditions for the film formations in all solution-processed TFT-FPD.

precursor gels when imprinted; it softens at a certain temperature during thermal imprinting so that the gel can be rheologically imprinted. The total process for the nRP method is illustrated in **Figure 12**. A thermal nanoimprinting machine (ST-50, Toshiba machine) has been used. First, a solution is coated and dried to make a semi-solid thin film (1). It is then loaded onto the heating stage of the imprinting machine, after which a mold is set onto the semi-solid film and pressure is applied (2). At this point, almost no deformation occurs. When the temperature is increased, the semi-solid film will suddenly soften at a certain temperature (2–3). The imprinting temperature (T_{im}) is maintained to complete the imprinting (3). Next, the temperature is lowered, and then, the mold is detached (4). Although a small amount of residual film is remained, it can be easily removed by etching in atmospheric air and other such simple methods. The etching process slightly reduces the sharpness of the edge but has no other significant influence on the pattern geometry. Well-defined patterns of various functional oxide materials including insulators (ZrO_2, SiO_2), semiconductors (IGZO, SnO_2), and conductors (La-Ru-O, ITO) with the size down to several 10 nm were successfully produced by the nRP method [33, 107]. The imprinted patterns show very small shrinkage during post annealing, thereby achieving a high-shape fidelity to the mold. This results from the metal-oxide condensation at imprinting. The viscoelastic transformation and metal-oxide condensation at imprinting constitute the basics for the nRP method, which is closely related to the cluster structure in the precursor gel. Operation of totally rheology-printed oxide TFT (nRP-TFT) has been demonstrated (**Figure 13**) [94]. The obtained "on/off" ratio, *SS*-factor, and μ were 10^7, 80 mV/decade, and 8.4 $cm^2\ V^{-1}\ s^{-1}$, respectively. Furthermore, excellent TFTs with sub-micron channel length could be completely printed by this method in an air ambient [33]. Active-matrix oxide-TFT array totally printed by the nRP method for display application has also been demonstrated [108], indicating great promise for large-area low-cost printed electronics application.

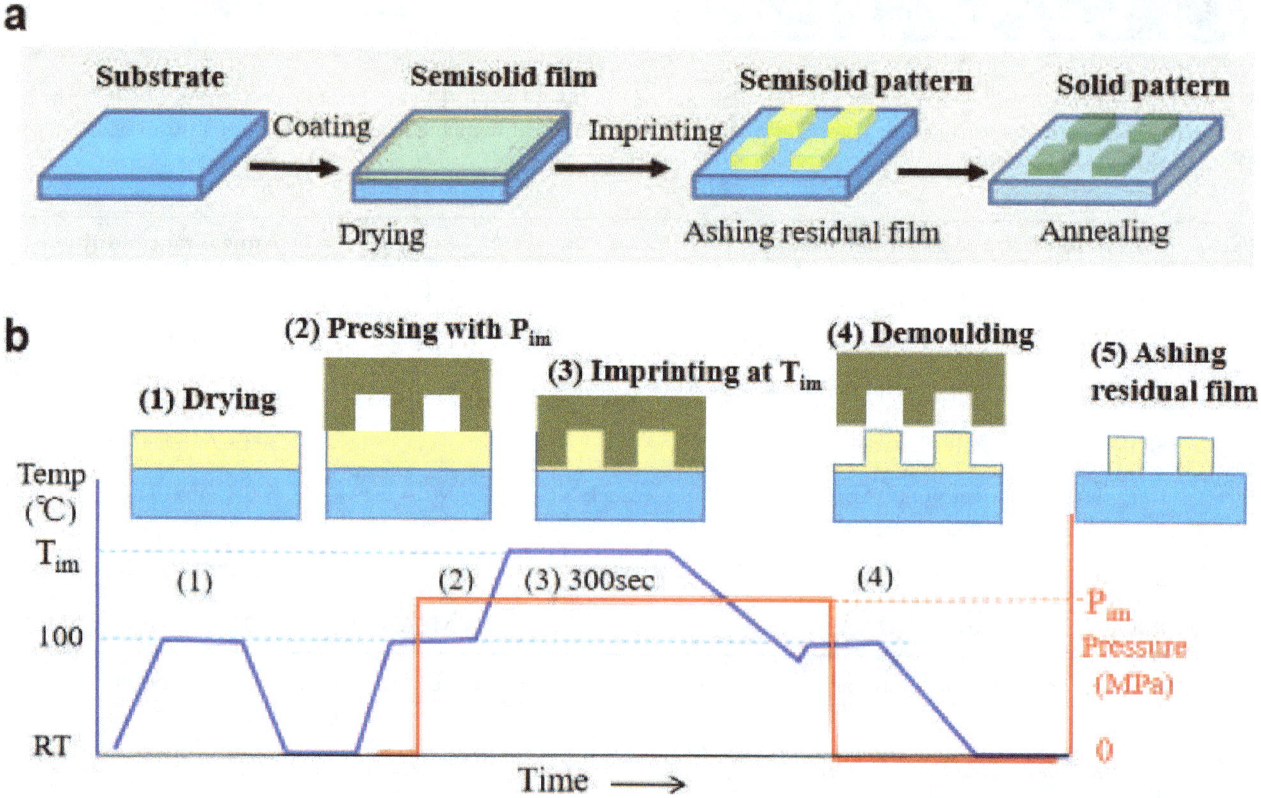

Figure 12. Process of the rheology printing. (a) Total process of rheology printing to form a one-layered pattern. (b) The profile of temperature and pressure during the imprinting process together with a schematic illustration of configurations of the imprinted film and the mold.

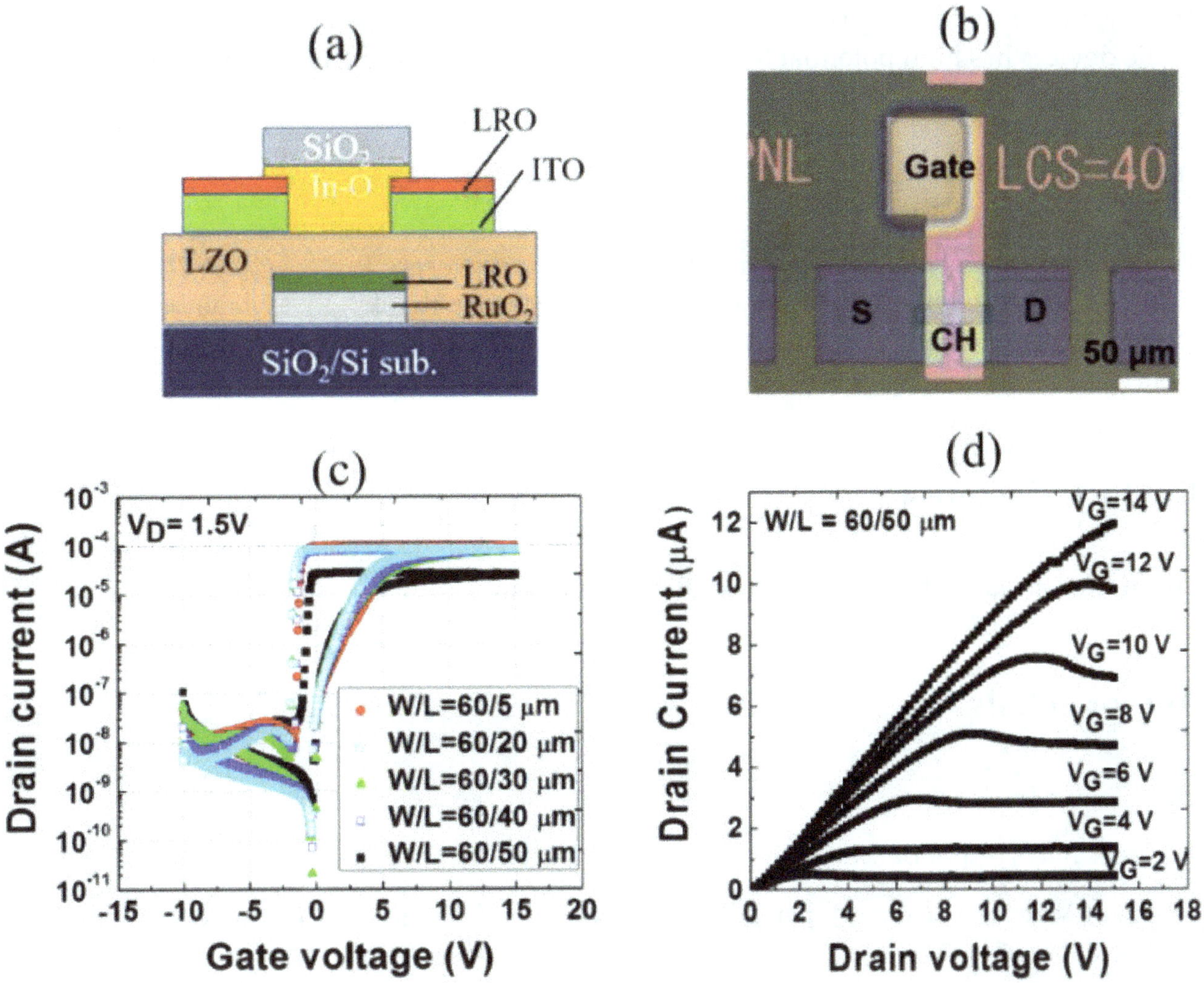

Figure 13. (a) Cross-sectional structure, (b) optical microscope image, (c) transfer curve, and (d) output curve of the nRP-TFT.

6. Conclusion

Solution-processed functional oxide materials deposited at a plastic-compatible temperature play a central role in the emerging printed electronic devices. The chapter critically reviews recent advances in the development of low-temperature solution-processable functional oxide materials including dielectric, semiconducting, and conductive thin-film oxides for printed electronics and highlights the challenges for the fundamental understanding and practical implementation of complex oxides in devices. Properties of oxide films and devices prepared from solution-based processes at a low temperature are now approaching those of vacuum-based counterparts. In addition, ultrahigh resolution printing technologies have been introduced enabling a feasible low-cost fabrication of electronic devices on a large area without the need of sophisticated lithography and vacuum systems. The remaining challenge lies in the production of p-type semiconducting oxide thin film whose performance must be improved to reach that of the n-type counterpart. Once both n-type and p-type oxide TFTs are ready, all-oxide CMOS, a building block of many electronic devices, can be realized.

Together with advanced printing technologies, it is expected that a truly printing of electronic devices has high potential to replace conventional production line in the near future.

Acknowledgements

The author thanks JST-ERATO Shimoda Nano-Liquid Process Project (2007–2014) and JST-CREST (2014–2019) for their financial supports. Special thanks go to members of Green Devices Research Center-JAIST and Center for Single Nano Innovative Devices-JAIST for their fruitful discussions.

Conflict of interest

The author declares that there is no conflict of interest.

Author details

Phan Trong Tue

Address all correspondence to: phan-tt@jaist.ac.jp

School of Materials Science, Japan Advanced Institute of Science and Technology, Nomi-shi, Ishikawa, Japan

References

[1] Yang JJ et al. Memristive switching mechanism for metal/oxide/metal nanodevices. Nature Nanotechnology. 2008;**3**:429

[2] Chen C-H, Suib SL. Control of catalytic activity via porosity, chemical composition, and morphology of nanostructured porous manganese oxide materials. Journal of the Chinese Chemical Society. 2012;**59**(4):465-472

[3] Gich M et al. Multiferroic iron oxide thin films at room temperature. Advanced Materials. 2014;**26**(27):4645-4652

[4] Vrejoiu I et al. Functional perovskites—From epitaxial films to nanostructured arrays. Advanced Functional Materials. 2008;**18**(24):3892-3906

[5] Hiroyuki A. Recent advances and future prospects in functional-oxide nanoelectronics: The emerging materials and novel functionalities that are accelerating semiconductor device research and development. Japanese Journal of Applied Physics. 2013;**52**(10R):100001

[6] Yu X, Marks TJ, Facchetti A. Metal oxides for optoelectronic applications. Nature Materials. 2016;**15**:383

[7] Glynn C, O'Dwyer C. Solution processable metal oxide thin film deposition and material growth for electronic and photonic devices. Advanced Materials Interfaces. 2017;**4**(2):1600610

[8] Nikam V, Bhuwalka KK, Kottantharayil A. Optimization of n-channel tunnel FET for the sub-22nm gate length regime. In: 2008 Device Research Conference. 2008

[9] Le Royer C. Interfaces and performance: What future for nanoscale Ge and SiGe based CMOS? Microelectronic Engineering. 2011;**88**(7):1541-1548

[10] Hu C. 3D FinFET and other sub-22nm transistors. In: 2012 19th IEEE International Symposium on the Physical and Failure Analysis of Integrated Circuits. 2012

[11] Hoffman RL, Norris BJ, Wager JF. ZnO-based transparent thin-film transistors. Applied Physics Letters. 2003;**82**(5):733-735

[12] Chiang HQ et al. High mobility transparent thin-film transistors with amorphous zinc tin oxide channel layer. Applied Physics Letters. 2005;**86**(1):013503

[13] Carcia PF, McLean RS, Reilly MH. High-performance ZnO thin-film transistors on gate dielectrics grown by atomic layer deposition. Applied Physics Letters. 2006;**88**(12):123509

[14] Yabuta H et al. High-mobility thin-film transistor with amorphous InGaZnO4 channel fabricated by room temperature rf-magnetron sputtering. Applied Physics Letters. 2006;**89**(11):112123

[15] Norris BJ et al. Spin-coated zinc oxide transparent transistors. Journal of Physics D: Applied Physics. 2003;**36**(20):L105

[16] David B, Mitzi MC, Murray CE, Kosbar L, Afzali A. Thin-film transistors based on spin-coated chalcogenide semiconductor channels. Proceedings-Electrochemical Society. 2005;**15**:10

[17] Lee DH et al. A general route to printable high-mobility transparent amorphous oxide semiconductors. Advanced Materials. 2007;**19**(6):843-847

[18] Srinivasan R et al. Crystallization and phase transformation process in zirconia: An in situ high-temperature X-ray diffraction study. Journal of the American Ceramic Society. 1992;**75**(5):1217-1222

[19] Ushakov SV, Brown CE, Navrotsky A. Effect of La and Y on crystallization temperatures of hafnia and zirconia. Journal of Materials Research. 2004;**19**(3):693-696

[20] John R. High dielectric constant gate oxides for metal oxide Si transistors. Reports on Progress in Physics. 2006;**69**(2):327

[21] Wilk GD, Wallace RM, Anthony JM. High-κ gate dielectrics: Current status and materials properties considerations. Journal of Applied Physics. 2001;**89**(10):5243-5275

[22] Wang L et al. High-performance transparent inorganic–organic hybrid thin-film n-type transistors. Nature Materials. 2006;**5**:893

[23] Ju S et al. Low operating voltage single ZnO nanowire field-effect transistors enabled by self-assembled organic gate nanodielectrics. Nano Letters. 2005;**5**(11):2281-2286

[24] Kukli K et al. Dielectric properties of zirconium oxide grown by atomic layer deposition from iodide precursor. Journal of the Electrochemical Society. 2001;**148**(12):F227-F232

[25] Park YM et al. Solution-processable zirconium oxide gate dielectrics for flexible organic field effect transistors operated at low voltages. Chemistry of Materials. 2013; **25**(13):2571-2579

[26] Yamada H et al. MOCVD of high-dielectric-constant lanthanum oxide thin films. Journal of the Electrochemical Society. 2003;**150**(8):G429-G435

[27] Yang C, Hang L, Xu J. Structural, optical and dielectric properties of HfSiO films prepared by co-evaporation method. Materials Science in Semiconductor Processing. 2015;**29**(Supplement C):321-325

[28] Talbot E et al. Atomic scale microstructures of high-k HfSiO thin films fabricated by magnetron sputtering. Materials Science and Engineering: B. 2012;**177**(10):717-720

[29] Yusuke O et al. HfO_2/Si and HfSiO/Si structures fabricated by oxidation of metal thin films. Japanese Journal of Applied Physics. 2009;**48**(5S1):05DA01

[30] Sokolov AS et al. Comparative study of Al_2O_3, HfO_2, and HfAlOx for improved self-compliance bipolar resistive switching. Journal of the American Ceramic Society. 2017;**100**(12):5638-5648

[31] Mikhelashvili V et al. Characteristics of metal-insulator-semiconductor capacitors based on high-k HfAlO dielectric films obtained by low-temperature electron-beam gun evaporation. Applied Physics Letters. 2004;**85**(24):5950-5952

[32] Li J et al. Hybrid cluster precursors of the LaZrO insulator for transistors: Properties of high-temperature-processed films and structures of solutions, gels, and solids. Scientific Reports. 2016;**6**:29682

[33] Kaneda T et al. Rheology printing for metal-oxide patterns and devices. Journal of Materials Chemistry C. 2014;**2**(1):40-49

[34] Tue PT et al. Low-temperature all-solution-derived amorphous oxide thin-film transistors. IEEE Electron Device Letters. 2013;**34**(12):1536-1538

[35] Tue PT et al. High-performance solution-processed ZrInZnO thin-film transistors. IEEE Transactions on Electron Devices. 2013;**60**(1):320-326

[36] Kim Y-H et al. Flexible metal-oxide devices made by room-temperature photochemical activation of sol–gel films. Nature. 2012;**489**:128

[37] Ridley BA, Nivi B, Jacobson JM. All-inorganic field effect transistors fabricated by printing. Science. 1999;**286**(5440):746-749

[38] Byrne PD, Facchetti A, Marks TJ. High-performance thin-film transistors from solution-processed cadmium selenide and a self-assembled multilayer gate dielectric. Advanced Materials. 2008;**20**(12):2319-2324

[39] Nomura K et al. Room-temperature fabrication of transparent flexible thin-film transistors using amorphous oxide semiconductors. Nature. 2004;**432**:488

[40] Mott NF. Silicon dioxide and the chalcogenide semiconductors; similarities and differences. Advances in Physics. 1977;**26**(4):363-391

[41] Narushima S et al. A p-type amorphous oxide semiconductor and room temperature fabrication of amorphous oxide p–n heterojunction diodes. Advanced Materials. 2003;**15**(17):1409-1413

[42] Hosono H. Ionic amorphous oxide semiconductors: Material design, carrier transport, and device application. Journal of Non-Crystalline Solids. 2006;**352**(9):851-858

[43] Kenji N et al. Amorphous oxide semiconductors for high-performance flexible thin-film transistors. Japanese Journal of Applied Physics. 2006;**45**(5S):4303

[44] Jeong S et al. Stable aqueous based cu nanoparticle ink for printing well-defined highly conductive features on a plastic substrate. Langmuir. 2011;**27**(6):3144-3149

[45] Jeong S et al. Printed Cu source/drain electrode capped by CuO hole injection layer for organic thin film transistors. Journal of Materials Chemistry. 2011;**21**(29):10619-10622

[46] Faber H et al. Low-temperature solution-processed memory transistors based on zinc oxide nanoparticles. Advanced Materials. 2009;**21**(30):3099-3104

[47] Schneider JJ et al. Synthesis, characterization, defect chemistry, and FET properties of microwave-derived nanoscaled zinc oxide. Chemistry of Materials. 2010;**22**(7):2203-2212

[48] Dasgupta S et al. Inkjet printed, high mobility inorganic-oxide field effect transistors processed at room temperature. ACS Nano. 2011;**5**(12):9628-9638

[49] Jeong S et al. Impact of metal salt precursor on low-temperature annealed solution-derived Ga-doped In_2O_3 semiconductor for thin-film transistors. The Journal of Physical Chemistry C. 2011;**115**(23):11773-11780

[50] Jeong WH, Bae JH, Kim HJ. High-performance oxide thin-film transistors using a volatile nitrate precursor for low-temperature solution process. IEEE Electron Device Letters. 2012;**33**(1):68-70

[51] Jeong S et al. Role of gallium doping in dramatically lowering amorphous-oxide processing temperatures for solution-derived indium zinc oxide thin-film transistors. Advanced Materials. 2010;**22**(12):1346-1350

[52] Banger KK et al. Low-temperature, high-performance solution-processed metal oxide thin-film transistors formed by a 'sol–gel on chip' process. Nature Materials. 2010;**10**:45

[53] Meyers ST et al. Aqueous inorganic inks for low-temperature fabrication of ZnO TFTs. Journal of the American Chemical Society. 2008;**130**(51):17603-17609

[54] Lin Y-H et al. High-performance ZnO transistors processed via an aqueous carbon-free metal oxide precursor route at temperatures between 80-180°C. Advanced Materials. 2013;**25**(31):4340-4346

[55] Kim M-G et al. Low-temperature fabrication of high-performance metal oxide thin-film electronics via combustion processing. Nature Materials. 2011;**10**:382

[56] Jun T et al. High-performance low-temperature solution-processable ZnO thin film transistors by microwave-assisted annealing. Journal of Materials Chemistry. 2011; **21**(4):1102-1108

[57] Carlos E et al. UV-mediated photochemical treatment for low-temperature oxide-based thin-film transistors. ACS Applied Materials & Interfaces. 2016;**8**(45):31100-31108

[58] Heo JS et al. Photochemically activated flexible metal-oxide transistors and circuits using low impurity aqueous system. IEEE Electron Device Letters. 2015;**36**(2):162-164

[59] Jaisutti R et al. Low-temperature photochemically activated amorphous indium-gallium-zinc oxide for highly stable room-temperature gas sensors. ACS Applied Materials & Interfaces. 2016;**8**(31):20192-20199

[60] Tsay C-Y, Liang S-C. Ultraviolet-assisted annealing for low-temperature solution-processed p-type gallium tin oxide (GTO) transparent semiconductor thin films. Materials Science in Semiconductor Processing. 2017;**71**:441-446

[61] Park SC et al. Advanced photo-annealing of indium zinc oxide films for thin-film transistors using pulse UV light. Journal of Information Display. 2016;**17**(1):1-7

[62] Rim YS et al. Simultaneous modification of pyrolysis and densification for low-temperature solution-processed flexible oxide thin-film transistors. Journal of Materials Chemistry. 2012;**22**(25):12491-12497

[63] Chen C et al. Solution-processed metal oxide arrays using femtosecond laser ablation and annealing for thin-film transistors. Journal of Materials Chemistry C. 2017;**5**(36):9273-9280

[64] Sun B, Sirringhaus H. Solution-processed zinc oxide field-effect transistors based on self-assembly of colloidal nanorods. Nano Letters. 2005;**5**(12):2408-2413

[65] Park SY et al. Low-temperature, solution-processed and alkali metal doped ZnO for high-performance thin-film transistors. Advanced Materials. 2012;**24**(6):834-838

[66] Han S-Y, Herman GS, Chang C-h. Low-temperature, high-performance, solution-processed indium oxide thin-film transistors. Journal of the American Chemical Society. 2011;**133**(14):5166-5169

[67] Oh H, Yang J-H, Hwang C-S. Ultralow-temperature sol-gel route to metal oxide semiconductors for soft platforms. Advanced Materials Interfaces. 2016;**3**(20):1600664

[68] Leppäniemi J et al. Rapid low-temperature processing of metal-oxide thin film transistors with combined far ultraviolet and thermal annealing. Applied Physics Letters. 2014;**105**(11):113514

[69] Minami T. New n-type transparent conducting oxides. MRS Bulletin. 2000;**25**(8):38-44

[70] Fraser DB, Cook HD. Highly conductive, transparent films of sputtered $In_{2-x}Sn_xO_{3-y}$. Journal of the Electrochemical Society. 1972;**119**(10):1368-1374

[71] Kim N-R et al. Enhanced conductivity of solution-processed indium tin oxide nanoparticle films by oxygen partial pressure controlled annealing. Journal of Materials Chemistry C. 2013;**1**(37):5953-5959

[72] Chen Z et al. Fabrication of highly transparent and conductive indium–tin oxide thin films with a high figure of merit via solution processing. Langmuir. 2013;**29**(45):13836-13842

[73] Kanehara M et al. Indium tin oxide nanoparticles with compositionally tunable surface plasmon resonance frequencies in the near-IR region. Journal of the American Chemical Society. 2009;**131**(49):17736-17737

[74] Lee J et al. A facile solution-phase approach to transparent and conducting ITO nanocrystal assemblies. Journal of the American Chemical Society. 2012;**134**(32):13410-13414

[75] Toki M, Aizawa M. Sol-gel formation of ITO thin film from a sol including ITO powder. Journal of Sol-Gel Science and Technology. 1997;**8**(1):717-720

[76] Ito T, Uchiyama H, Kozuka H. Evaporation-driven deposition of ITO thin films from aqueous solutions with low-speed dip-coating technique. Langmuir. 2017;**33**(21):5314-5320

[77] Tue PT et al. Combustion synthesized indium-tin-oxide (ITO) thin film for source/drain electrodes in all solution-processed oxide thin-film transistors. Applied Physics A. 2016;**122**(6):623

[78] Ban S-G et al. Low-temperature postfunctionalization of highly conductive oxide thin-films toward solution-based large-scale electronics. ACS Applied Materials & Interfaces. 2017;**9**(31):26191-26200

[79] Koo B-R, Bae J-W, Ahn H-J. Low-temperature conducting performance of transparent indium tin oxide/antimony tin oxide electrodes. Ceramics International. 2017;**43**(8):6124-6129

[80] Asakuma N et al. Low-temperature synthesis of ITO thin films using an ultraviolet laser for conductive coating on organic polymer substrates. Journal of Sol-Gel Science and Technology. 2003;**27**(1):91-95

[81] Ota R et al. Fabrication of indium-tin-oxide films by dip coating process using ethanol solution of chlorides and surfactants. Thin Solid Films. 2002;**411**(1):42-45

[82] Seki S et al. Highly conducting indium-tin-oxide transparent films prepared by dip-coating with an indium carboxylate salt. Surface and Coatings Technology. 2003;**169-170**:525-527

[83] Alam MJ, Cameron DC. Preparation and properties of transparent conductive aluminum-doped zinc oxide thin films by sol-gel process. Journal of Vacuum Science & Technology A: Vacuum, Surfaces, and Films. 2001;**19**(4):1642-1646

[84] Nanto H et al. Electrical and optical properties of zinc oxide thin films prepared by rf magnetron sputtering for transparent electrode applications. Journal of Applied Physics. 1984;**55**(4):1029-1034

[85] Fortunato E, Barquinha P, Martins R. Oxide semiconductor thin-film transistors: A review of recent advances. Advanced Materials. 2012;**24**(22):2945-2986

[86] Liu H et al. Transparent conducting oxides for electrode applications in light emitting and absorbing devices. Superlattices and Microstructures. 2010;**48**(5):458-484

[87] Beyer W, Hüpkes J, Stiebig H. Transparent conducting oxide films for thin film silicon photovoltaics. Thin Solid Films. 2007;**516**(2):147-154

[88] Kamiya T et al. Electrical properties and structure of p-type amorphous oxide semiconductor $xZnO\cdot Rh_2O_3$. Advanced Functional Materials. 2005;**15**(6):968-974

[89] Kawazoe H et al. P-type electrical conduction in transparent thin films of $CuAlO_2$. Nature. 1997;**389**:939

[90] Kawazoe H et al. Transparent p-type conducting oxides: Design and fabrication of p-n heterojunctions. MRS Bulletin. 2000;**25**(8):28-36

[91] Kim S et al. Room temperature deposited oxide p-n junction using p-type zinc-cobalt-oxide. Journal of Applied Physics. 2010;**107**(10):103538

[92] Aksit M et al. Misfit layered $Ca_3Co_4O_9$ as a high figure of merit p-type transparent conducting oxide film through solution processing. Applied Physics Letters. 2014;**104**(16):161901

[93] Li J et al. P-type conductive amorphous oxides of transition metals from solution processing. Applied Physics Letters. 2012;**101**(5):052102

[94] Tue PT, Fukada K, Shimoda T. High-performance oxide thin film transistor fully fabricated by a direct rheology-imprinting. Applied Physics Letters. 2017;**111**(22):223504

[95] Li J et al. Highly conductive p-type amorphous oxides from low-temperature solution processing. Applied Physics Letters. 2012;**101**(13):132104

[96] Götzendörfer S, Bywalez R, Löbmann P. Preparation of p-type conducting transparent $CuCrO_2$ and $CuAl_{0.5}Cr_{0.5}O_2$ thin films by sol-gel processing. Journal of Sol-Gel Science and Technology. 2009;**52**(1):113-119

[97] Chen H-Y, Wu J-H. Transparent conductive $CuFeO_2$ thin films prepared by sol–gel processing. Applied Surface Science. 2012;**258**(11):4844-4847

[98] Chiu T-W et al. Preparation of p-type conductive transparent $CuCrO_2$:Mg thin films by chemical solution deposition with two-step annealing. Ceramics International. 2012;**38**:S673-S676

[99] Wang J et al. Combustion synthesis of p-type transparent conducting $CuCrO_{2+x}$ and Cu:CrO_x thin films at 180 °C. In: ACS Applied Materials & Interfaces. 2018

[100] Farrell L et al. Spray pyrolysis growth of a high figure of merit, nano-crystalline, p-type transparent conducting material at low temperature. Applied Physics Letters. 2015;**107**(3):031901

[101] Liu Y, Pollaor S, Wu Y. Electrohydrodynamic processing of p-type transparent conducting oxides. Journal of Nanomaterials. 2015;**2015**:14

[102] Inoue S et al. Electrophoretic displays driven by all-oxide thin-film transistor backplanes fabricated using a solution process. Physica Status Solidi (a). 2015;**212**(10):2133-2140

[103] Sirringhaus H et al. High-resolution inkjet printing of all-polymer transistor circuits. Science. 2000;**290**(5499):2123-2126

[104] Wang JZ et al. Dewetting of conducting polymer inkjet droplets on patterned surfaces. Nature Materials. 2004;**3**:171

[105] Shimoda T et al. Solution-processed silicon films and transistors. Nature. 2006;**440**:783

[106] Kim D et al. Inkjet-printed zinc tin oxide thin-film transistor. Langmuir. 2009;**25**(18): 11149-11154

[107] Nagahara K et al. Rheology printing of an ultra-fine conductive Ru–La–O line. Ceramics International. 2016;**42**(6):7730-7741

[108] Hirose D et al. All-solution-printed oxide thin-film transistors by direct thermal nano-imprinting for use in active-matrix arrays. Physica Status Solidi (a). 2017;**214**(1):1600397

9

Study of Heat Dissipation Mechanism in Nanoscale MOSFETs Using BDE Model

Houssem Rezgui, Faouzi Nasri,
Mohamed Fadhel Ben Aissa and
Amen Allah Guizani

Abstract

In this chapter, we report the nano-heat transport in metal-oxide-semiconductor field effect transistor (MOSFET). We propose a ballistic-diffusive model (BDE) to inquire the thermal stability of nanoscale MOSFET's. To study the mechanism of scattering in the interface oxide-semiconductor, we have included the specularity parameter defined as the probability of reflection at boundary. In addition, we have studied the effective thermal conductivity (ETC) in nanofilms we found that ETC depend with the size of nanomaterial. The finite element method (FEM) is used to resolve the results for a 10 nm channel length. The results prove that our proposed model is close to those results obtained by the Boltzmann transport equation (BTE).

Keywords: nano-devices, thermal conductivity, BDE model, temperature jump, heat dissipation

1. Introduction

Considerable exploration focused on the fast progress of nanodevice. In recent years, the thermal analysis is important to inquire the phonon transport in nano-materials. The study of the heat conduction in nano-electronics lead to compare the thermal stability of nano-transistors [1–3]. The increase of the heat dissipation has been owned by the miniaturization and the reduction of the thermal conductivity [3]. The smaller channel device was assumed to

(13) nm for the current (2018) and less than (6) nm for long-term (2026) [4]. The Fourier's law has generally used to predict the diffusive heat conduction [5]. In nanoscale the characteristic time and size of nanodevice was smaller than the mean free path (MFP). The classical heat conduction based on local equilibrium lead to the linear equation

$$q = -\kappa \nabla T \tag{1}$$

where q is the heat flux, ∇T is the temperature gradient and κ is the bulk thermal conductivity. Actually, the BTE is an effective method to study the non-continuous temperature and heat flux in nanosystems [6, 7]. Many transport models have been derived from the BTE used to investigate the thermal transport in solid interface [8], nano-transistor [9–12] and carbon nanotubes [13].

Alvarez et al. [14] have studied the nonlocal effect in nanoscale devices. They inquired the heat transport in ballistic regime. They found that the thermal conductivity bank on the Knudsen number.

Nano-heat transport includes both temporal aspects and spatial aspects. More elaborated model have been developed to describe the nanoheat conduction [15]. The phonon hydrodynamic model [16–19], single- phase-lag model [2] and dual-lag phase model [1] have been applied in modeling thermal transport in nanostructures. Nasri et al. [2, 3] have been investigated the heat transfer in many architecture of nano-MOSFET. They found that the Tri-gate SOI-MOSFET with a wideness of 20 nm is more thermally stable than the device having a length of 10 nm. To study the nature of collision, it is found that the temperature jump boundary condition was an accurate approach to explore the heat transport in interfaces [2]. The ballistic-diffusive equation (BDE) was used to explain the temperature dependence in nano-structure [20, 21]. Humian et al. [22] proposed the BDE to evaluate the heat transport in two-dimensional domain. They have been used the finite element analysis to validate the BDE model. Yang et al. [23] solved the BDE model to access the heat transfer in two-dimensional conventional MOSFET. In this work, we have been developed the BDE model to address the phonon transport in nanodevice. Due to the miniaturization the thermal conductivity, reduce by scattering [24, 25]. The scattering mechanism induced to the use of the ETC [26–29]. We have proposed a theoretical approach, which describe the nature of phonon collision with boundary. The specularity parameter defined as the probability of reflection at boundary [25, 30]. We include this parameter in the ETC to portend the rise of the temperature in nano-structure. To validate our results, the proposed model is tested with results obtained by Yang et al. [23] and a previous work [2, 9]. The proposed (ETC) will be compared with the results obtained by McGaughey et al. [31]. To compute the proposed BDE model depended with the temperature jump boundary condition, we have used the FEM. This method is a useful procedure to model the thermal properties of nanodevice [2].

2. Computer model

The phonon BTE can be defined as [7]:

$$\frac{\partial f(r,v,t)}{\partial t}+v\nabla f(r,v,t)=-\frac{f-f_0}{\tau_R} \tag{2}$$

where f is the distribution function, f^0 is the equilibrium distribution function, v is the group velocity, and τ_R is the relaxation time related to resistive collisions written as

$$\tau_R=\frac{3\times k}{C\times v^2} \tag{3}$$

where κ is the thermal conductivity written as

$$\kappa=\frac{C\times v\times \Lambda}{3} \tag{4}$$

where Λ is the mean free path defined as $\Lambda = v\tau_R$ and C is the volumetric heat capacity [20]. The ballistic-diffusive approximation is to divide the distribution function into a diffusive term f_m and ballistic term f_b [20]:

$$f=f_m+f_b \tag{5}$$

where f_b arise from the boundary scattering [20], defined as

$$\frac{\partial f_b(r,v,t)}{\partial t}+v\nabla f_b(r,v,t)=-\frac{f_b}{\tau_R} \tag{6}$$

The second part grouped into f_m. The basic equation for f_m is defined as:

$$\frac{\partial f_m(r,v,t)}{\partial t}+v\nabla f_m(r,v,t)=-\frac{f_m-f_0}{\tau_R} \tag{7}$$

We can calculate the diffusive flux [7, 20].

$$q_m(t,r)=\int_{\varepsilon} v(r,t)f_m(r,\varepsilon,t)\varepsilon \mathrm{D}(\varepsilon)d\varepsilon \tag{8}$$

where ε is the kinetic energy and $D(\varepsilon)$ is the density of states. By the development in Taylor series to the first order of Eq. (7), we obtain [7]:

$$\tau_R\frac{\partial q_m(r,t)}{\partial t}+q_m(r,t)=-\kappa\nabla T_m(r,t) \tag{9}$$

We use the energy conservation equation to eliminate q_m

$$-\nabla.q(r,t)+\dot{q}_h=\frac{\partial u(r,t)}{\partial t} \tag{10}$$

where $\dot{q}_h$ is the volumetric heat generation, q is the heat flux and u is the internal energy defined as [20].

$$q(t,r) = q_b(t,r) + q_m(t,r)$$
$$u(t,r) = u_b(t,r) + u_m(t,r) \quad (11)$$

We can rewritten two temperature T_b and T_m such that [20, 21].

$$\frac{\partial u}{\partial t} = C\frac{\partial T}{\partial t} = \frac{\partial u_m}{\partial t} + \frac{\partial u_b}{\partial t} = C\left(\frac{\partial T_m}{\partial t} + \frac{\partial T_b}{\partial t}\right) \quad (12)$$

where $T = T_m + T_b$

T_m is the temperature to the diffusive part and T_b arise from the ballistic parts.

By using the same reasoning of Eq. (9), Eq. (6) becomes [20].

$$\tau_R \frac{\partial^2 T_b(r,t)}{\partial t^2} + C\frac{\partial T_b}{\partial t}(r,t) = -\tau_R \times \frac{\partial(\nabla . q_b(r,t))}{\partial t} \quad (13)$$

Substituting Eqs. (9) and (13) into Eq. (10) we obtain the ballistic-diffusive-equation [20, 23].

$$\tau_R \frac{\partial^2 T_m(r,t)}{\partial t^2} + \frac{\partial T_m(r,t)}{\partial t} = \frac{\kappa}{C}\nabla\nabla T_m(r,t) - \frac{\nabla q_m(r,t)}{C} + \frac{\dot{q}_h}{C} + \frac{\tau_R}{C}\frac{\partial \dot{q}_h}{\partial t} \quad (14)$$

The conventional Fourier heat conduction equation cannot predict the heat transport in nanostructure. Hua et al. [26] studied the ETC in nanostructure. They derived a model for the ETC based on the phonon BTE written as [26]:

$$\kappa_{eff} = \kappa/(1 + \alpha \times Kn) \quad (15)$$

where $Kn = \frac{\Lambda}{L}$ is the Knudsen number, L is the length of nanofilms and α is a coefficient depend with the geometries.

For $Kn = 0$, Eq. (15) becomes $\kappa_{eff} = \kappa$ (diffusive regime).

For $Kn > 1$, the thermal conductivity reduce due to the ballistic transport. Using the Fourier's law the ETC defined as [26]:

$$\kappa_{eff} = \frac{q \times L}{\Delta T} \quad (16)$$

where L is length of the nanostructure q is the heat flux and ΔT is the temperature difference. Kaiser et al. [28] proposed a non-Fourier heat conduction at the nanoscale. They recently derived an analytic expression for the ETC. In addition, they proved the impact of the temperature jump in nanostructure. In this work, we propose a theoretical model for the ETC [19], defined as

$$k_{eff}(Kn) = \kappa\left[1 - \frac{2Kn \times \tanh(1/2Kn)}{1 + C_W \times \tanh(1/2Kn)}\right] \quad (17)$$

where $C_W = 2 \times \left(\frac{1+p}{1-p}\right)$ is a constant related to the properties of the walls [19] and p is the specularity parameter [30].

For high values of Kn, where the regime ballistic is dominant, the thermal conductivity reduced by scattering (reflection at boundary) [24, 25], Eq. (17) predict that the ETC behaves [19]:

$$\kappa_{eff}(Kn) = \kappa\left(\frac{C_W}{2Kn}\right) \tag{18}$$

In our case, the BDE model rewritten as:

$$\tau_b \frac{\partial^2 T_m(r,t)}{\partial t^2} + \frac{\partial T_m(r,t)}{\partial t} = \frac{\kappa_{eff}}{C} \nabla\nabla T_m(r,t) - \frac{\nabla q_b(r,t)}{C} + \frac{\dot{q}_h}{C} + \frac{\tau_b}{C}\frac{\partial \dot{q}_h}{\partial t} \tag{19}$$

where τ_b is the relaxation time related to the phonon scattering at boundary [7] defined as:

$$\tau_b = \frac{3 \times \kappa_{eff}}{C \times v^2} \tag{20}$$

Substituting Eq. (18) into Eq. (20), we obtain the Ziman formula [24]

$$\frac{1}{\tau_b} = \left(\frac{1-p}{1+p}\right) \times \left(\frac{v}{L}\right) \tag{21}$$

where $\frac{1}{\tau_b}$ is the collision rate.

3. Boundary and initial condition

When the ballistic transport appear the temperature jump at boundary occur and cause the reduction of the thermal conductivity [27, 32]. Ben Aissa et al. [12] have been explored a nano-heat conduction in cylindrical surrounding-gate (SG) MOSFET. They used the DPL with the temperature jump applied in the interface oxide-semiconductor defined as

$$\Delta T_{Jump} = -d \times Kn \times L \times \nabla T \tag{22}$$

where d is an adjustable coefficient, the ETC defined as [12]:

$$d = \frac{R \times \kappa_{eff}}{Kn \times L_c} \tag{23}$$

where R is the thermal boundary resistance. The proposed ETC given by Ben Aissa et al. [12, 32] is written as:

$$\kappa_{eff} = \kappa/(1 + (4 \times Kn)) \tag{24}$$

Hua et al. [33, 34] discussed the temperature jump in nanofilms. They have studied the phonon transport in interfaces. They derived a boundary temperature jump defined as

Work	Adjustable coefficient d
Ben Aissa et al. [12]	0.05 For $Kn = 3.33$
Hua and Cao [33]	0.66 For $Kn < 5$
Sing et al. [36]	$\frac{1.25}{\mathrm{Pr}}$ For $Kn < 0.1$
Present work	0.09 For $Kn = 10$

Table 1. First-order temperature jump condition.

$$T - T_W = -d \times \Lambda \times \frac{\partial T}{\partial x} \tag{25}$$

where T_W is the temperature jump at the wall.

Yang et al. [35] explained the impact of the temperature jump in the continuum flow and slip flow. They have investigated the heat transfer in nanofluids. At the wall, the temperature jump lead to the following expression by Gad-el-Hak [35]

$$T_S - T_W = \frac{2\beta}{\beta + 1} \frac{2 - \sigma_T}{\sigma_T} \frac{\Lambda}{\mathrm{Pr}} \frac{dT}{dy}\bigg|_{Wall} \tag{26}$$

where T_S is the system temperature and T_W is the wall temperature, σ_T is the thermal accommodation coefficient, β is the ratio of specific heats, and Pr is the gas Prandtl number. Singh et al. [36] noted that in ideal monoatomic gas ($Kn < 0.1$), the temperature jump at the solid interface rewritten as

$$T_S - T_W = \frac{1.25 \times \Lambda}{\mathrm{Pr}} \frac{dT}{dy}\bigg|_{Wall} \tag{27}$$

Due to the utility of the temperature jump, the following expressions are summarized in **Table 1**.

4. Structure to model and numerical method

The architecture used in this present work is the two-dimensional conventional MOSFET. The proposed structure shown in **Figure 1**. The substrate is compound by Silicon (Si). The Si-MOSFETs thickness used in this model is 50 nm. The channel length is L_c = 10 nm. In order to compare our results with similar works, the reference temperature is T_0 = 300 K and the maximal power generation is $\dot{q}_h = 10^{19} w/m^3$ [23]. The right and left boundaries are assumed to be adiabatic. The temperature jump boundary condition is applied in the interface (Si-SiO_2). In this side the phenomena of collision phonon-wall is more frequent. The MFP used in this proposed work is $\Lambda = 100\, nm$ [23]. Using Eq. (17), we found that the thermal conductivity reaches 18 $Wm^{-1}\,k^{-1}$. In this case, the adjustable coefficient d attains 0.09 for R = 0.503 K m^2 W^{-1} [37].

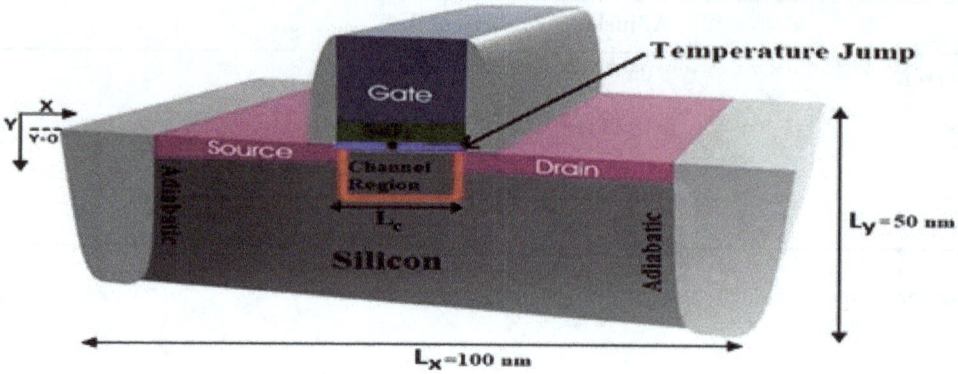

Figure 1. Schematic geometries of the MOSFET transistor.

Symbol	V (m s^{-1})	K (Wm^{-1} K^{-1})	C (J m^{-3} k^{-1})	Λ (nm)	Kn
Si	3000	150	1.5×10^6	100	10
SiO_2	5900	1.4	1.75×10^6	0.4	0.04

Table 2. Thermal properties of Silicon and Silicon dioxide.

To solve the BDE model coupled with the temperature jump at boundary we use the FEM [4]. The finite-element approximation used in the BDE model can be defined as:

$$[B]\{T_t\} + [B_1]\{T_{tt}\} - [D]\{T\} = \{m\} \tag{28}$$

where $[B]$, $[B_1]$ and $[D]$ is a matrix valued, $\{T_t\}$, $\{T_{tt}\}$ and $\{T\}$ represent the nodal temperature, $\{m\}$ is the matrix vector.

The discretization of Eq. (25) leads to

$$\{T_{p+1}\}\left(\frac{[B]}{\Delta t}+\frac{[B]}{\Delta t^2}\right) = \{T_p\}\left(\frac{[B]}{\Delta t}+2\frac{[B]}{\Delta t^2}+[D]\right) - \{T_{p-1}\}\left(\frac{[B]}{\Delta t^2}\right) + \{m\}_p \tag{29}$$

where Δt is the time step and $\{T_p\}$ is the nodal temperature at the time t_p.

The materials used in our simulation are Silicon [2] and Silicon dioxide [2] and their thermal properties are illustrated in **Table 2**.

5. Results and discussion

The reduction of the thermal conductivity have a strong dependence with the Knudsen number. We take account the specularity parameter and Knudsen number because we studied the mechanism of boundary scattering.

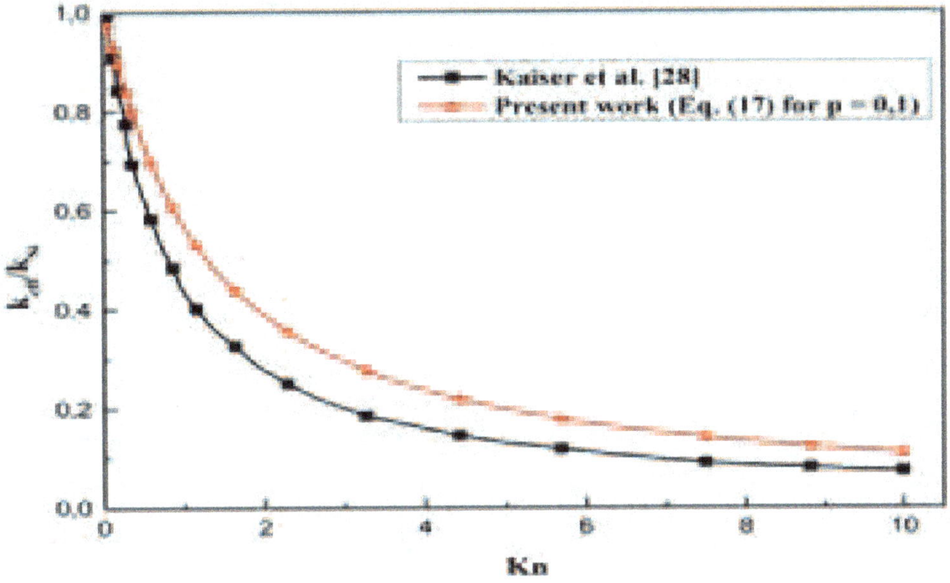

Figure 2. Effective thermal conductivity vs. Knudsen number.

In **Figure 2** we shows the impact of the thermal conductivity which depend on the specularity parameter. In this case, we use Eq. (17) to inquire the ETC. It is obvious that the thermal conductivity reduce when the Knudsen number increase. In similarity to the analytical model proposed by Hua and Cao et al. [32], it is found that the thermal conductivity reaches 62% of the bulk value for Kn = 1.

Figure 3 plot the ETC for various the length of nanofilms. For low thin films (L = 10 nm) the thermal conductivity attains 10–20% of the bulk value. The ballistic transport involve the rapid increase of the ETC. For p = 0.25 we shows the same shape obtained by Ma [29]. For thin films (L > 1000 nm), p = 0.25 is a good approximation.

The advantage of our proposed model is the capture of the increase of the temperature better than the other transport model (DPL, SPL and classical BDE). We associated the BDE model with the temperature jump. The obtained results are presented along the centerline ($L_x/2$, Y = 0) at the time t = 30 ps. In the ballistic regime (Kn = 10) we use Eq. (18). For high Knudsen number the heat transport influenced by the mechanism of scattering related to the boundary. This type of collision was examined by Guo et al. [38] they deduced a discrete-ordinate-method (DOM) derived by the Callaway's model [38]. They found acceptable results to determinate the ETC in a rectangular graphene ribbon. The Callaway's model based on a simple boundary scattering.

Figure 4 illustrate the comparison of the peak temperature rise in the nano-transistor at t = 30 ps. The classical BDE, BTE, DPL, SPL, Fourier law and our proposed model reaches respectively 318.7, 327, 320.5, 318.9, 305 and 322.7 K. The new BDE model capture the increase of the temperature near the BTE. For low thin-film (L_c = 10 nm) one can see that the temperature attaints the maximal at short time. The saturation of the temperature varied to 12–15 ps for all model transport. The classical Fourier law cannot predict the temperature profile due the nature of phonon thermal transport [39].

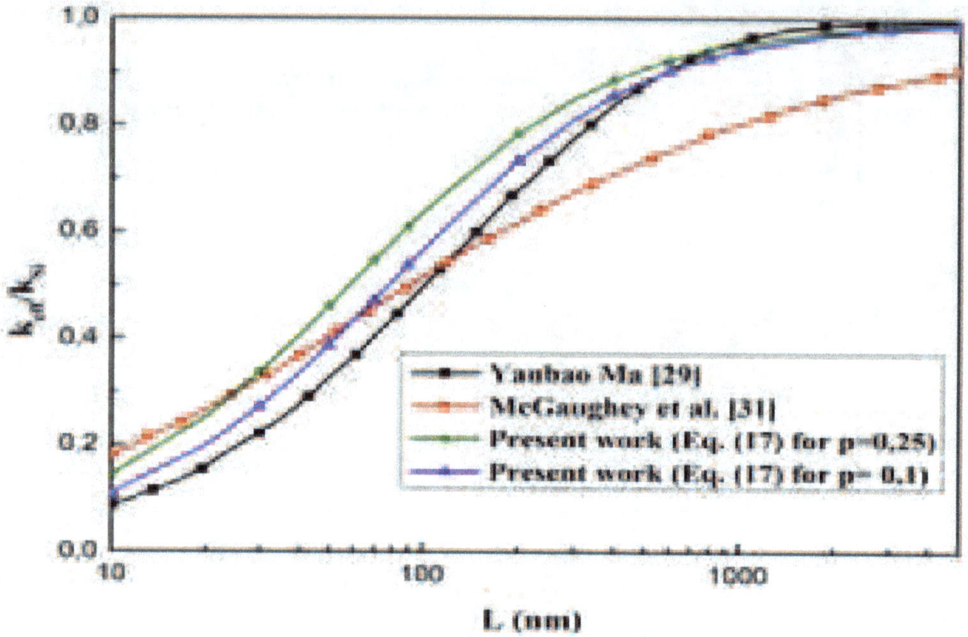

Figure 3. Effective thermal conductivity in Silicon thin film at room temperature.

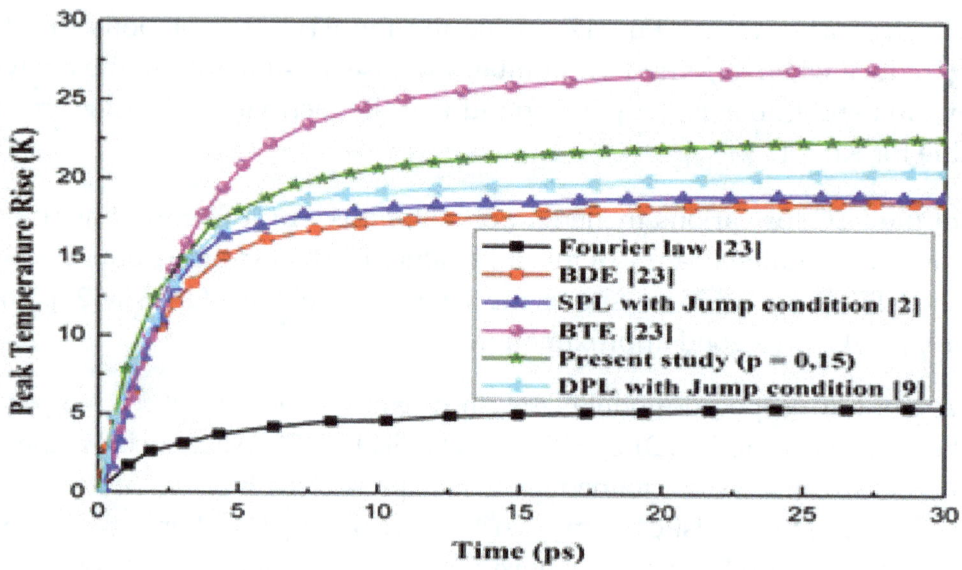

Figure 4. Comparison of the peak temperature at the centerline.

The temperature peak rise in the Y-direction at the centerline of the nanodevice is demonstrated in **Figure 5**. The decrease of the temperature is owned to the reduction of the thermal conductivity. Our present model has the same form with the classical BDE model. The difference appear in low temperature due to the collision rate, which depend on the specularity.

Figure 6 illustrate the 2D distribution of the temperature at t = 30 ps. In a short time, the temperature increase in the left and right side of the channel region. This side known as the heat zone of the nanodevice. A self-heating processes appear due to nature of the phonon

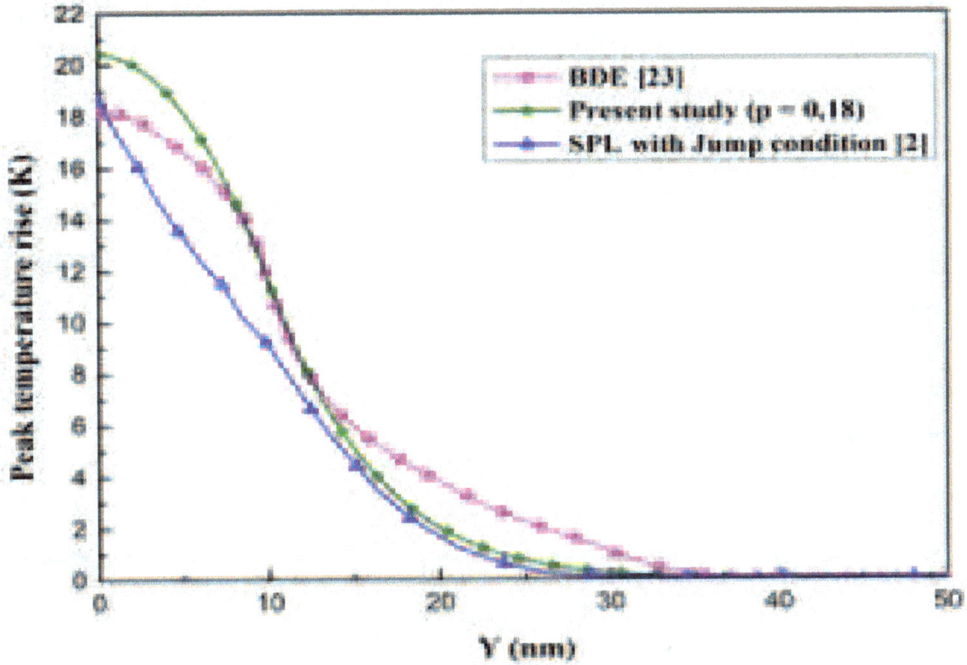

Figure 5. Peak temperature rise versus Y-axis at the centerline of the MOSFET at t = 10 ps.

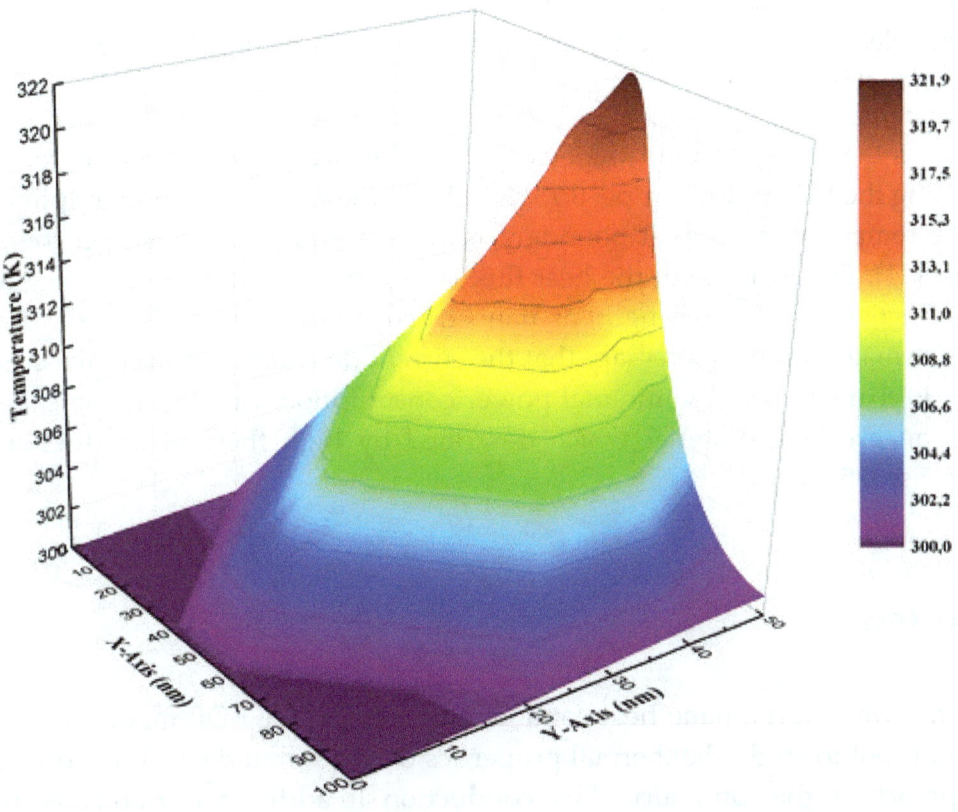

Figure 6. A 2D temperature distribution for p = 0.18 at t =30 ps.

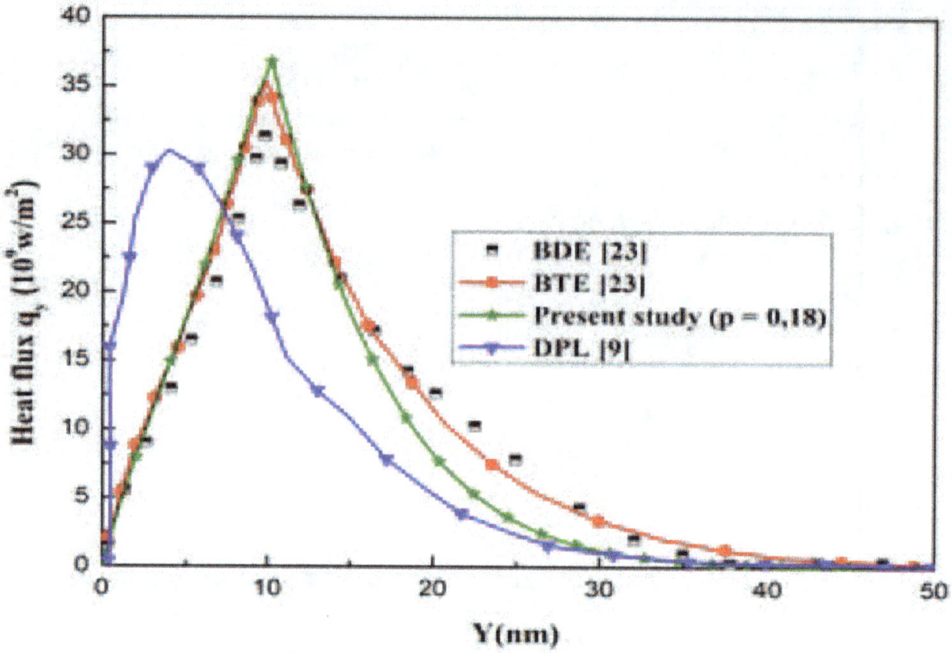

Figure 7. Comparison of the heat flux in the Y direction at t = 10 ps.

collision which is characterized by a frequent scattering at boundary. The augmentation of the temperature cause an important dissipation of energy, which affect the environment. In the last years, organic electronic was made to reduce the energy consumption and the thermal resistance between materials [40].

Figure 7 shows the comparison of the heat flux in the Y-direction at the centerline of nano-MOSFET. Using the BDE model for p = 0.18, we obtain the same shape and amplitude given by the BTE. The temperature jump at boundary is a good argument to predict the non-Fourier heat transfer [41]. The increase of the heat flux is caused by two raison: the reduction of the thermal conductivity and the length of nanostructure. To reduce the heat dissipation in nanoelectronic materials, it is necessary that the current densities was minimized [9]. A preferment devise is characterized by minimal power consumption. In a technological concept, the graphene is an excellent material which described by high thermal conductivity and low temperature rise [25, 42].

6. Conclusions

In this chapter, we report a nano-heat conduction based on the BDE model. The temperature jump is good proof to study the thermal properties of nano-materials. Our proposed model is efficient approach for the non-Fourier heat conduction. In addition, our obtained results agree with other transport model. In nanostructure the reduction of the thermal conductivity and phonon collision mechanism. Our study explain the distribution of the temperature in 10 nm

MOSFET. The maximal temperature is located in the interface oxide-semiconductor. To reduce the effect of the thermal transport in nano-electronic materials it is obvious that we should replace Si based materials by organic technologies (carbon based). Green electronics are involved in recent integrated circuits, solar cell and high-speed processors. In addition, green materials has a wide biocompatibility and a safe impact on the environment [43].

Author details

Houssem Rezgui, Faouzi Nasri*, Mohamed Fadhel Ben Aissa and Amen Allah Guizani

*Address all correspondence to: nasrifaouzi90@yahoo.fr

Laboratory of Thermal Processes, Research and Technology Centre of Energy, Hammam-Lif, Tunisia

References

[1] Nasri F, Ben Aissa MF, Belmabrouk H. Effect of second-order temperature jump in metal-oxide-semiconductor field effect transistor with dual-phase-lag model. Microelectronics Journal. 2015;**46**(1):67-74

[2] Nasri F, Ben Aissa MF, Gazzah MH, Belmabrouk H. 3D thermal conduction in nanoscale tri-gate MOSFET based on single-phase-lag model. Applied Thermal Engineering. 2015; **91**:647-653

[3] Nasri F, Ben Aissa MF, Belmabrouk H. Microscale thermal conduction based on Cattaneo-Vernotte model in silicon on insulator and double gate MOSFETs. Applied Thermal Engineering. 2015;**76**:206-211

[4] Fiori G, Bonaccorso F, lannaccone G, Palacios T, Neumaier D, Seabaugh A, Banerjee S, Colombo L. Electronic based on two-dimensional materials. Nature Nanotechnologie. 2014;**9**. DOI: 10.1038/NNANO.2014.207

[5] Dong Y, Cao BY, Guo ZY. Thermomass theory: A mechanical pathway to analyse anomalous heat conduction in nanomaterials. In: Vakhrushev A, editor. Nanomechanics. Rijeka: InTech; 2017. DOI: 10.5772/67780

[6] Xu M, Li X. The modeling of nanoscale heat conduction by Boltzmann transport equation. International Journal of Heat and Mass Transfer. 2012;**55**:1905-1910

[7] Xu M, Hu H. A ballistic-diffusive heat conduction model extracted from Boltzmann transport equation. Proceedings of the Royal Society A. 2011;**467**:1851-1864

[8] Pisipati S, Chen C, Geer J, Sammakia B, Murray BT. Multiscale thermal device modeling using diffusion in the Boltzmann transport equation. International Journal of Heat and Mass Transfer. 2013;**64**:286-303

[9] Nasri F, Echouchene F, Ben Aissa MF, Graur I, Belmabrouk H. Investigation of Self-heating effects in a 10-nm SOI-MOSFET with an insulator region using electrothermal modeling. IEEE (Institute of Electrical and Electronics Engineers) Transaction on Electron Devices. 2015;**62**:2410-2415

[10] Nasri F, Ben Aissa MF, Belmabrouk H. Nanoheat conduction performance of black phosphorus filed-effect transistor. IEEE (Institute of Electrical and Electronics Engineers) Transaction on Electron Devices. 2017;**64**:2765-2769

[11] Nasri F, Ben Aissa MF, Belmabrouk H. Nonlinear electrothermal model for investigation of heat transfer process in a 22-nm FD-SOI MOSFET. IEEE (Institute of Electrical and Electronics Engineers) Transaction on Electron Devices. 2017;**4**:1461-1466

[12] Ben Aissa MF, Nasri F, Belmabrouk H. Multidimensional nano heat conduction in cylindrical transistors. IEEE (Institute of Electrical and Electronics Engineers) Transaction on Electron Devices. 2017;**64**:5236-5241

[13] Wang HD, Cao BY, Guo ZY. Heat flow in carbon nanotubes. International Journal of Heat and Mass Transfer. 2010;**53**:1796-1800

[14] Alvarez FX, Jou D. Memory and nonlocal effects in heat transport: From diffusive to ballistic regimes. Applied Physics Letters. 2007;**90**:083109

[15] Tzou DY. Macro-to Microscale Heat Transfer: The Lagging Behavior. Chichester: John Wiley & Sons; 2014

[16] Alvarez FX, Jou D, Sellitto A. Phonon hydrodynamics and phonon-boundary scattering in nanosystems. Journal of Applied Physics. 2009;**105**(1):014317

[17] Jou D, Criado-Sancho M, Casas-Vázquez J. Heat fluctuations and phonon hydrodynamics in nanowires. Journal of Applied Physics. 2010;**107**(8):084308

[18] Guo Y, Wang M. Phonon hydrodynamics for nanoscale heat transport at ordinary temperatures. Physical Review B. 2018;**97**(3):035421

[19] Sellito A, Carlomango I, Jou D. Two-dimensional phonon hydrodynamics in narrow strips. Proceedings of the Royal Society A. 2015;**471**:20150376

[20] Chen G. Ballistic-diffusive heat-conduction equations. Physical Review Letters. 2001;**86**: 2297-2300

[21] Chen G. Ballistic-diffusive equations for transient heat conduction from nano to macroscales. ASME (The American Society of Mechanical Engineers) Journal Heat Transfer. 2002;**124**:320-328. DOI: 10.115/1.1447938

[22] Hamian S, Yamada T, Faghri M, Park K. Finite element analysis of transient ballistic-diffusive phonon heat transport in two-dimensional domains. International Journal of Heat and Mass Transfer. 2015;**80**:781-788

[23] Yang R, Chen G, Laroche M, Taur Y. Simulation of nanoscale multidimensional transient heat conduction problems using ballistic-diffusive equations and phonon Boltzmann

equation. ASME (The American Society of Mechanical Engineers) Journal Heat Transfer. 2005;**127**:298-306

[24] Ghosh S, Bao W, Nika DL, Subrina S, Pokatilov EP, Lau CN, Balandin AA. Dimensional crossover of thermal transport in few-layer graphene. Nature Materials. 2010;**9**:555-558

[25] Balandin AA. Thermal properties of graphene and nanostructured carbon materials. Nature Materials. 2011;**10**:569-581

[26] Hua YC, Cao BY. An efficient two-step Monte Carlo method for heat conduction in nanostructures. Journal of Computational Physics. 2017;**342**:253-266

[27] Sobolev SL. Discrete space-time model for heat conduction: Application to size dependent thermal conductivity in nano-films. International Journal of Heat and Mass Transfer. 2017; **108**:933-939

[28] Kaiser J, Feng T, Maassen J, Wang X, Ruan X, Lundstrom M. Thermal transport at the nanoscale: A Fourier's law vs. phonon Boltzmann equation study. Journal of Applied Physics. 2017;**121**:044302. DOI: 10. 1063/1.4974872

[29] Ma Y. Size-dependent thermal conductivity in nanosystems based on non-Fourier heat transfer. Applied Physical Letters. 2012;**101**(21):211905

[30] Ziman JM. Electrons and Phonons: The Theory of Transport Phenomena in Solids. Oxford: Oxford University Press; 2001

[31] McGaughey AJ, Landry ES, Sellan DP, Amon CH. Size-dependent model for thin film and nanowire thermal conductivity. Applied Physical Letters. 2011;**99**(13):131904

[32] Hua YC, Cao BY. The effective thermal conductivity of ballistic-diffusive heat conduction in nanostructures with internal heat source. International Journal of Heat and Mass Transfer. 2016;**92**:995-1003

[33] Hua YC, Cao BY. Slip boundary conditions in ballistic-diffusive heat transport in nanostructures. Nanoscale and Microscale Thermophysical Engineering. 2017;**21**(3):159-176

[34] Hua YC, Cao BY. Phonon ballistic-diffusive heat conduction in silicon nanofilms by Monte Carlo simulations. International Journal of Heat and Mass Transfer. 2014;**78**:755-759

[35] Yang C, Wang Q, Nakayama A, Qiu T. Effect of temperature jump on forced convective transport of nanofluids in the continuum flow and slip flows regimes. Chemical Engineering Science. 2015;**137**:730-739

[36] Singh D, Guo X, Alexeenko A, Murthy JY, Fisher TS. Modeling of subcontinuum thermal transport across semiconductor-gas interfaces. Journal of Applied Physics. 2009;**106**:024314

[37] Mahajan SS, Subbarayan G. Estimating Kapitza resistance between Si-SiO2 Interface using molecular dynamics simulations. IEEE (Institute of Electrical and Electronics Engineers) Transactions on components, Packaging and Manufacturing Technology. 2011;**1**(8):1132-1139

[38] Guo Y, Wang M. Heat transport in two-dimensional materials by directly solving the

phonon Boltzmann equation under Callaway's dual relaxation model. Physical Review B. 2017;**96**:134312

[39] Nie BD, Cao BY. Reflection and refraction of a thermal wave at an ideal interface. International Journal of Heat and Mass Transfer. 2018;**116**:314-328

[40] Irimia-Vladu M. Green electronics: Biodegradable and biocompatible materials and devices for sustainable future. The Royal Society of Chemistry. 2014;**43**:588-610. DOI: 10.1039/c3cs60235d

[41] Criado-Sancho M, Jou D. A simple model of thermoelastic heat switches and heat transistors. Journal of Applied Physics. 2017;**121**:024503. DOI: 10.1063/1.4974011

[42] Pop E, Varshney V, Roy AK. Thermal propertis of graphene: Fundamentals and applications. MRS Bulletin. 2012;**37**:1273-1281. DOI: 10.1557/mrs.2012.203

[43] Irimia-Vladu M, Glowacki E, Voss G, Bauer S, Sariciftci NS. Green and biodegradable electronics. Materials Today. 2012;**15**:340-346. DOI: 10.1016/S1369-7021(12)70139-6

10

Conducting Polymers as Elements of Miniature Biocompatible Sensor

Joanna Cabaj and Jadwiga Sołoducho

Abstract

Conducting polymers (CPs), the so-called "fourth generation of polymeric materials", can solve essential problems in biosensing technologies due to their unique material properties and implementation in innovative device systems. CPs have excellent biocompatibility. They can provide advantageous interfaces for bioelectrodes owing to their hybrid conducting mechanics, combining both electron and ionic charge carriers. Many (i.e. glucose) biosensors use immobilized enzymes to form a selective layer on CP structure. Miniaturization of sensors is a new requirement. Mini sensors are portable and wearable with low utilization of sample and cost-effective technology of production.

Keywords: conducting polymers, nanomaterials, sensing devices, biosensors, miniaturization, biocompatibility

1. Introduction

Green electronics represents an occurring area of research aimed at identifying compounds of natural origin and establishing cost-reasonable ways for the synthetic materials that have utility in environmentally safe and/or biocompatible devices. There are several biocompatible sensing technologies that can perform innumerable physical and physiological measurements. Apart from carbon-based nanomaterials, other active sensing components are widely reported. These materials include polymers, semiconductors and metallic conductor-based nanomaterials as well as ionic and metallic liquids.

Carbon-based technologies are meant to address the energy and cost inefficiency issues posed by their inorganic counterparts. Organic electronics (based on i.e. conjugated polymers, CPs)

entered the research field in the 1970s holding the high promise of delivering cost-reasonable and energy-efficient materials and devices. Despite intense effort of the scientific community during the past 30 years, the efficiency and stability of organic semiconductors endure at current times' major obstacles in their development as solid challengers of the inorganic materials [1–3]. Consequently, the large-scale rapid replacement of hard core inorganic counterparts, like the ones active in high-speed processors, integrated circuits, and solar cell modules, with organic components is not immediately expected [1–3]. However, the "soft" nature of carbon-based components confers them a serious benefit over the inorganic materials, enabling production of flexible, conformable and even extremely thin electronic equipment [4].

Conducting polymers, the so-called "fourth generation of polymeric materials", can provide effective methods for the diagnosis and treatment of different disorders, that is, diabetes. Conducting polymers have often excellent biocompatibility. They can provide favorable interfaces for bioelectrodes owing to their hybrid conducting processes, combining both electron and ionic charge carriers. Many (i.e. glucose) biosensors use immobilized enzymes to construct a selective layer on CP structure. Miniaturization of sensors is a new demand. Mini sensors are portable and wearable with low utilization of sample. New biosensors with a market size of a US$13 billion annual turnover have quickly become valuable instruments in the healthcare. Actually, glucose biosensors (accounting for 85% of the total biosensor market) have notably mended the quality of life of diabetics [5].

2. Green electronic materials: conducting polymers

Conducting polymers (CPs) have occurred as competitive sensing materials for biological sensing applications. Their convenience of synthesis by chemical or electrochemical routes at ambient conditions, functionalization with monomer, dopant, monomer/dopant ratios and oxidation state to enhance the conductivities over 15 orders of magnitude, biocompatibility and low energy optical transitions have caused a significant concern. CPs have been synthesized by differing procedures, namely, electrochemical dip-pen lithography, mechanical stretching, electrospinning and template-directed electrochemical synthesis [6].

Since CPs were discovered, they have found many utilization. The swift progress in conductive polymer technologies is a significant motivating force for utilization of these materials as alternatives [7, 8] to conventional conductors, such as copper and gold [9, 10], as elements in the construction of electromagnetic devices. Regardless of that, the conductivity of polymers is lower than in metals, it has been presented to be adequate to construct antennas [11, 12]. Simultaneously, appearing green materials are contemplated to achieve a more aspiring purpose, designated by the integration of biocompatible, biodegradable and cost-reasonable electronics, such as the monitoring or the diagnosis of humans with environmentally benign technologies. Examples of the most common conductive polymers are shown in **Tables 1** and **2** [11–14] and **Figure 1**. Among the different conducting polymers, poly(3,4-ethylenedioxythiophene) (PEDOT) is one of the most encouraging materials for bioelectronics due to its relatively high conductivity, stability and more importantly its organic nature and good compatibility with bioorganic molecules such as enzymes compared to other CPs. Nevertheless, because of

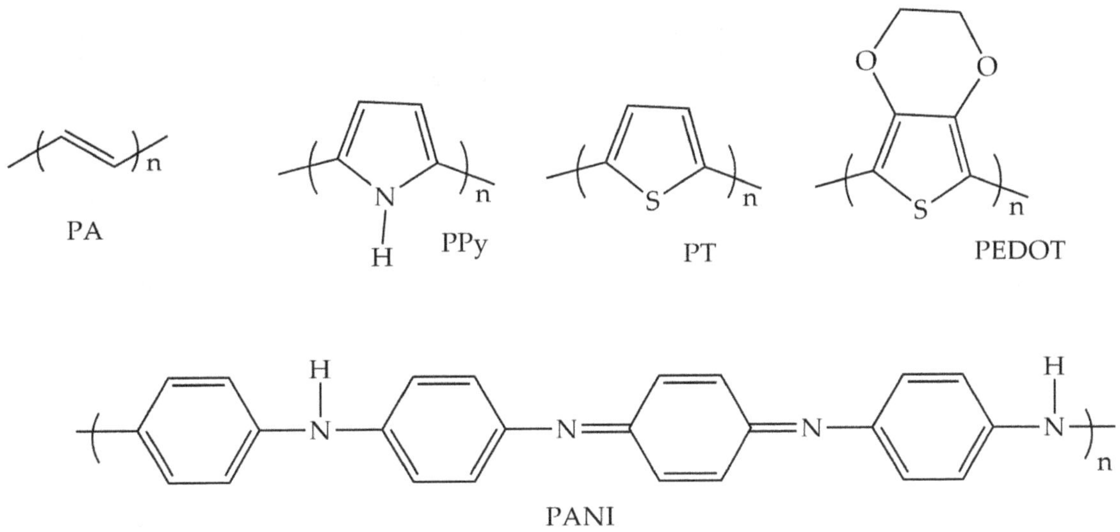

Figure 1. The most common CPs as sensor elements.

its poor solubility and processability, PEDOT is often mixed with PSS to generate water soluble poly(3,4-ethylenedioxythiophene):poly(styrene sulphonate)—PEDOT:PSS, which was developed and patented in 1988 by Bayer AG [15]. Doping of PSS could disrupt the combination of the PEDOT chains and lower their electrochemical efficiency [16]. Despite the fact that water soluble PEDOT:PSS can be applied to generate conducting thin films, the resulting planar layer lacks morphological benefit, having only restricted accessible surface in comparison with processable colloidal interface materials [16].

To date, some electrical conductors have been applied in implantable biological interfaces such as cardiac patches [17], neural interfaces [18], electroceuticals [19] and on-command drug delivery platforms [20]. Although classic inorganic electrical materials (i.e. metals and semiconductors) are not appropriate for a seamless biointerface because of the need for extracellular functionalities, the semiconducting polymers seem to be valuable alternatives for these applications [21]. Controlling this limited biocompatibility in flexible electronic materials endures a challenge, and is recently drawing fair research efforts in the bioelectronics group.

Conducting polymer	Maximum conductivity (S/CM)	Type of doping
Polyacetylene (PA)	200–1000	n, p
Polyparaphenylene (PPP)	500	n,p
Polyaniline (PANI)	5	n, p
Polyparavinylene (PPV)	1–1000	p
Polyparaphenylene sulphide (PPS)	3–300	p
Polypyrrole (PPY)	40–200	p
Polythiophene (PT)	10–100	p
PEDOT:PSS	100–1500	p

Table 1. The conductivity of the most common CPs.

CPs	Synthesis	Properties	Application
Polypyrrole (PPy)	Electrochemical and chemical synthesis	High conductivity (up to 160 S cm^{-1}) when doped with iodine; opaque, brittle, amorphous material	Biosensors, antioxidants, drug delivery, bioactuators, neural prosthetics, cardiovascular application
Polythiophenes (PT)	Electrochemical and chemical synthesis	Good electrical conductivity and optical property	Biosensors, food industry
Polyaniline (PANI)	Electrochemical and chemical synthesis	Belongs to the semi-flexible rod polymer family; requires simple doping/dedoping chemistry; exists as bulk films or dispersions; high conductivity up to 100 S cm^{-1}	Biosensors, antioxidants, drug delivery, bioactuators, food industry, cardiovascular application
Poly(3,4-ethylenedioxythiophene) (PEDOT)	Electrochemical and chemical synthesis	High temperature stability; ability to suppress the so-called "thermal runaway" of the capacitor; transparent conductor; moderate band gap and low redox potential; conductivity up to 210 S cm^{-1}	Biosensors, antioxidants, drug delivery, neural prosthetics

Table 2. Properties of the CPs.

2.1. Surface modification of conducting polymers

Surface modification of the CPs for incorporating biomolecules has been obtained by both physical and chemical moderations. Such modifications can be applied to create both physical and chemical guidance cues, which can be adapted for the required biomedical utilization [22]. Chemical modification has been extensively studied using biomolecules as dopants (bio-specific dopants such as peptides, proteins and neurotrophins) [23, 24] or by immobilizing bioactive moieties on the surface of the material [24].

For example, neural microelectrodes are commonly used in chronic, long-term implantations. Due to the fact, highly stable materials are needed that can tolerate the implantation procedure as well as the presence of biochemical environment in living tissue. Polypyrrole, however, has a poorly defined chemical structure in which there is a notable amount of α-β' coupling. The presence of these defect sites along the polymer chain induces structural disorder, limits the electrochemical response and has been implicated as the primary site of polymer breakdown due to over-oxidation [23]. Moreover, oxidized polypyrrole is unstable to reduction by relatively weak, but biologically relevant, reducing agents such as dithiothreitol and glutathione [23], which act as a p-dopants.

Physical modification has been investigated by enlarging surface roughness by different processes such as generating microporous layers using polystyrene sphere templates, creating composites of nanoparticles and polylactide [25], growing CPs within hydrogels [18] and mixing with biomolecules to yield "fuzzy" structures.

The electrode coatings used are rather soft [26] and can be tailored at the micrometer, nanometer and molecular scale to have fibrillary, nodular, fuzzy, tubular [27] or porous surface morphologies [25]. As a consequence, most tissue- and device-compatible surface tempering of the electrode would be bringing electrical activity, bioactivity, mechanical softness and architectural properties on a similar scale to that of cells in tissues. The surface roughness character of the conducting polymer layers can be tailored by modification of the conducting polymer synthesis temperature [22]. Exceptional adaptation of the surface roughness characteristics is important because rougher topology corresponds with increasing of surface area, which would expand the signal conduction by growing the interface with neurons. For example, polypyrrole films synthesized at a lower temperature (418°C) were rougher than the same layers achieved at 2518°C [28]. Kmecko et al. [29] presented that the introduction of carbon nanotubes as dopants to PPy and PEDOT prefers the creation of bumps and grows the surface roughness. Functionalization of CPs with biomolecules has permitted engineers to modify CPs with biological sensing elements, and to turn ON and OFF different signaling routes demanded for several cellular mechanisms to form conducting polymers that extend cell proliferation/differentiation. Moreover, dopants can be applied as intermediates to allow further modification of CPs, that is, doping with poly(glutamic acid) supplies a carboxylic acid pendent group, which can be functionalized further by covalent binding to any amino group, such as those found in polylysine and laminin [30].

The electrochemical character of the CPs can be varied by modifying the dopant concentration. Electrical conductivities can be varied by as much as 15 orders of magnitude by changing the dopant concentrations so that control is feasible over the all ranging from insulator to semiconductor and then to metal [31]. The usually used dopants contain aromatic sulphonate variants such as *p*-toluenesulphonate, styrene, sodium benzenesulphonate to dope the polymers [32]. Other appropriate dopants for oxidation polymerization contain buffer salts, I_2, BF_4, perchlorates and $FeCl_3$ (**Table 3**). Biological dopants include laminin peptide sequences, hydroxyapatite or a silk-like polymer with fibronectin units and polysaccharides [22]. Nevertheless, a main disadvantage of introducing dopants is the possible diffusion of the dopant into the culture medium with effects on cytotoxicity and deterioration of the electrical characteristics of the CP layer itself. For example, dodecyl sulphate-doped PPy layers undergo structural modifications after 7 days of soaking in deionized water [23].

The scope of possible dopants is huge as long as the selected dopant is charged. On the other hand, covalent methods can be used to more constantly functionalize CPs. The monomer can be synthesized with required functional groups and then polymerized, or post-polymerization covalent modification is also possible. It is crucial to note that the steric effects of any introduced functional group may interrupt the planarity of the conjugated arrangement, which could in turn lower the conductivity [22].

Polymer	Conductivity (S cm^{-1})	Dopant	Conductivity after doping (S cm^{-1})	Ref.
PEDOT	6×10^{-4}	polystyrene sulphonate (PSS)	10	[33]
PPY	10^{-2}–10^{-3}	ClO_4^-	10	[34]
PT	10–10^{-3}	BF_4	10–20	[34]
PT	10–10^{-3}	SO_3^-	10–20	[34]

Table 3. Examples of common CP dopants in biological use.

2.2. Conducting polymers as effective electron relays in sensor devices

Effective electron transfer between the biorecognition species (e.g. an enzyme) and the electrode is challenging element when creating enzymatic biosensors. Classically, the distance among the active centre of the enzyme and the electrode surface is too long for direct electron transfer (DET) owing to the protective disk of the enzyme. Because electron transfer (ET) *via* a tunneling mechanism is rarely observed in classic electrodes, establishing electron relays that allow for fast ET, thus avoiding free-diffusing redox species between the electrode and the enzyme, is vital [35]. Due to the fact, organic electronic materials present very attractive expectants for molecular wiring owing to their polymeric essence and conducting character [36].

Doped PPy was the first CP presented to provide an electron relay among the surface of the electrode and the active centre of an enzyme [37, 38]. Nevertheless, owing to deficient electrochemical stability (potentially affecting long-term functionality) [39], attempts moved to other materials such as PEDOT, a polythiophene derivative which appeared as a more stable expectant owing to its low bandgap and high electrochemical stability in the oxidized state [40]. The first example of a PEDOT-based glucose sensor with potential for long-term measurements was presented by Kros et al. [41]. They physically introduced a positively charged polymer in the conducting substrate of the biosensor, permitting more effective ET as a result of the grown electrostatic interaction between the positively charged entrapped polymer and the negatively charged enzyme (**Figure 2A**). An optional procedure to enhance the electron relay in CPs post-synthesis involves intermixing with redox hydrogels, which have been presented to reveal rapid substrate and counter-ion diffusion effects with high flexibility and quick electron transfer rates. The non-conducting nature of such hydrogels hinders their effective and spatially placed immobilization on the active electrode surface, and mixing them with CPs can thus overcome this issue resulting in an ideal electron-transfer strategy. PEDOT:PSS was used to improve the deficient performance of a mediator-based biosensor by its introduction into nanocomposite enzyme electrodes, emerging in enhanced electron hopping in terms of the electron diffusion coefficient and charge transfer resistance (**Figure 2B**) [42]. Going one step further, Bao et al. investigated intrinsically conducting nanostructured polyaniline (PANI)-redox hydrogels [43]. In another strategy, a CP-based glucose-permeable redox hydrogel was created by crosslinking polymer acid-templated PANI together with glucose oxidase, leading to the electrical wiring of the enzyme and permitting electrocatalytic oxidation of glucose at low oxidation potentials (**Figure 2C**) [44]. Recently, CP hydrogels with high permeability to enzymes were utilized to produce metabolite biosensors with excellent sensing character without the need for a mediator (**Figure 2D**) [45].

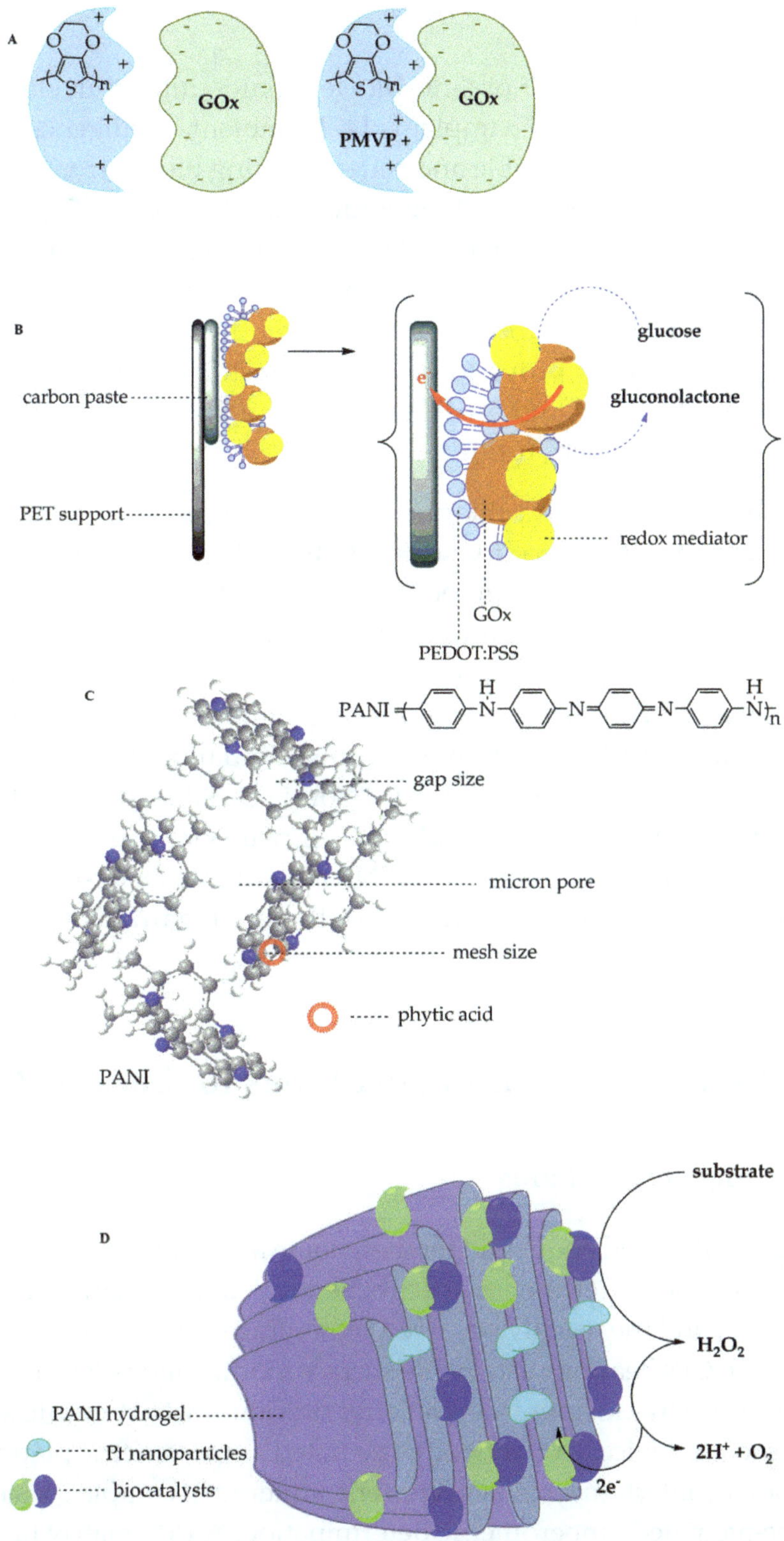

Figure 2. Organic conducting materials as effective enzyme immobilization supports and transducers for different body metabolite determination. (A) Electrostatic binding (weak and strong) of glucose oxidase (GOx) onto a PEDOT and PEDOT supplied with positively charged PMVP, poly(*N*-methyl-4-pyridine), according to [41]. (B) Schematic diagram of the working electrode coated with ferrocene-branched polyethylenimine, PEDOT:PSS and GOx for glucose detection, according to [42]. (C) Phytic acid gelated and doped PANI hydrogel according to [44]. (D) The PANI hydrogel matrix including Pt nanoparticles and the proper biocatalysts for the detection of uric acid, cholesterol and triglyceride, according to [45].

2.3. Biocompatibility of CPs

Among the effective presentations of organic bioelectronics, ultra-thin electronic systems for surgical, point-of-care [46], diagnostic implants [47], ambient intelligence for daily-life assistance [48], soft robotics [49], conformable and self-sustaining bioelectronic elements for sports and recreation [48], or even disposable (biodegradable) electronics [50] for food packaging [51] or throw-away applications [52] can be listed. The connection of novel electronic elements with biosensing constituents will open the possibility for investigating disposable diagnostic and drug delivery platforms. This topic has been recently reviewed in detail [53, 54]. The organic bioelectronics field may prove to be the satisfactory host for greeting natural and nature-inspired carbon materials and a perfect base for achieving the ambitious purpose of "green" and sustainable electronics future.

Conducting polymers of pyrrole and thiophene connected by ester linkages have been considered for the generation of temporary scaffolds for cell attachment and proliferation for tissue engineering applications [22]. Moreover, these scaffolds are biodegradable [55]. The possibility of growing cells on CPs has proven the biocompatibility of these polymers [21, 56]. Additionally, recently the biocompatibility of PPy and PEDOT layers and PPy and PEDOT nanotubes was estimated by utilizing a dorsal root ganglion model [57]. The implantation of CPs *in vivo* for several weeks has led to only minimal inflammation, again pointing to low toxicities and good tissue compatibility [21, 55]. Moreover, Abidian and Martin [18] successfully presented that PEDOT nanotubes could record neuronal spikes about 30% more than control sites with a high signal-to-noise ratio (SNR) for 7 weeks post-implantation *in vivo*. In addition, there have been a number of reviews on CPs with regard to biomedical applications [22, 27, 55, 58–63].

3. Organic electronic-based sensing platforms for body metabolites

3.1. Metabolite sensing of body fluids

Blood is the most generally used body fluid for metabolite level monitoring. Nevertheless, owing to the wealth of electroactive elements, electrochemical determination procedures become somewhat challenging, and the usually observed biofouling of the sensing electrodes poses further restrictions [5, 64]. CPs bearing sustainable surface modifications (i.e. incorporation of electron mediators, permselective membranes) can offer precious instruments towards modern and more accurate diagnostic devices. An antibody-mediated amperometric platform was designed by Wei et al. to avoid the interfering signals often encountered in complex systems (blood) when utilizing enzymatic-mediated amperometric determination. A PPy matrix favored for immobilization of the capture antibody, on top of a 16-array gold electrochemical sensor, which could therefore determine creatinine fast and accurately in whole blood, resulting in a point-of-care (POC) assay for allograft dysfunction (**Figure 3A**) [65]. Liao et al. recently investigated a flexible organic electrochemical transistor (OECT) platform based on PEDOT:PSS to selective detection of urea and glucose in saliva samples [66]. To exclude electrochemical interference in saliva, thus increasing sensitivity and selectivity, the gate electrodes were modified with oppositely charged bilayer polymeric layers for both anionic and cationic charge exclusion of interferes. Moving towards multiplexing, a PEDOT:PSS-based OECT biosensing platform integrated with microfluidics was investigated for contemporary screening of glucose, lactate and cholesterol in human saliva samples [5]. The final tool was tested with human volunteers before and after exercise to present comparative differences in their metabolite profiles under stimuli (**Figure 3B**) [67]. In a similar procedure, contemporary sensing of lactate and glucose was presented by integrating two OECT-based tools, each with a separate microfluidic channel. They created a prototype portable glucose sensor by linking a smartphone with the sensing platform through Bluetooth connection, highlighting the ease of integration of such devices for POC systems [68].

3.2. Metabolite sensing from whole cells

Determination of cellular metabolites under different stimuli or environmental conditions can give useful prospects for drug discovery and toxicology. Larsen et al. used PEDOT:tosylate microelectrodes as an all polymer electrochemical chip for the determination of potassium-induced transmitter release from neuron-like cells, presenting the potential of the procedure

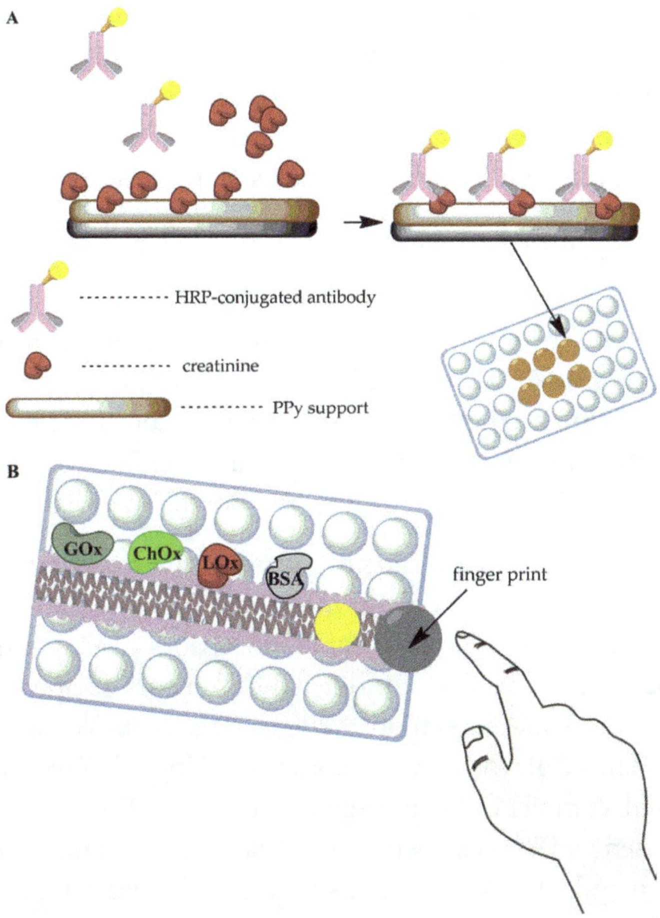

Figure 3. Integrated point-of-care systems based on organic electronics. (A) Scheme of the conducting polymer electrochemical sensor for the direct measurement of creatinine from serum, according to [65]. (B) Schematic demonstrating the OECT-based multianalyte system, according to [67]; BSA, bovine serum albumin; ChOx, cholesterol oxidase; GOx, glucose oxidase; HRP, horseradish peroxidase; lox, lactate oxidase.

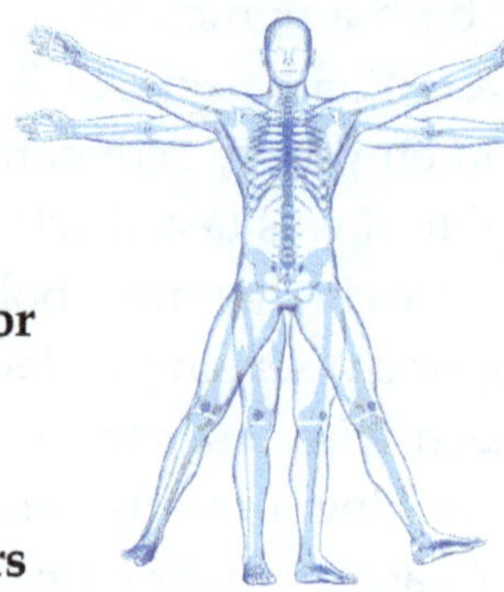

Figure 4. Overview of the swiftly increasing field of wearable biosensors.

for drug screening applications [69]. To enhance the electrocatalytic effect of the sensing electrode, the PEDOT:PSS gate can also be supplied with electrodeposited Pt nanoparticles [70]. Owing to the high surface area of the nanoparticles and the high specificity of the biocatalyst, the authors obtained very sensitive detection of the crucial metabolites such as glucose and lactate from live cells. Lactate production in tumor cell cultures derived from real patients was also studied using an OECT circuit. Lactate production could be measured from a few cells, underlining the sensitivity of the tool in a highly complex milieu, thus shown its potential for utilization in *in vivo* applications for cancer diagnostics [71]. Also, recently, Curto et al. presented a multiparametric on-chip platform integrated with microfluidics for cell cultures, using among other in-line methods the OECT-based detection of glucose produced by the cells as a measure to validate their improved differentiation under stimuli conditions [72].

3.3. Wearable metabolite biosensors

An indisputable trend in biosensor technology is on-body continuous monitoring of metabolites using wearable devices (**Figure 4**) [1]. Wearable biosensor applications aim to transform centralized hospital-based care systems to home-based personal medicine, reducing healthcare cost and time for diagnosis. Electrochemical transducers offer many benefits as wearable sensors for physiological monitoring, and can be easily integrated onto textile materials or directly on the skin.

Sweat-based wearable sensors, although mostly focused on a small number of physical or electrophysiological parameters, can yield crucial information about the health status of a patient based on levels of vital metabolites [73]. Wearable biosensors can be either textile/plastic-based or epidermal (tattoo)-based systems [74]. Epidermal biosensors supply better contact with skin but commonly exhibit shorter lifetimes than the textile-based tools. Such biosensors were first developed in 2009 by Kim et al. for continuous monitoring of physical parameters [75] and, shortly thereafter, Jia et al. combined this route with biorecognition elements to generate the first printed tattoo-based biosensor [76]. A screen-printed electrode on no permanent tattoo paper was investigated with carbon and silver (Ag)/AgCl serving as the working and reference electrodes, respectively. The working electrode was also modified with carbon nanotubes carrying a mediator together with lactate oxidase for endlessly monitoring lactate in sweat during exercise [76].

CPs are specially beneficial for wearable sensor technology owing to their compatibility with production on flexible solids [77]. In a very interesting way, Pal et al. investigated PEDOT:PSS electrodes on flexible fully biodegradable silk protein fibroin supports using a simple photolithographic process and an aqueous ink composed of the CP and carrier proteins (**Figure 4B**) [78]. In an almost identical route by the same scientific group, silk proteins including fibroin and sericin were modified with photoreactive methacrylate groups for use as substrate inks for water-dispersible PEDOT:PSS that was micropatterned to investigate a biodegradable bioelectrode for glucose sensing *in vitro* [79]. This pathway presents a new trend for generating an entirely organic and free-standing system with controllable biodegradability including scalability and processability, leading to applications in wearable or implantable bioelectronics [80].

3.4. Miniaturization: implantable devices

The perspectives of implantable instruments and especially home-based metabolic monitoring can only be reached if they can be simply implanted and explanted (i.e. needle-assisted) without the necessity of complicated surgery [81]. Due to that, the implantable tool should be small, which calls for novel miniaturization of different functional elements such as electrodes, power sources, signal processing systems and sensory components. In addition, miniaturized biosensors implanted by ultrafine needles induce less tissue damage and then less inflammation and foreign body response [82]. Miniaturization of implantable instruments and particularly biosensors can be listed under: (1) miniaturization of sensing electrodes and elements and (2) miniaturization of driving electronics for power, communication and their subsequent integration/packaging. Referring to the production of miniaturized electrodes for analyte sensing, immobilization of biocatalyst onto an ultra-thin Pt wire (diameters less than 50 μm) or carbon nanofibres has been substantial [83]. The latter is convenient for generating nerve stimulating microelectrodes because of the possibility of ultra-fine dimensions and flexibility [84]. Due to subsequent improvement of the electrocatalytic feature of carbon nanofibres, these were modified with different metal nanoparticles without compromising their flexibility [85]. Recently, the advent of sub-micron lithography and its further use to produce miniaturized transistors has encouraged investigators to develop solid state electrochemical sensing systems in a transistor order [86]. Biosensors based on classic Si-based transistors as well as the incipient organic thin film transistors are being investigated for a scope of analytes. The unique electrical character of 1-D nanomaterials (CPs) [86] has led researchers to use them as channel materials and investigates sensors based on modifications induced in either gate conductance, modulation, transconduction, hysteresis or threshold voltage.

The flexible nature of polymers together with their low-temperature processing and demonstrated biocompatibility with enzymes renders them beneficial over classic Si- and glass-based materials [81]. Additionally, the soft and flexible character of polymers could reduce the possibility of tissue damage to the body during implantation and can be beneficial for applications where the instrument has to be able to adjust itself to the shape of the human body.

4. Discussion

Green electronics constitutes not only a novel term but also twenty-first century's slogan; it means an emerging area of research covered the identifying compounds of natural origin and

Device	CPS	Analyte	Ref.
Fluorescence-based biosensor	Poly(dithienotetraphenylsilane)	Dopamine	[87]
Electrochemical biosensor	Poly(bis-selenophene)-*N*-nonyl carbazole	Phenolic compounds	[88]
Optical LTCC biosensor	Poly(bis-thiophene) acridone	Phenolic compounds	[89]
Glucose biosensor	Langmuir-Schaefer film of *N*-hexadecyl-2,8-bis(thianthrene)phenothiazine	Glucose	[90]

Table 4. Miniature sensors based on CPs.

determining economically efficient ways for the fabrication of materials that have applicability in environmentally friendly technologies and devices. The key factor of this chapter is to generate routes for the production of human- and environmentally safe electronics in the main and the integration of such electronic circumferences with biological tissue.

Scientific researches into the class of green electronics may implement not only the original assurance of organic electronics that is to carry cost-reasonable and energy efficient materials but also achieve inconceivable functionalities for electronics, for example, benign integration into life and environment. Modern electronics technology has turned the relationship energy consumed during fabrication versus energy consumed during exploitation of the product to a complete imbalance [48]. A key prerequisite for achieving sustainability in the electronics industry is the usage of materials and technologies that have low embodied energy. In this context, it is worth to emphasize miniaturization procedures and alternative conducting materials as—CPs, which our group successfully implemented in miniature sensor devices (**Table 4**).

5. Concluding remarks

Organic electronics recently has swiftly gone to the forefront of biological applications, regardless its beginnings in wide, flexible applications realized by tools such as the organic light-emitting diode or organic photovoltaics. One of the agents responsible for the favorable outcome of organic electronics in biosensing applications is the accurate flexibility and tunability of the materials to suit the requirements in biological environment. Organic electronics can then demonstrate the technology to meet the requirements of the biosensor market. With regard to classical raised issues in terms of stability and lifetime of the biocatalyst and other biological elements, the trends in this dynamic field of organic electronic sensors are foreseen to contain the investigation of biomimetic architectures (i.e. molecular imprinted structures), harnessing the versatility in synthesis of such electronic materials. Transducer and biorecognition elements can be met as a single active agent that combines electronic functionalities and the best properties of the biological element in a more stable support, therefore, opening up modern prospects in sensor technologies in both fundamental and practical aspects.

Acknowledgements

The authors gratefully acknowledge the financial support of Wroclaw University of Science and Technology (10401/0194/17).

Author details

Joanna Cabaj* and Jadwiga Sołoducho

*Address all correspondence to: joanna.cabaj@pwr.edu.pl

Faculty of Chemistry, Wroclaw University of Science and Technology, Wroclaw, Poland

References

[1] Ma H, Acton O, Hutchins DO, Cernetic N, AK-Y J. Multifunctional phosphonic acid self-assembled monolayers on metal oxides as dielectrics, Interface modification layers and semiconductors for low-voltage high-performance organic field-effect transistors. Physical Chemistry Chemical Physics. 2012;**14**:14110-14126

[2] Mei J, Diao Y, Appleton AL, Fang L, Bao Z. Integrated materials design of organic semiconductors for field-effect transistors. Journal of the American Chemical Society. 2013;**135**:6724-6746

[3] Jorgensen M, Norrman K, Gevorgyan SA, Tromholt T, Andreasen B, Krebs FC. Stability of polymer solar cells. Advanced Materials. 2012;**24**:580-612

[4] Kaltenbrunner M, White MS, Głowacki ED, Sekitani T, Someya T, Sariciftci NS, Bauer S. Ultrathin and lightweight organic solar cells with high flexibility. Nature Communications. 2012;**3**:770

[5] Pappa A-M, Parlak O, Scheiblin G, Mailley P, Salleo A, Owens RM. Organic electronics for point-of-care metabolite monitoring. Trends in Biotechnology. 2018;**36**:45-59

[6] Ahmad R, Mahmoudi T, Ahn M-S, Hahn Y-B. Recent advances in nanowires-based field-effect transistors for biological sensor applications. Biosensors and Bioelectronics. 2018;**100**:312-325

[7] Peisert H, Liu X, Olligs D, Petr A, Dunsch L, Schmidt T, Chass T, Knupfer M. Highly ordered phthalocyanine thin films on a technically relevant polymer substrate. Journal of Applied Physics. 2004;**96**(7):4009-4011

[8] Kang HS, Lee JW, Kim MK, Joo J, Ko JM, Lee JY. Electrical characteristics of pentacene-based thin film transistor with conducting poly(3,4-ethylenedioxythiophene) electrodes. Journal of Applied Physics. 2006;**100**(6):064508

[9] Herth E, Rolland N, Lasri T. Circularly polarized millimeter-wave antenna using 0-level packaging. IEEE Antennas and Wireless Propagation Letters. 2010;**9**:934-937

[10] Herth E, Seok S, Rolland N, Lasri T. Wafer level packaging compatible with millimeter-wave antenna. Sensors and Actuators A: Physical. 2012;**173**(1):238-243

[11] Pudas M, Halonen N, Granat P, Vhkangas J. Gravure printing of conductive particulate polymer inks on flexible substrates. Progress in Organic Coating. 2005;**54**(4):310-316

[12] Bai H, Shi G. Gas sensors based on conducting polymers. Sensors. 2007;**7**(3):267-307

[13] Herth E, Guerchouche K, Rousseau L, Calvet LE, Loyez C. A biocompatible and flexible polyimide for wireless sensors. Microsystem Technologies. 2017;**23**:5921-5929

[14] Woods M, Batchelor J. Low-profile slot antenna integrated with a thin polymer non-metallic battery. In: Antennas and Propagation Conference (LAPC); Loughborough. 2013. pp. 441-445

[15] Jonas FD, Heywang GD, Schmidtberg W. AgB Festst off-Elektrolyte und diese enthaltende Elektrolyt-Kondensatoren. 1989

[16] Alemu D, Wei H, Ho K, Chu C. Highly conductive PEDOT:PSS electrode by simple film treatment with methanol for ITO-free polymer solar cells. Energy & Environmental Science. 2012;**5**:9662-9671

[17] Shin SR, Jung SM, Zalabany M, Kim K, Zorlutuna P, Kim SB, Nikkhah M, Khabiry M, Azize M, Kong J, Wan T, Palacios T, Dokmeci MR, Bae H, Tang X, Khademhosseini A. Carbon-nanotube-embedded hydrogel sheets for engineering cardiac constructs and bioactuators. ACS Nano. 2013;**7**:2369-2380

[18] Abidian MR, Martin DC. Multifunctional nanobiomaterials for neural interfaces. Advanced Functional Materials. 2009;**19**:573-585

[19] Famm K. A jump-start for electroceuticals. Nature. 2013;**2013**(496):159-161

[20] Wu CH, Cao C, Kim JH, Hsu CH, Wanebo HJ, Bowen WD, Xu J, Marshall J. Trojan-horse nanotube on-command intracellular drug delivery. Nano Letters. 2012;**12**:5475-5480

[21] Schmidt CE, Shastri VR, Vacanti JP, Langer R. Stimulation of neurite outgrowth using an electrically conducting polymer. Proceedings of the National Academy of Sciences of the United States of America. 1997;**94**:8948-8953

[22] Ravichandran R, Sundarrajan S, Reddy Venugopa J, Mukherjee S, Ramakrishna S. Applications of conducting polymers and their issues in biomedical engineering. Journal of the Royal Society Interface. 2010;**7**:S559-S579

[23] Cui XY, Martin DC. Electrochemical deposition and characterization of poly (3,4-ethylenedioxythiophene) on neural microelectrode arrays. Sensors and Actuators B: Chemical. 2003;**89**:92-102

[24] Zhong YH, Yu XJ, Gilbert R, Bellamkonda RV. Stabilizing electrode-host interfaces: A tissue engineering approach. Journal of Rehabilitation Research and Development. 2001;**38**:627-632

[25] Yang JY, Martin DC. Microporous conducting polymers on neural microelectrode arrays. I. Electrochemical deposition. Sensors and Actuators B: Chemical. 2004;**101**:133-142

[26] Yang JY, Martin DC. Impedance spectroscopy and nanoindentation of conducting poly(3,4-ethylenedioxythiophene) coatings on microfabricated neural prosthetic devices. Journal of Materials Research. 2006;**21**:1124-1132

[27] Abidian MR, Kim DH, Martin DC. Conducting polymer nanotubes for controlled drug release. Advanced Materials. 2006;**18**:405-409

[28] Guimard NK, Gomezb N, Schmidt CE. Conducting polymers in biomedical engineering. Progress in Polymer Science. 2007;**32**:876-921

[29] Kmecko T, Hughes G, Cauller L, Jeong-Bong L, Romero-Ortega M. Nanocomposites for neural interfaces. Materials Research Society Symposia Proceedings. Materials Research Society. 2006;**926E**:0926-CC04-06

[30] Song HK, Toste B, Ahmann K, Hoffman-Kim D, Palmore GTR. Micropatterns of positive guidance cues anchored to polypyrrole doped with polyglutamic acid: A new platform for characterizing neurite extension in complex environments. Biomaterials. 2006;**27**:473-484

[31] Malhotra BD, Singhal R. Conducting polymer based biomolecular electronic devices. Pramana. 2003;**61**:331-343

[32] Cui X, Lee VA, Raphael Y, Wiler JA, Hetke JF, Anderson DJ. Surface modification of neural recording electrodes with conducting polymer/biomolecule blends. Journal of Biomedical Materials Research. 2001;**56**:261-272

[33] Crispin X, Jakobbson FLE, Crispin A, Grim PCM, Andersson P, Volodin A, van Haesendonck C, van der Auweraer M, Solaneck WR, Berggren M. The origin of the high conductivity of PEDOT-PSS plastic electrode. Chemistry of Materials. 2006;**18**:4354-4360

[34] Tourillon G, Garnier F. Effect of dopant on the physicochemical and electrical properties of organic conducting polymers. The Journal of Physical Chemistry. 1983;**87**:2289-2292

[35] Sekretaryova AN, Vagin M, Beni V, Turner APF, Karyakin AA. Unsubstituted phenothiazine as a superior water-insoluble mediator for oxidases. Biosensors & Bioelectronics. 2014;**53**:275-282

[36] Vallejo-Giraldo C, Kelly A, Biggs MJ. Biofunctionalisation of electrically conducting polymers. Drug Discovery Today. 2014;**19**:88-94

[37] Schuhmann W, Zimmermann H, Habermüller K, Laurinavicius V. Electron-transfer pathways between redox enzymes and electrode surfaces: Reagentless biosensors based on thiol-monolayer-bound and polypyrrole-entrapped enzymes. Faraday Discussions. 2000;**2000**:245-255

[38] Swann MJ, Bloor D, Haruyama T, Aizawa M. The role of polypyrrole as charge transfer mediator and immobilization matrix for d-fructose dehydrogenase in a fructose sensor. Biosensors & Bioelectronics. 1997;**12**:1169-1182

[39] Pud AA. Stability and degradation of conducting polymers in electrochemical systems. Synthetic Metals. 1994;**66**:1-18

[40] Kros A, Sommerdijk NAJM, Nolte RJM. Poly(pyrrole) versus poly(3,4-ethylenedioxythiophene): Implications for biosensor applications. Sensors and Actuators B: Chemical. 2005;**106**:289-295

[41] Kros A, van Hovell WFM, Sommerdijk NAJM, Nolte RJM. Poly(3,4-ethylenedioxythiophene)-based glucose biosensors. Advanced Materials. 2001;**13**:1555-1557

[42] Wang J-Y, Chen L-C, Ho K-C. Synthesis of redox polymer nanobeads and nanocomposites for glucose biosensors. ACS Applied Materials & Interfaces. 2013;**5**:7852-7861

[43] Pan L, Yu G, Zhai D, Lee HR, Zhao W, Liu N, Wang H, Tee BC, Shi Y, Cui Y, Bao Z. Hierarchical nanostructured conducting polymer hydrogel with high electrochemical activity. Proceedings of the National Academy of Sciences. 2012;**109**:9287-9292

[44] Mano N, Yoo JE, Tarver J, Loo YL, Heller A. An electron-conducting cross-linked polyaniline-based redox hydrogel, formed in one step at pH 7.2, wires glucose oxidase. Journal of the American Chemical Society. 2007;**129**:7006-7007

[45] Li L, Wang Y, Pan L, Shi Y, Cheng W, Shi Y, Yu G. A nanostructured conductive hydrogels-based biosensor platform for human metabolite detection. Nano Letters. 2015;**15**:1146-1151

[46] Pollock NR, Rolland JP, Kumar S, Beattie PD, Jain S, Noubary F, Wong VL, Pohlmann RA, Ryan US, Whitesides GM. A paper-based multiplexed transaminase test for low-cost, point-of-care liver function testing. Science Translational Medicine. 2012;**4**:152ra129

[47] Tybrandt K, Forchheimer R, Berggren M. Logic gates based on ion transistors. Nature Communications. 2012;**3**:871

[48] Irimia-Vladu M. "Green" electronics: Biodegradable and biocompatible materials and devices for sustainable future. Chemical Society Reviews. 2014;**43**:588-610

[49] Martinez RV, Branch JL, Fish CR, Jin L, Shepherd RF, Nunes RMD, Suo Z, Whitesides GM. Robotic tentacles with three-dimensional mobility based on flexible elastomers. Advanced Materials. 2013;**25**:205-212

[50] Irimia-Vladu M, Troshin PA, Reisinger M, Shmygleva L, Kanbur Y, Schwabegger M, Bodea R, Schwödiauer A, Mumyatov JW, Fergus V, Razumov G, Sitter H, Sariciftci NS, Bauer S. Biocompatible and biodegradable materials for organic field-effect transistors. Advanced Functional Materials. 2010;**20**:4069-4076

[51] Tao H, Brenckle MA, Yang M, Zhang J, Liu M, Siebert SM, Averitt RD, Mannoor MS, McAlpine MC, Rogers JA, Kaplan DL, Omenetto FG. Silk-based conformal, adhesive, edible food sensors. Advanced Materials. 2012;**24**:1067-1072

[52] Tobjork D, Osterbacka R. Paper electronics. Advanced Materials. 2011;**23**:1935-1961

[53] Angione MD, Pilolli R, Cotrone S, Magliulo M, Mallardi A, Palazzo G, Sabbatini L, Fine D, Dodabalapur A, Cioffiand N, Torsi L. Carbon based materials for electronic bio-sensing. Materials Today. 2011;**14**:424-433

[54] Svennersten K, Larsson KC, Berggren M, Richter-Dahlfors A. Organic bioelectronics in nanomedicine. Biochimica et Biophysica Acta. 2011;**1810**:276-285

[55] Rivers TJ, Hudson TW, Schmidt CE. Synthesis of a novel, biodegradable electrically conducting polymer for biomedical applications. Advanced Functional Materials. 2002;**12**:33-37

[56] Garner B, Georgevich A, Hodgson AJ, Liu L, Wallace GG. Polypyrrole–heparin composites as stimulus-responsive substrates for endothelial cell growth. Journal of Biomedical Materials Research. 1999;**44**:121-129

[57] Abidian MR, Corey JM, Kipke DR, Martin DC. Conducting-polymer nanotubes improve electrical properties, mechanical adhesion, neural attachment, and neurite outgrowth of neural electrodes. Small. 2010;**6**:421-429

[58] Ahuja T, Mir IA, Kumar D. Biomolecular immobilization on conducting polymers for biosensing applications. Biomaterials. 2007;**28**:791-805

[59] Kenry YJC, Lim CT. Emerging flexible and wearable physical sensing platforms for healthcare and biomedical applications. Microsystems & Nanoengineering. 2016;**2**:1-19

[60] Gong S, Schwalb W, Wang Y, Chen Y, Tang Y, Si KJ, Shirinzadeh B, Cheng W. A wearable and highly sensitive pressure sensor with ultrathin gold nanowires. Nature Communications. 2014;**5**:1-8

[61] Kim SY, Park S, Park HW, Park DH, Jeong Y, Kim DH. Highly sensitive and multimodal all-carbon skin sensors capable of simultaneously detecting tactile and biological stimuli. Advanced Materials. 2015;**27**:4178-4185

[62] Pang C, Lee G-Y, Kim T-I, Kim SM, Kim HN, Ahn SH, Suh KY. A flexible and highly sensitive strain-gauge sensor using reversible interlocking of nanofibres. Nature Materials. 2012;**11**:795-801

[63] Segev-Bar M, Landman A, Nir-Shapira M, Shuster G, Haick H. Tunable touch sensor and combined sensing platform: Toward nanoparticle-based electronic skin. ACS Applied Materials & Interfaces. 2013;**5**:5531-5541

[64] Krishnan S, Weinmanb CJ, Oberb CK. Advances in polymers for anti-biofouling surfaces. Journal of Materials Chemistry. 2008;**18**:3405-3413

[65] Wei F, Cheng S, Korin Y, Reed EF, Gjertson D, Ho CM, Gritsch HA, Veale J. Serum creatinine detection by a conducting-polymer-based electrochemical sensor to identify allograft dys-function. Analytical Chemistry. 2012;**84**:7933-7937

[66] Liao C, Mak C, Zhang M, Chan HL, Yan F. Flexible organic electrochemical transistors for highly selective enzyme biosensors and used for saliva testing. Advanced Materials. 2015;**27**:676-681

[67] Pappa AM, Curto VF, Braendlein M, Strakosas X, Donahue MJ, Fiocchi M, Malliaras GG, Owens RM. Organic transistor arrays integrated with finger-powered microfluidics for multianalyte saliva testing. Advanced Healthcare Materials. 2016;**5**:2295-2302

[68] Ji X, Lau HY, Ren H, Peng B, Zhai P, Feng S-P, Chan PK. Highly sensitive metabolite bio-

sensor based on organic electrochemical transistor integrated with microfluidic channel and poly(N-vinyl-2-pyrrolidone)-capped platinum nano-particles. Advanced Materials Technologies. 2016;**1**:1600042(1-8)

[69] Larsen ST, Vreeland RF, Heien ML, Taboryski R. Characterization of poly(3,4-ethylenedioxythiophene): Tosylate conductive polymer microelectrodes for transmitter detection. Analyst. 2012;**137**:1831-1836

[70] Strakosas X, Huerta M, Donahue MJ, Hama A, Pappa AM, Ferro M, Ramuz M, Rivnay J, Owens RM. Catalytically enhanced organic tran-sistors for in vitro toxicology monitoring through hydrogel entrap-ment of enzymes. Journal of Applied Polymer Science. 2017;**134**(1-7):44483

[71] Braendlein M, Pappa AM, Ferro M, Lopresti A, Acquaviva C, Mamessier E, Malliaras GG, Owens RM. Lactate detection in tumor cell cultures using organic transistor circuits. Advanced Materials. 2017;**29**:1605744

[72] Curto VF, Marchiori B, Pappa AM, Ferro MP, Braendlein M, Rivnay J, Fiocchi M, Malliaras GG, Ramuz M, Owens RM. Organic transistor platform with integrated microfluidics for in-line multi-paramentric in vitro cell monitoring. Microsyst. Nano. 2017;**17028**(1-12):3

[73] Heikenfeld J. Technological leap for sweat sensing. Nature. 2016;**529**:475-476

[74] Parrilla M, Cánovas R, Jeerapan I, Andrade FJ, Wang J. A textile-based stretchable multi-ion potentiometric sensor. Advanced Healthcare Materials. 2016;**5**:996-1001

[75] Kim DH, Lu N, Ma R, Kim YS, Kim RH, Wang S, Wu J, Won SM, Tao H, Islam A, Yu KJ, Kim TI, Chowdhury R, Ying M, Xu L, Li M, Chung HJ, Keum H, McCormick M, Liu P, Zhang YW, Omenetto FG, Huang Y, Coleman T, Rogers JA. Epidermal electronics. Science. 2011;**333**:838-843

[76] Jia WZ, Bandodkar AJ, Valdés-Ramírez G, Windmiller JR, Yang Z, Ramírez J, Chan G, Wang J. Electrochemical tattoo biosensors for real-time noninvasive lactate monitoring in human perspiration. Analytical Chemistry. 2013;**85**:6553-6560

[77] Khodagholy D, Curto VF, Fraser KJ, Gurfinkel M, Byrne R, Diamond D, Malliaras GG, Benito-Lopez F, Owens RM. Organic electrochemical transistor incorporating an ionogel as a solid state electrolyte for lactate sensing. Journal of Materials Chemistry. 2012;**22**:4440-4443

[78] Pal RK, Farghaly AA, Wang C, Collinson MM, Kundu SC, Yadavalli VK. Conducting polymer-silk biocomposites for flexible and biodegradable electrochemical sensors. Biosensors & Bioelectronics. 2016;**81**:294-302

[79] Kurland NE, Dey T, Kundu SC, Yadavalli VK. Precise patterning of silk microstruc-tures using photolithography. Advanced Materials. 2013;**25**:6207-6212

[80] Kurland NE, Dey T, Wang C, Kundu SC, Yadavalli VK. Silk protein lithography as a route to fabricate sericin microarchitectures. Advanced Materials. 2014;**26**:4431-4437

[81] Vaddiraju S, Tomazos I, Burgess DJ, Jain FC, Papadimitrakopoulos F. Emerging synergy between nanotechnology and implantable biosensors: A review. Biosensors & Bioelectronics. 2010;**25**:1553-1565

[82] Kvist PH, Iburg T, Aalbaek B, Gerstenberg M, Schoier C, Kaastrup P, Buch-Rasmussen T, Hasselager E, Jensen HE. Biocompatibility of an enzyme-based, electrochemical glucose sensor for short-term implantation in the subcutis. Diabetes Technology & Therapeutics. 2006;**8**(5):546-559

[83] Crespi F. In vivo voltammetric detection of neuropeptides with micro carbon fiber biosensors: Possible selective detection of somatostatin. Analytical Biochemistry. 1991;**194**(1):69-76

[84] Receveur RAM, Lindemans FW, De Rooij NF. Microsystem technologies for implantable applications. Journal of Micromechanics and Microengineering. 2007;**17**(5):R50-R80

[85] Tamiya E, Sugiura Y, Takeuchi T, Suzuki M, Karube I, Akiyama A. Ultra-micro glutamate sensor using platinized carbon Fiber electrode integrated counter electrode. Sensors and Actuators B: Chemical. 1993;**10**(3):179-184

[86] Forzani ES, Zhang H, Nagahara LA, Amlani I, Tsui R, Tao N. A conducting polymer nanojunction sensor for glucose detection. Nano Letters. 2004;**4**(9):1785-1788

[87] Baluta S, Malecha K, Zając D, Sołoducho J, Cabaj J. Dopamine sensing with fluorescence strategy based on lowtemperature co-fired ceramic technology modified with conduct-ingpolymers. Sensors and Actuators B: Chemical. 2017;**252**:803-812

[88] Cabaj J, Jędrychowska A, Świst A, Sołoducho J. Tyrosinase biosensor for antioxidants based on semiconducting polymer support. Electroanalysis. 2016;**28**:1383-1390

[89] Jędrychowska A, Malecha K, Cabaj J, Sołoducho J. Laccase biosensor based on low temperature co-fired ceramics for the permanent monitoring of water solutions. Electrochimica Acta. 2015;**165**:372-382

[90] Cabaj J, Sołoducho J, Jędrychowska A, Zając D. Biosensing invertase-based Langmuir-Schaefer films: Preparation and characteristic. Sensors and Actuators B: Chemical. 2012;**166-167**:75-82

Permissions

All chapters in this book were first published in GE, by InTech Open; hereby published with permission under the Creative Commons Attribution License or equivalent. Every chapter published in this book has been scrutinized by our experts. Their significance has been extensively debated. The topics covered herein carry significant findings which will fuel the growth of the discipline. They may even be implemented as practical applications or may be referred to as a beginning point for another development.

The contributors of this book come from diverse backgrounds, making this book a truly international effort. This book will bring forth new frontiers with its revolutionizing research information and detailed analysis of the nascent developments around the world.

We would like to thank all the contributing authors for lending their expertise to make the book truly unique. They have played a crucial role in the development of this book. Without their invaluable contributions this book wouldn't have been possible. They have made vital efforts to compile up to date information on the varied aspects of this subject to make this book a valuable addition to the collection of many professionals and students.

This book was conceptualized with the vision of imparting up-to-date information and advanced data in this field. To ensure the same, a matchless editorial board was set up. Every individual on the board went through rigorous rounds of assessment to prove their worth. After which they invested a large part of their time researching and compiling the most relevant data for our readers.

The editorial board has been involved in producing this book since its inception. They have spent rigorous hours researching and exploring the diverse topics which have resulted in the successful publishing of this book. They have passed on their knowledge of decades through this book. To expedite this challenging task, the publisher supported the team at every step. A small team of assistant editors was also appointed to further simplify the editing procedure and attain best results for the readers.

Apart from the editorial board, the designing team has also invested a significant amount of their time in understanding the subject and creating the most relevant covers. They scrutinized every image to scout for the most suitable representation of the subject and create an appropriate cover for the book.

The publishing team has been an ardent support to the editorial, designing and production team. Their endless efforts to recruit the best for this project, has resulted in the accomplishment of this book. They are a veteran in the field of academics and their pool of knowledge is as vast as their experience in printing. Their expertise and guidance has proved useful at every step. Their uncompromising quality standards have made this book an exceptional effort. Their encouragement from time to time has been an inspiration for everyone.

The publisher and the editorial board hope that this book will prove to be a valuable piece of knowledge for researchers, students, practitioners and scholars across the globe.

List of Contributors

Arijit Banerjee
University of Virginia, Charlottesville, Virginia, USA

Ruby Srivastava
CSIR-Indian Institute of Chemical Technology, Hyderabad, India

Xiaojuan Cui, Zhirong Zhang and Hua Xia
Anhui Provincial Key Laboratory of Photonic Devices and Materials, Anhui Institute of Optics and Fine Mechanics, Chinese Academy of Sciences, Hefei Anhui, China
Key Laboratory of Environmental Optics and Technology, Anhui Institute of Optics and Fine Mechanics, Chinese Academy of Sciences, Hefei, China

Fengzhong Dong
Anhui Provincial Key Laboratory of Photonic Devices and Materials, Anhui Institute of Optics and Fine Mechanics, Chinese Academy of Sciences, Hefei Anhui, China
Key Laboratory of Environmental Optics and Technology, Anhui Institute of Optics and Fine Mechanics, Chinese Academy of Sciences, Hefei, China
University of Science and Technology of China, Hefei, China

Tao Pang, Pengshuai Sun and Bian Wu
Anhui Provincial Key Laboratory of Photonic Devices and Materials, Anhui Institute of Optics and Fine Mechanics, Chinese Academy of Sciences, Hefei Anhui, China

Shuo Liu, Luo Han, Zhe Li and Runqing Yu
Anhui Provincial Key Laboratory of Photonic Devices and Materials, Anhui Institute of Optics and Fine Mechanics, Chinese Academy of Sciences, Hefei Anhui, China
University of Science and Technology of China, Hefei, China

Golebiowski Wlodzimierz, Kubiak Przemysław and Szymon Madziara
Department of Vehicles and Fundamentals of Machine Design, Faculty of Mechanical Engineering, Technical University of Lodz, Poland

Chang Wook Han, Hong-Seok Choi, Chanki Ha, Hongjae Shin, Hyun Chul Choi and In Byeong Kang
LG Display, Seoul, Korea

Wenquan Du, Zixin Wang and Dihu Chen
Sun Yat-Sen University, Guangzhou, China

Cristian Ravariu
Department of Electronic Device, Circuits and Architectures, Faculty of Electronics Telecommunications and Information Technology, "Politehnica" University of Bucharest, Bucharest, Romania

Elena Manea, Alina Popescu and Cătălin Pârvulescu
National Institute for Research and Development of Microtechnology - IMT Bucharest, Voluntari, Ilfov, Romania

Phan Trong Tue
School of Materials Science, Japan Advanced Institute of Science and Technology, Nomi-shi, Ishikawa, Japan

Houssem Rezgui, Faouzi Nasri, Mohamed Fadhel Ben Aissa and Amen Allah Guizani
Laboratory of Thermal Processes, Research and Technology Centre of Energy, Hammam-Lif, Tunisia

Joanna Cabaj and Jadwiga Sołoducho
Faculty of Chemistry, Wroclaw University of Science and Technology, Wroclaw, Poland

Index

www.ingramcontent.com/pod-product-compliance
Lightning Source LLC
LaVergne TN
LVHW060604210926
848485LV00005B/1370
* 9 7 8 1 6 3 9 8 9 5 0 1 4 *